# Climate Change and Insect Biodiversity

This up-to-date reference book discusses the effects of climate change on the biodiversity of insect pests. The changing climate and agricultural intensification practices impact negatively on insect biodiversity. The book explains the significance of insect pests for evaluating climatic impacts on a wide range of ecological systems. It covers the effect of climate change on pollinators and household and agricultural insect pests. It explains how climate-smart agriculture can enhance productivity and food security.

**FEATURES**

- Reviews the effects of climate change on plant-insect interactions
- Includes topics such as insect biodiversity informatics and conservation
- Discusses food security, pest management, and beneficial and social insects
- Covers topics such as precision agriculture and climate-smart agriculture
- Provides insights on the relation between agriculture intensification and insect biodiversity

This book is meant for scientists, researchers, and students working in the fields of agriculture, entomology, ecology, plant science, environmental biology, and biotechnology.

# Climate Change and Insect Biodiversity

## Challenges and Implications

Edited by

Habib Ali
Youming Hou
Muhammad Bilal Tahir

CRC Press
Taylor & Francis Group
Boca Raton London New York

CRC Press is an imprint of the
Taylor & Francis Group, an **informa** business

Designed cover image: Shutterstock

First edition published 2024
by CRC Press
2385 NW Executive Center Drive, Suite 320, Boca Raton FL 33431

and by CRC Press
4 Park Square, Milton Park, Abingdon, Oxon, OX14 4RN

*CRC Press is an imprint of Taylor & Francis Group, LLC*



ISBN: 978-1-032-44835-0 (hbk)
ISBN: 978-1-032-46379-7 (pbk)
ISBN: 978-1-003-38208-9 (ebk)

DOI: 10.1201/9781003382089

Typeset in Times
by Apex CoVantage, LLC

# Contents

# Editors

**Habib Ali, PhD,** is an esteemed researcher specializing in the field of entomology and insect pest control. With a strong academic background and a passion for scientific inquiry, he has made significant contributions to the field through his research work. He earned a doctorate in agricultural entomology and insect pest control, providing him with a solid foundation for his future endeavors. Ali is a highly accomplished researcher in entomology and insect pest control, with an extensive publication record of over 100 international research papers in renowned journals. He has been recognized for his exceptional research capabilities and dedication to academic excellence through awards such as Best International Student of the Year and the Best Thesis Award. Ali's prominence in the scientific community is further evident through nominations for the World Scientist Index and the Council of Asian Science Editorship Award. He actively participates in conferences, workshops, and training programs to stay abreast of the latest advancements in his field. Additionally, Ali contributes to the scientific community as an editor and reviewer for national and international journals, ensuring the maintenance of high standards in scientific publications. His multidisciplinary approach and focus on sustainable pest management underscore his significance in the field of entomology and insect pest control.

**Youming Hou, PhD,** is a Distinguished Professor in the Plant Protection Department, Fujian Agriculture and Forestry University, Fuzhou, China. He also serves as Dean of the College of Plant Protection, Deputy Director of State Key Laboratory, Director of Key Laboratory of Ministry of Education and Fujian Entomological Ecology, and Innovation Chief Scientist of the Ministry of Agriculture in Fujian-Taiwan Crop pest disaster mechanism and control.

**Muhammad Bilal Tahir, PhD,** is a researcher at Khawaja Fareed University of Engineering and Technology, Rahim Yar Khan (KFUEIT), and is currently working on nanotechnology for material processing, photocatalysis, energy and environmental applications, colloids, and surfactants. His initial work was on photocatalysis-based processes for decontamination of pollutants, adsorption of heavy metals in water and oil, and energy production. He has more than 500 publications in reputed International Scientific Index (ISI) journals and serves as Editor for numerous journals. He was the recipient of Best University Teacher Award, 2015–2016, and numerous awards from 2017 until now for his academic, research, and administrative performance in teaching and research.

# Contributors


**Asim Abbasi**
Department of Environmental Sciences
Kohsar University Murree
Murree, Punjab, Pakistan

**Mariam Khalid Abbasi**
Department of Botany
Kohsar University Murree
Murree, Punjab, Pakistan

**Usama Abdullah**
Department of Life Sciences
Khwaja Fareed University of Engineering and Information Technology
Rahim Yar Khan, Punjab, Pakistan

**Mohammad Abbas Ahmad**
Department of Entomology
Dr. Rajendra Prasad Central Agricultural University
Pusa, Bihar, India

**Taswar Ahsan**
Institute of Plant Protection
Liaoning Academy of Agricultural Sciences
Shenyang, China

**Muhammad Akmal**
Department of Entomology
Bahauddin Zakariya University
Multan, Punjab, Pakistan
and
Department of Evolutionary Biology and Environmental Studies
University of Zurich
Zurich, Switzerland

**Azad Ali**
Department of Agricultural Engineering
Khwaja Fareed University of Engineering and Information Technology
Rahim Yar Khan, Punjab, Pakistan

**Basharat Ali**
Department of Agricultural Engineering
Khawaja Fareed University of Engineering and Information Technology
Rahim Yar Khan, Punjab, Pakistan

**Habib Ali**
Department of Agricultural Engineering
Khawaja Fareed University of Engineering and Information Technology
Rahim Yar Khan, Punjab, Pakistan

**Kashif Ali**
Department of Zoology
Division of Science and Technology
University of Education
Lahore, Punjab, Pakistan

**Mohsin Ali**
Institute of Biological Sciences
University of Talca
Talca, Chile

**Sajjad Ali**
Department of Entomology
Faculty of Agriculture and Environment
The Islamia University of Bahawalpur
Bahawalpur, Punjab, Pakistan

**Shahbaz Ali**
Department of Agricultural Engineering
Khwaja Fareed University of Engineering and Information Technology
Rahim Yar Khan, Punjab, Pakistan

**Yasir Ali**
Khwaja Fareed University of Engineering and Information Technology
Rahim Yar Khan, Pakistan

**Muhammad Anjum Aqueel**
Department of Entomology
Faculty of Agriculture and Environment
The Islamia University of Bahawalpur
Bahawalpur, Punjab, Pakistan

**Muhamamd Arshad**
Department of Entomology
University of Agriculture Faisalabad
Faisalabad, Punjab, Pakistan

**Muhammad Asad**
Department of Zoology
Division of Science and Technology
University of Education
Lahore, Punjab, Pakistan

**Asma Ashraf**
Department of Zoology
Division of Science and Technology
University of Education
Lahore, Punjab, Pakistan

**Tanmaya Kumar Bhoi**
Forest Protection Division
Indian Council of Forestry Research and Education (ICFRE)
Arid Forest Research Institute (AFRI)
Jodhpur, Rajasthan, India

**Kinza Bibi**
Department of Botany
Kohsar University Murree
Murree, Punjab, Pakistan

**Sadiqa Bibi**
Department of Botany
Kohsar University Murree
Murree, Punjab, Pakistan

**Noman Ali Buttar**
Department of Agricultural Engineering
Khawaja Fareed University of Engineering and Information Technology
Rahim Yar Khan, Punjab, Pakistan

**Sanhita Chowdhury**
Department of Entomology
Post Graduate College of Agriculture
Dr. Rajendra Prasad Central Agricultural University
Pusa, Bihar, India

**Vinod Kumar Dubey**
Institute of Zoology
University of the Punjab
Lahore, Punjab, Pakistan

**Shoaib Fareed**
Department of Agricultural Entomology
Faculty of Agriculture Science and Technology
Bahauddin Zakariya University
Multan, Punjab, Pakistan

**Akhilesh Gupta**
Department of Agricultural Statistics
College of Agriculture
Odisha University of Agriculture and Technology (OUAT)
Bhubaneswar, Odisha, India

**Inzamam Ul Haq**
College of Plant Protection
Gansu Agricultural University
Lanzhou, China

**Kamran Ikram**
Department of Agricultural Engineering
Khawaja Fareed University of Engineering and Information Technology
Rahim Yar Khan, Punjab, Pakistan

**Hifza Inam**
Department of Botany
Kohsar University Murree
Murree, Punjab, Pakistan

**Neha Jamwal**
Department of Zoology
University of Jammu
Jammu, Jammu and Kashmir, India

**Raman Jasrotia**
Department of Zoology
University of Jammu
and
Department of Zoology
Cluster University of Jammu
Jammu, Jammu and Kashmir, India

**Adnan Khalil**
Department of Physics
Khawaja Fareed University of Science and Technology
Rahim Yar Khan, Punjab, Pakistan

**J. Komal**
Department of Entomology
Navsari Agricultural University
Navsari, Gujarat, India

**Laiba Kousar**
Department of Botany
Kohsar University Murree
Murree, Punjab, Pakistan

**Seema Langer**
Department of Zoology
University of Jammu
Jammu, Jammu and Kashmir, India

**Chunhao Liang**
Institute of Plant Protection
Liaoning Academy of Agricultural Sciences
Shenyang, China

**Deepak Kumar Mahanta**
Department of Entomology
Dr. Rajendra Prasad Central Agricultural University
Pusa, Bihar, India

**Waqar Majeed**
Department of Zoology, Wildlife and Fisheries
University of Agriculture Faisalabad
Faisalabad, Punjab, Pakistan

**Prasanta Kumar Majhi**
Department of Plant Breeding and Genetics
Odisha University of Agriculture and Technology (OUAT)
Bhubaneswar, Odisha, India

**Saif ul Malook**
Entomology and Nematology Department
University of Florida
Gainesville, Florida, USA

**Saiqa Maqbool**
Department of Chemistry
University of Agriculture Faisalabad
Faisalabad, Punjab, Pakistan

**Khalid Naveed**
University of Agriculture Faisalabad, Sub-Campus Depalpur Okara
Okara, Punjab, Pakistan

**Muhammad Nawaz**
Department of Agricultural Engineering
Khawaja Fareed University of Engineering and Information Technology
Rahim Yar Khan, Punjab, Pakistan

**Yasir Niaz**
Department of Agricultural Engineering
Khwaja Fareed University of Engineering and Information Technology
Rahim Yar Khan, Punjab, Pakistan

**Prachi Pattnaik**
Department of Horticulture
Institute of Agricultural Sciences
Banaras Hindu University (BHU)
Varanasi, Uttar Pradesh, India

**Chandini Pradhan**
Department of Soil Science and Agricultural Chemistry
Institute of Agricultural Sciences
Banaras Hindu University (BHU)
Varanasi, Uttar Pradesh, India

**Saima Qadeer**
Department of Zoology
Division of Science and Technology
University of Education
Lahore, Punjab, Pakistan

**Mirza Abdul Qayyum**
Institute of Plant Protection
Muhammad Nawaz Sharif University of Agriculture
Multan, Punjab, Pakistan

**Mawra Rasheed**
Department of Life Sciences
Khwaja Fareed University of Engineering and Information Technology
Rahim Yar Khan, Punjab, Pakistan

**Ali Raza**
Department of Microbiology and Molecular Genetics
University of Okara
Okara, Punjab, Pakistan

**Abu Bakar Muhammad Raza**
Department of Entomology
Faculty of Agriculture and Environment
The Islamia University of Bahawalpur
Bahawalpur, Punjab, Pakistan

**Maqsood Razzaq**
Department of Zoology
University of Sialkot
Sialkot, Pakistan

**Aamir Riaz**
Department of Genetics
University of Okara
Okara, Punjab, Pakistan

**Ammara Riaz**
Department of Life Sciences
Faculty of Natural and Applied Sciences
Khwaja Fareed University of Engineering & Information Technology
Rahim Khan, Punjab, Pakistan

**Aqsa Riaz**
Department of Zoology, Wildlife and Fisheries
University of Agriculture Faisalabad
Faisalabad, Punjab, Pakistan

**Saba**
Department of Molecular Biology
University of Okara
Okara, Punjab, Pakistan

**Laiba Sadaqat**
Department of Botany
Kohsar University Murree
Murree, Punjab, Pakistan

**Asif Sajjad**
Department of Entomology
Faculty of Agriculture and Environment
The Islamia University of Bahawalpur
Bahawalpur, Punjab, Pakistan

**Muhammad Saleem**
Department of Microbiology and Molecular Genetics
The Women University Multan
Multan, Punjab, Pakistan

**Ipsita Samal**
Department of Entomology
Faculty of Agriculture
Sri Sri University
Cuttack, Odisha, India

**Nafiu Bala Sanda**
Department of Crop Protection
Faculty of Agriculture
Bayero University
Kano, Nigeria

**Zahid Mehmood Sarwar**
Department of Agricultural Entomology
Bahauddin Zakariya University
Multan, Pakistan

**Jaffer Sattar**
Department of Agricultural Engineering
Khawaja Fareed University of Engineering and Information Technology
Rahim Yar Khan, Punjab, Pakistan

**Subrat Kumar Senapati**
Department of Floriculture and Landscaping
Odisha University of Agriculture and Technology (OUAT)
Bhubaneswar, Odisha, India

**Farah Rauf Shakoori**
Institute of Zoology
University of the Punjab
Lahore, Punjab, Pakistan

**Vyshnavi Sunil**
Department of Entomology
College of Agriculture
Kerala Agricultural University
Thiruvananthapuram, Kerala, India

**Muhammad Tahir**
Department of Zoology, Wildlife and Fisheries
University of Agriculture Faisalabad
Faisalabad, Punjab, Pakistan

**Roughaina Tahir**
Department of Entomology
Faculty of Agriculture and Environment
The Islamia University of Bahawalpur
Bahawalpur, Punjab, Pakistan

**Sana Ullah**
Department of Zoology
Division of Science and Technology
University of Education
Lahore, Punjab, Pakistan

**Muhammad Yasin**
Department of Entomology
Faculty of Agriculture and Environment
The Islamia University of Bahawalpur
Bahawalpur, Punjab, Pakistan

**Hafiz Muhammad Bilal Yousuf**
Department of Entomology
Faculty of Agriculture and Environment
The Islamia University of Bahawalpur
Bahawalpur, Punjab, Pakistan

**Jannat Zaka**
Department of Botany
Kohsar University Murree
Muree, Punjab, Pakistan

**Hina Zawar**
Department of Zoology
The Islamia University of Bahawalpur
Bahawalpur, Punjab, Pakistan

# 1 Overview of Precision and Climate-Smart Agriculture

## *Challenges and Solutions*

*Noman Ali Buttar, Yasir Niaz, Kamran Ikram, Muhammad Nawaz, Zahid Mehmood Sarwar, Khalid Naveed, Basharat Ali, Habib Ali, and Jaffer Sattar*

## 1.1 INTRODUCTION

Climate change has an impact on agricultural output and food systems, creating uncertainty and vulnerability for farmers and policymakers (Vermeulen et al., 2013). There is no doubt that climate change will pose enormous threats to global food security, and this will probably worsen in years to come due to pressures on livelihoods such as population expansion, economic development, urbanization, and the regular occurrence of natural catastrophes such as excessive heat, drought, and flooding, among other things (Bank et al., 2015). Some world-renowned organizations like the Food and Agricultural Organization (FAO) and the World Bank are partnering to develop agricultural systems that will improve and grow food production at all levels, from global to local. The adoption of climate-smart agriculture (CSA) has since received support from a number of organizations, including those involved in research, policymaking, and investments, as well as from the corporate, public, and civil society sectors (Taylor, 2018). FAO explicitly recommended CSA and discussed climate change in 2010 at the Hague Conference on Agriculture. CSA is a concept that addresses production and development of agricultural output while also improving adaptation, lowering greenhouse gas (GHG), and improving food security ("State of Food and Agriculture 2020," 2020). To secure food security and promote sustainable agricultural development, three major components of society, economy, and atmosphere work together to solve the difficulties posed by climate change. The specific goals of CSA are primarily presented in three areas: increasing the agricultural system's production efficiency to support farmers' incomes, guaranteeing food security, and improving GHG emissions. Finally, the triple goals of increased productivity, stress resilience, and decreased GHG emissions in agricultural systems are attained (Lipper et al., 2014; Palombi et al., 2013). Even though CSA strives to accomplish all three goals, not every strategy adopted in every place will necessarily lead to three successes. CSA must take into account these three objectives from short- to long-term perspectives at local to global scales in order to find solutions that are acceptable for the local environment. Depending on the environment and circumstances, each objective will have a different relative relevance (Figure 1.1). Finding a balance between the three is necessary, with the three CSA objectives to be realized as a priority (Lipper et al., 2014).

The available literature suggests that the CSA approach proposes the integration of climate change considerations into the planning and execution of sustainable agriculture policies, which involves identifying interconnections and potential conflicts among the three CSA pillars. This would ultimately aid in climate change–focused policy development and decision-making (Nagothu et al., 2016). It offers enabling methods for evaluating various technologies and practices' effects, particularly national development and food security goals. Additionally, CSA incorporates expertise and knowledge in sustainable agriculture development as well as

DOI: 10.1201/9781003382089-1

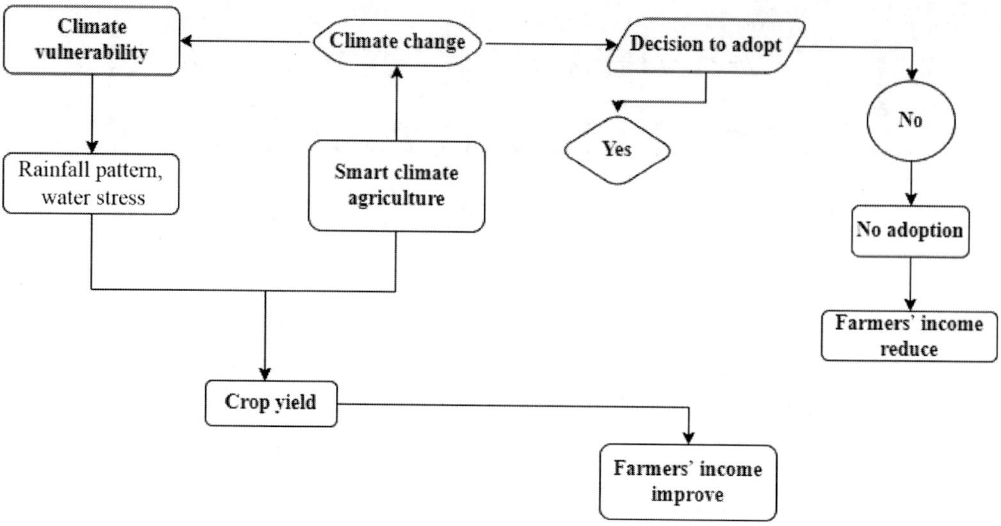


**FIGURE 1.1** The relation between climate-smart agriculture practices and the impact of climate change on farmers' income.

community-driven participatory methods (Nagothu et al., 2016). Integrating existing agricultural land conservation measures with sustainable intensification is the primary driver of on-farm production and profitability. To enhance food productivity and cope with climate change, CSA highlights the importance of adopting low-cost agricultural systems like conservation agriculture, agroecology, ecosystem services, small-scale irrigation, aquaculture, and agroforestry, in addition to integrated approaches that consider crops, livestock, landscapes, grasslands, and forestry management, CSA promotes the adoption of best practices for reducing tillage and breeding (Chandra et al., 2018).

The entire agricultural sector is directly impacted by climate change, which can have either beneficial or bad effects (e.g., expanding agricultural production zones or a rise in natural disasters like heat stress). Due to the increased proportion of agriculture, developing nations are disproportionately impacted. In addition, compared to developed nations, they are frequently more susceptible to such changes (Long et al., 2016). Smallholder farms provide food and income to the vast majority of the world's most impoverished, which form the backbone of the global food chain (Bogdanski, 2012). Small-scale farmers (less than 2 hectares) account for 84% of all farmers worldwide, with developing nations housing the majority of them ("State of Food and Agriculture 2020," 2020).

Sustainable intensification (SI) is frequently used in conjunction with CSA. SI stands for more sustainable, more intensive production. The complementarity between them, since they share so many components, like diversification, was highlighted (Campbell et al., 2014). They included instances to highlight the significance of SI in CSA, such as the intensification of cattle diets through agroforestry or the intercropping of bananas and coffee. However, the same guideline that applies to CSA also applies to CSA-SI: implementation should always take into account spatial variance. A local response is necessary to the global goal of CSA. Conservation agriculture (CA) is a related concept that is frequently linked to CSA. The least amount of soil disturbance, a continuous organic cover, and crop variety serve as the foundation for this. These values and the CSA pillars have a lot in common. So, one may say that CSA includes CA (Branca et al., 2021).

## 1.2 CONSTRAINTS IN THE ADOPTION OF CSA

CSA has been a key factor in the development of agriculture and ultimately the economy of a country. However, many barriers remain in the adoption of CSA in different countries, some of these countries along with the obstructions are discussed here (Table 1.1).

### 1.2.1 Market Inflation

Market incentives can be quite effective in encouraging the adoption of CSA practices. CSA entails applying practices that raise agricultural output, improve resistance to climate change, and minimize GHG. Following are some examples of market incentives that can promote the adoption of CSA:

1. Companies may offer higher prices for crops that are produced using CSA practices. This can encourage farmers to adopt these practices in order to boost their income.
2. Certification schemes can provide market recognition for sustainably produced crops. This can enable farmers to access markets that value sustainably produced crops and provide price premiums.
3. Climate financing systems can give farmers financial incentives to practice CSA. Subsidies, grants, and other financial instruments that enable farmers to invest in practices that improve their resilience to climate change are examples of this.
4. Carbon markets provide financial incentives for farmers to reduce their GHG emissions. This can include payments for carbon sequestration, emission reductions, or other climate-related benefits. Market incentives can overcome the financial barriers that farmers face in adopting CSA practices. By providing farmers with financial incentives to adopt these practices, market incentives can promote the adoption of practices that increase agricultural output (Bogdanski, 2012; Brunke et al., 2014; Security, 2014; Westermann et al., 2018).

### 1.2.2 Institutional Support

Institutional support is essential for the successful adoption of CSA, which involves implementing practices. Following are some examples of institutional support that can facilitate the adoption of CSA:

1. Extension services can provide farmers with the latest information on CSA practices, including technical assistance, training, and demonstrations.
2. Research institutions can develop and promote new CSA technologies and practices that are adapted to local conditions.
3. Governments can provide policy support for CSA by developing and implementing policies and regulations that incentivize the adoption of these practices. This can include subsidies, tax incentives, and market-based mechanisms.
4. Institutional support can help overcome the difficulties that farmers face in adopting CSA practices. Through the provision of financial incentives to farmers, market-based incentives can stimulate the adoption of practices that enhance agricultural productivity (Chandra et al., 2018; Harvey et al., 2014; Neufeldt et al., 2013; Scherr et al., 2012).

### 1.2.3 Organizational Barriers

Organizational barriers can significantly impact the adoption of CSA. Following are some examples of organizational barriers that can impede the adoption of CSA:

1. There may be a lack of access to information and knowledge. Farmers may not have access to the latest information on CSA practices, or they may not be aware of the benefits of these

**TABLE 1.1**
**Significant Constraints and Measures Related to the Adoption of Climate-Smart Agriculture in Different Countries**

| Countries | Significant Constraints | Adopted Measures for CSA | | | | | | | | | | | | |
|---|---|---|---|---|---|---|---|---|---|---|---|---|---|---|
| | | Nutrient Management | Controlled Chemical Application | Agroforestry | Water Savings | Inventive Seed Cultivars | Modern Sowing and Harvesting Approaches | Surface and Ground Water Management | Hydroponic-Based Agriculture | Planting Mushroom in Deserts | Improving Soil Biology and Physical Properties | Fields Land Leveling | Drainage Management | Crop Rotations |
| India | Climate threats, Bone-dry irrigation, GHG Emissions, Marginal yields | ✓ | | | | | | ✓ | | | | | | |
| Vietnam | | | ✓ | | | ✓ | ✓ | ✓ | | | | | | |
| Nepal | | | | | | | | ✓ | | | ✓ | | | |
| Pakistan | | | | | | ✓ | | ✓ | | | | ✓ | ✓ | |
| Zambia | | | ✓ | | | ✓ | | | | | | | | ✓ |
| Malawi | | | | ✓ | | | | | | | | | | ✓ |
| Namibia | | | | | ✓ | | | | ✓ | ✓ | | | | |

practices. This can be due to a lack of extension services or inadequate communication channels between researchers and farmers.
2. CSA practices may require additional investment or up-front costs, which can be a barrier for farmers who have limited financial resources. This may include the cost of equipment, seeds, and training.
3. Government policies and regulations can play a critical role in promoting CSA practices. However, policies may not exist or may not be implemented effectively, which can create a barrier for farmers.
4. Farmers may not have access to markets that provide incentives for CSA practices. This can be due to a lack of infrastructure, poor transportation networks, or inadequate pricing mechanisms.
5. Farmers may not have adequate social networks that provide support for CSA practices. This can be due to a lack of farmer organizations or inadequate communication between farmers. Addressing these organizational barriers will require a coordinated effort between government, civil society, and private sector actors. This can involve providing access to information and knowledge, improving access to finance, strengthening policy support, building market linkages, and supporting the formation of farmer organizations (Brunke et al., 2014; Johnson, 2010; Montalvo, 2008).

### 1.2.4 Economic Barriers

Economic constraints can have a substantial impact on the adoption of CSA. Following are some examples of economic barriers that can impede the adoption of CSA:

1. Farmers may have limited access to finances to invest in CSA practices. This may include the cost of equipment, seeds, and training.
2. CSA practices may require additional inputs, such as organic fertilizers or improved seed varieties. These inputs may be more expensive than traditional inputs, which can create a barrier for farmers who have limited financial resources.
3. Farmers may face low prices for their agricultural products, which can reduce their income and limit their ability to invest in CSA practices.
4. Market access may be lacking. Farmers may have limited access to markets that provide incentives for CSA practices. This can be due to a lack of infrastructure, poor transportation networks, or inadequate pricing mechanisms.

Addressing these economic barriers will require a coordinated effort between government, civil society, and private sector actors. This can involve providing access to finance, developing and promoting low-cost inputs, improving market access and pricing mechanisms, and supporting the formation of farmer organizations to negotiate better prices for their products. By addressing these economic barriers, farmers will be better positioned to adopt CSA practices, which can enhance their productivity, improve their resilience to climate change, and reduce GHG emissions (Cullen et al., 2013; Faber et al., 2013; Montalvo, 2008; Westermann et al., 2018).

## 1.3 CONSTRAINTS IN THE IMPLEMENTATION OF CSA

### 1.3.1 Lack of Water Resources for Agriculture

Food security is based on the availability of water. Currently, the availability of agricultural water resources, as well as the sustainable development of CSA, are severely curtailed. There will be a 55% increase in the world's water demand. Water use by agriculture is being impacted globally by the current water deficit, which is escalating quickly (Salehi, 2022). According to one study, water scarcity in agriculture has reduced sweet corn production and protein in Turkey (Oktem, 2008). The possibility of agricultural water resource shortages was expected to climb drastically to a medium-high risk level by 2030. The Loess Plateau, a typical dry land agricultural region in China, has shown that as rainfall became scarce, grain productivity and water-use efficiency fell (Zhang et al., 2022). The danger of agricultural water resource constraints in arid Northwest China in 2010 was moderate to low (Roy et al., 2021). Furthermore, some coal-mining areas are associated with hostile settings in which a lack of water and critical nutrients has a negative impact on plant growth and development (Roy et al., 2021). Agriculture water resource evaluation is crucial for long-term planning and management. Water scarcity has a significant impact on South Korea's irrigated agriculture due to seasonal changes in rainfall and water quality (Lee et al., 2019; Nam et al., 2017). The amount of agricultural water resources accessible has recently been reduced due to a scarcity of water resources and their unequal distribution at spatiotemporal scales induced by climate variability, such as drought and heat waves (Kim et al., 2018). As a result, anticipating the supply and demand for agricultural water resources under warmer climatic conditions became increasingly critical for CSA expansion.

### 1.3.2 Impact of Climate Change and Variability

The production of food around the world and the sustainability of CSA are threatened by climate change brought on by human activity. The worldwide community frequently debates ways to address climate change actively, increase agricultural output, and cut GHG emissions more sustainably. The availability and distribution of vital agricultural climate resources, such as light, heat, and water, have been impacted by climate variability and change. Small-scale farmers are particularly susceptible to the adverse consequences of climate change, including diminished crop yields, reduced income, and food insecurity (McKinley et al., 2021). According to a survey conducted in Ghana, climate change has a significant impact on subsistence farming, causing anxiety over food in 58% of families and making it difficult for 62% of families to obtain the quantity and quality of food they prefer (Asare-Nuamah, 2021). Climate change has a direct impact on agriculture by extending the growing season, changing precipitation patterns, and raising $CO_2$ concentrations (Janowiak et al., 2016). It is generally acceptable for some places to have an increase in expected yield while experiencing a decrease in predicted yield. The regional climates that result from temperature and precipitation variations brought on by climate change may also be significantly altered, which may result in changes to how crops are distributed (Zhao et al., 2015). Climate change has the potential to alter the specific nutrients, carbohydrates, proteins, and other ingredients when combined with income increases. Climate change has a direct impact on agriculture by extending the growing season, varying precipitation patterns, and raising $CO_2$ concentrations (Nelson et al., 2018).

### 1.3.3 GHG Emission Effect

The obstacles to CSA's sustainable development are exacerbated by agricultural GHG emissions. As a result of the use of fossil fuels, changes in land use, and deforestation, anthropogenic GHG emissions have increased dramatically. The Intergovernmental Panel on Climate Change (IPCC) has underlined the consequences of GHG emissions on climate change (Buckle, 2009). In terms of anthropogenic GHG emissions, the agricultural ecosystem ranks second, contributing 56% of all non-$CO_2$ emissions to the atmosphere. Agriculture, while a major contributor to GHG emissions, also has considerable promise for reducing them. To attain the goals of increasing agricultural productivity and efficiency and ensuring food security while simultaneously mitigating agricultural GHG emissions, a range of technologies can be implemented. These technologies include GHG emission reduction strategies, such as GHG absorption enhancement and adaptation to climate change.

### 1.3.4 Integrated Resource Management

Product information and data support are key components in the development, administration, management, shipping, and sale of CSA. To ensure the smooth development of CSA, information and data security must be maintained. However, there are several concerns with CSA information and security from the aspect of the development process, such as low standardization, insufficient data collection and sorting, and a lack of efficacy in agricultural data gathering. It is currently difficult to exchange real-time information technology resources, and the CSA information security issue is pressing. The fundamental security of agricultural data has evolved to be a crucial issue in the long-term development of CSA. The issue of too much agricultural information will also be a challenge for CSA. The long-term growth of CSA will be hampered if erroneous information cannot be appropriately identified.

## 1.4 FUTURE DIRECTIONS OF CSA

### 1.4.1 Involvement of Internet of Things

The internet of things (IoT) is a vast network of linked equipment, sensors, and computing devices. Every gadget has a distinct identification and the capacity to use remote sensing and monitoring (Pylianidis et al., 2021). In the agricultural sector, IoT is primarily utilized for data collection via a diverse range of sensors, crop factors such as temperature, humidity, pH levels, leaf color, and more. IoT potential in agricultural sectors has been explored from a variety of perspectives, including analyzing IoT applications and developing IoT food control systems (Dias et al., 2021; Jusoh et al., 2021). The following are a number of areas that need to be strengthened in the IoT's future development for CSA: First, because farmer demands vary greatly, the IoT system needs to be very adaptable and able to be customized to local conditions. Second, each system's IoT deployment must be effective and configured. Reliable network connections and farm infrastructure are required, as well as the allocation of sufficient human and financial resources. As a result, the IoT sensor network utilized in agriculture must be equipped with security protocols that align with cloud databases and current operational computing networks (Kleinschmidt et al., n.d.).

### 1.4.2 Crop Production and Management Techniques

Crop diversification techniques, no-till farming, and rice-wheat rotation are a few examples of diverse cropping patterns that can boost agricultural production and lower GHG emissions. By introducing suitable dry-land crops, the yearly planting cycle's submergence duration can be shortened (Datta et al., 2011). To enhance soil quality and crop yield, particularly for low fertility soils, it is common practice to combine inorganic fertilizers with organic improvers (Mi et al., 2018). For instance, biochar is regarded as a beneficial soil additive because it has the ability to fix carbon, rebuild soil, and enhance crop output (El-Naggar et al., 2019). These feasible methods can assist in further conserving water, improving soil structure, strengthening the element cycle, raising agricultural production, and reducing GHG emissions. It is critical to remember that the rates of carbon sequestration and carbon release from soil differ significantly between agricultural systems. Additionally, the growing method known as CA is recognized as a significant sustainable agricultural technique and has been employed extensively in agricultural production due to its distinctive ecological preservation role. Future CSA expansion will advance agriculture's sustainable development and the various objectives of ensuring food security, facilitating climate adaptation, and cutting GHG emissions.

### 1.4.3 Importance of Remote Sensing Applications

Remote sensing technology is widely used in a variety of industries because of its quick, wide-area observation and macro, real-time, dynamic, and straightforward economic access (Meier et al.,

2020). The sensor's ability to identify targets in the electromagnetic spectrum is determined by the spectral sampling, size, and amount. Data acquisition frequency determines the temporal resolution. Because of continual advancements in the spatial, temporal, and spectral resolutions of remote sensing data, as well as the methods and products utilized for them, remote sensing has evolved into a key instrument with various potential applications in regional-scale CSA (Meier et al., 2020). For many years, image-based remote sensing has been deployed for precision crop management. However, hyperspectral images have considerably enhanced the identification and separation of agricultural nutrients, pathogens, and canopy structures (Mulla, 2013). Yuan et al. were able to identify the spectral differences between background and rice, using the spectral index and the wealth of information from the combined components of rice (Yuan et al., 2021). The use of a wireless sensor network on the ground in conjunction with a remotely controlled aerial vehicle can improve fertilizer and pesticide management in the field (Costa et al., 2012). The spatiotemporal resolution of observation data from remote sensing currently contains significant discrepancies. It is difficult to capture the dynamic changes of crops during periods of rapid development since the data have a low temporal resolution despite having a high spatial resolution. Future sustainable development of CSA will heavily rely on the fusion of data from multiple remote sensing sources. The increased accessibility of remote sensing data has made it feasible to acquire precise, real-time insights into the evolving progress of crops. This is accomplished through the amalgamation of data from multiple sources with different temporal and geographical resolutions.

### 1.4.4 Involvement of Artificial Intelligence

Future applications of artificial intelligence (AI) are another CSA focus. AI is the simulation, extension, and expansion of human intelligence through the use of digital computers or other controlled machines that can also observe their environment and learn pertinent information. AI has already proven to have significant advantages in a variety of study areas. The most recent developments in computer technology and large data (data storage capacity) have made room for the use of AI in agriculture. The use of AI in agriculture is widespread. AI has the capability to process and consolidate data from various agricultural domains for a multitude of purposes, including plant identification, weed detection, crop yield forecasting, GHG emission estimation, climate projection, pest management, crop planting risk evaluation, and more (Hamrani et al., 2020). AI can boost agricultural productivity by precisely forecasting optimal sowing and harvesting dates, assessing crop health, and minimizing the cost of agricultural inputs such as fertilizers, pesticides, and irrigation. As a result, agricultural risks can be reduced by finding solutions to issues like inadequate precipitation, weed growth, and disaster losses. The quantity and quality of the data, however, heavily influence the AI model's algorithms. AI technology should improve data collecting and analysis from several sources in the future, including soil, climate, illnesses, and pests. Crop quality and quantity are ensured by anticipating different factors that influence crop output, assessing crop health, and determining the risk that diseases and pests will arise.

### 1.4.5 Quality of Agricultural Service

Agriculture is currently moving toward mechanization and modernization. It is inevitable that it will advance stricter specifications for meteorological services during the course of future development. The demands of farmers should be successfully merged with meteorological services using internet technology. To establish a vertical meteorological service system for agriculture, the technology platform is employed to firmly integrate producers, transport houses. In the coming years, the incorporation of "internet + weather" services, coupled with the advancement of agricultural service standards and the integration of human-computer interaction with comprehensive meteorological observations and self-sufficient weather stations can effectively address the needs of rural and agricultural development.

## 1.5 CONCLUSION

CSA serves as a valuable tool for both stakeholders and policymakers in addressing the challenges at hand. To achieve the desired sustainability objectives, it is imperative to implement modifications and reforms to agricultural and food systems. Due to the increased emphasis on food security, the economy, and reductions in poverty, it is of prime importance to implement more sustainable agricultural solutions, livelihoods, and CSA. The majority of collaborations are between wealthy countries, which means that poor countries are lagging behind, despite the fact that the consequences of climate variability and change affect all countries regardless of national position. The adoption and integration of new CSA technologies and big data applications are still in their infancy. This study's main implications are as follows: (a) spreading the notion that there is evidence that implementing CSA throughout the agricultural economy might revolutionize the sector and (b) understanding that CSA technologies are likely to change and diversify, especially with the onset of the Fourth Industrial Revolution. The forthcoming developments in CSA technology are expected to radically impact the agricultural business, indicating that all stakeholders, including legislators, must be prepared to accept these disruptive advances. Small-scale CSA implementation faces comparable issues around the world that include low skills and knowledge as well as potentially high investment costs and delayed outcomes. The majority of constraints are frequently more challenging for women. It should be noted that the bare essentials for farming, such as officially recognized land rights and tenure security, are not always available. A fundamental requirement for smart farming is consistent and fast internet access. The study reviewed these challenges via various forms of information-sharing financial instruments and, ultimately, climate information services. Finally, considering that technical interventions alone may not be sufficient to achieve long-term agricultural transformation, the study emphasizes the importance of increasing the funding of new initiatives and investments to improve and boost CSA technical innovation.

## REFERENCES


Asare-Nuamah, P. (2021). Climate variability, subsistence agriculture and household food security in rural Ghana. *Heliyon*, *7*(4), e06928. doi:10.1016/j.heliyon.2021.e06928

Bank, W., President, G., Yong, J., & States, U. (2015). Global faith leaders and World Bank Group President Jim Yong Kim on call and commitment to end extreme poverty. *Documents1.Worldbank.Org*. Retrieved from http://documents1.worldbank.org/curated/en/306261467988936104/pdf/101975-WP-Box393267B-PUBLIC-2015-04-09-JK-Global-Faith-Leaders-Jim-Yong-Kim-on-Call-and-Commitment-to-End-Extreme-Poverty.pdf

Bogdanski, A. (2012). Integrated food-energy systems for climate-smart agriculture. *Agriculture and Food Security*, *1*(1), 1–10. doi:10.1186/2048-7010-1-9/TABLES/1

Branca, G., Arslan, A., Paolantonio, A., Grewer, U., Cattaneo, A., Cavatassi, R., Lipper, L., Hillier, J., & Vetter, S. (2021). Assessing the economic and mitigation benefits of climate-smart agriculture and its implications for political economy: A case study in Southern Africa. *Journal of Cleaner Production*, *285*. doi:10.1016/j.jclepro.2020.125161

Brunke, J. C., Johansson, M., & Thollander, P. (2014). Empirical investigation of barriers and drivers to the adoption of energy conservation measures, energy management practices and energy services in the Swedish iron and steel industry. *Journal of Cleaner Production*, *84*(1), 509–525. doi:10.1016/j.jclepro.2014.04.078

Buckle, S. (2009). Mitigation of climate change. *Weather*, *64*(6), 165–166. doi:10.1002/wea.422

Campbell, B. M., Thornton, P., Zougmoré, R., van Asten, P., & Lipper, L. (2014). Sustainable intensification: What is its role in climate smart agriculture? *Current Opinion in Environmental Sustainability*, *8*, 39–43. doi:10.1016/j.cosust.2014.07.002

Chandra, A., McNamara, K. E., & Dargusch, P. (2018). Climate-smart agriculture: Perspectives and framings. *Climate Policy*, *18*(4), 526–541. doi:10.1080/14693062.2017.1316968

Costa, F. G., Ueyama, J., Braun, T., Pessin, G., Osorio, F. S., & Vargas, P. A. (2012). The use of unmanned aerial vehicles and wireless sensor network in agricultural applications. *International Geoscience and Remote Sensing Symposium (IGARSS)*, 5045–5048. doi:10.1109/IGARSS.2012.6352477

Cullen, R., Forbes, S. L., & Grout, R. (2013). Non-adoption of environmental innovations in wine growing. *New Zealand Journal of Crop and Horticultural Science*, *41*(1), 41–48. doi:10.1080/01140671.2012.744760

Datta, A., Rao, K. S., Santra, S. C., Mandal, T. K., & Adhya, T. K. (2011). Greenhouse gas emissions from rice based cropping systems: Economic and technologic challenges and opportunities. *Mitigation and Adaptation Strategies for Global Change*, *16*(5), 597–615. doi:10.1007/s11027-011-9284-z

Dias, R. M., Marques, G., & Bhoi, A. K. (2021). Internet of things for enhanced food safety and quality assurance: A literature review. *Lecture Notes in Electrical Engineering*, *709*, 653–663. doi:10.1007/978-981-15-8752-8_66

El-Naggar, A., El-Naggar, A. H., Shaheen, S. M., Sarkar, B., Chang, S. X., Tsang, D. C. W., Rinklebe, J., & Ok, Y. S. (2019). Biochar composition-dependent impacts on soil nutrient release, carbon mineralization, and potential environmental risk: A review. *Journal of Environmental Management*, *241*, 458–467. doi:10.1016/j.jenvman.2019.02.044

Hamrani, A., Akbarzadeh, A., & Madramootoo, C. A. (2020). Machine learning for predicting greenhouse gas emissions from agricultural soils. *Science of the Total Environment*, *741*, 140338. doi:10.1016/j.scitotenv.2020.140338

Harvey, C. A., Chacón, M., Donatti, C. I., Garen, E., Hannah, L., Andrade, A., Bede, L., Brown, D., Calle, A., Chará, J., Clement, C., Gray, E., Hoang, M. H., Minang, P., Rodríguez, A. M., Seeberg-Elverfeldt, C., Semroc, B., Shames, S., Smukler, S., . . . Wollenberg, E. (2014). Climate-smart landscapes: Opportunities and challenges for integrating adaptation and mitigation in tropical agriculture. *Conservation Letters*, *7*(2), 77–90. doi:10.1111/conl.12066

Janowiak, M., Dostie, D., Wilson, M., Kucera, M., Howard Skinner, R., Hatfield, J., Hollinger, D., Swanston, C., & USDA. (2016). *Adaptation resources for agriculture: Responding to climate variability and change in the Midwest and Northeast*. 72. Retrieved from https://ageconsearch.umn.edu/record/320856/

Johnson, M. (2010). Barriers to innovation adoption: A study of e-markets. *Industrial Management & Data Systems*, *110*(2), 157–174. doi:10.1108/02635571011020287/FULL/XML

Jusoh, M. F., Muttalib, M. F. A., Krishnan, K. T., & Katimon, A. (2021). An overview of the internet of things (IoT) and irrigation approach through bibliometric analysis. *IOP Conference Series: Earth and Environmental Science*, *756*(1), 12041. doi:10.1088/1755-1315/756/1/012041

Kim, S. M., Kang, M. S., & Jang, M. W. (2018). Assessment of agricultural drought vulnerability to climate change at a municipal level in South Korea. *Paddy and Water Environment*, *16*(4), 699–714. doi:10.1007/s10333-018-0661-z

Kleinschmidt, J., Kamienski, C., Prati, R. C., Kolehmainen, K., & Aguzzi, C. (n.d.). End-to-end security in the IoT computing continuum: Perspectives in the SWAMP project. *Ieeexplore.Ieee.Org*. Retrieved from https://ieeexplore.ieee.org/abstract/document/8995724/

Lee, J. M., Kwon, E. H., & Woo, N. C. (2019). Natural and human-induced drivers of groundwater sustainability: A case study of the Mangyeong River Basin in Korea. *Sustainability (Switzerland)*, *11*(5). doi:10.3390/su11051486

Lipper, L., Thornton, P., Campbell, B., Baedeker, T., Braimoh, A., Bwalya, M., Caron, P., Cattaneo, A., Garrity, D., Henry, K., Hottle, R., Jackson, L., Jarvis, A., Kossam, F., Mann, W., McCarthy, N., Meybeck, A., Neufeldt, H., Remington, T., . . . Torquebiau, E. F. (2014). Climate-smart agriculture for food security. *Nature*, *4*. doi:10.1038/NCLIMATE2437

Long, T. B., Blok, V., & Coninx, I. (2016). Barriers to the adoption and diffusion of technological innovations for climate-smart agriculture in Europe: Evidence from the Netherlands, France, Switzerland and Italy. *Journal of Cleaner Production*, *112*, 9–21. doi:10.1016/j.jclepro.2015.06.044

McKinley, J. D., LaFrance, J. T., & Pede, V. O. (2021). Climate change adaptation strategies vary with climatic stress: Evidence from three regions of Vietnam. *Frontiers in Sustainable Food Systems*, *5*, 389. doi:10.3389/fsufs.2021.762650

Meier, J., Mauser, W., Hank, T., & Bach, H. (2020). Assessments on the impact of high-resolution-sensor pixel sizes for common agricultural policy and smart farming services in European regions. *Computers and Electronics in Agriculture*, *169*. doi:10.1016/j.compag.2019.105205

Mi, W., Sun, Y., Xia, S., Zhao, H., Mi, W., Brookes, P. C., Liu, Y., & Wu, L. (2018). Effect of inorganic fertilizers with organic amendments on soil chemical properties and rice yield in a low-productivity paddy soil. *Geoderma*, *320*, 23–29. doi:10.1016/j.geoderma.2018.01.016

Montalvo, C. (2008). General wisdom concerning the factors affecting the adoption of cleaner technologies: A survey 1990–2007. *Journal of Cleaner Production*, *16*(1 Suppl. 1), S7–S13. doi:10.1016/j.jclepro.2007.10.002

Mulla, D. J. (2013). Twenty five years of remote sensing in precision agriculture: Key advances and remaining knowledge gaps. *Biosystems Engineering*, *114*(4), 358–371. doi:10.1016/j.biosystemseng.2012.08.009

Nagothu, U. S., Kolberg, S., & Stirling, C. M. (2016). Climate smart agriculture; Is this the new paradigm of agricultural development. *Taylorfrancis.com*. Retrieved from www.taylorfrancis.com/chapters/edit/10.4324/9781315685953-9/climate-smart-agriculture-new-paradigm-agricultural-development-udaya-sekhar-nagothu-solveig-kolberg

Nam, W. H., Kim, T., Hong, E. M., & Choi, J. Y. (2017). Regional climate change impacts on irrigation vulnerable season shifts in agricultural water availability for South Korea. *Water (Switzerland)*, *9*(10). doi:10.3390/w9100735

Nelson, G., Bogard, J., Lividini, K., Arsenault, J., Riley, M., Sulser, T. B., Mason-D'Croz, D., Power, B., Gustafson, D., Herrero, M., Wiebe, K., Cooper, K., Remans, R., & Rosegrant, M. (2018). Income growth and climate change effects on global nutrition security to mid-century. *Nature Sustainability*, *1*(12), 773–781. doi:10.1038/s41893-018-0192-z

Neufeldt, H., Jahn, M., Campbell, B. M., Beddington, J. R., DeClerck, F., De Pinto, A., Gulledge, J., Hellin, J., Herrero, M., Jarvis, A., LeZaks, D., Meinke, H., Rosenstock, T., Scholes, M., Scholes, R., Vermeulen, S., Wollenberg, E., & Zougmoré, R. (2013). Beyond climate-smart agriculture: Toward safe operating spaces for global food systems. *Agriculture and Food Security*, *2*(1). doi:10.1186/2048-7010-2-12

Oktem, A. (2008). Effect of water shortage on yield, and protein and mineral compositions of drip-irrigated sweet corn in sustainable agricultural systems. *Agricultural Water Management*, *95*(9), 1003–1010. doi:10.1016/j.agwat.2008.03.006

Pylianidis, C., Osinga, S., & Athanasiadis, I. N. (2021). Introducing digital twins to agriculture. *Computers and Electronics in Agriculture*, *184*. doi:10.1016/j.compag.2020.105942

Roy, R., Wang, J., Mostofa, M. G., & Fornara, D. (2021). Optimal water and fertilizer applications improve growth of *Tamarix chinensis* in a coal mine degraded area under arid conditions. *Physiologia Plantarum*, *172*(2), 371–390. doi:10.1111/ppl.13147

Salehi, M. (2022). Global water shortage and potable water safety; today's concern and tomorrow's crisis. *Environment International*, *158*. doi:10.1016/j.envint.2021.106936

Scherr, S. J., Shames, S., & Friedman, R. (2012). From climate-smart agriculture to climate-smart landscapes. *Agriculture and Food Security*, *1*(1), 1–15. doi:10.1186/2048-7010-1-12/TABLES/2

Security, F. (2014). *Climate-smart agriculture: Smallholder adoption and implications for climate change adaptation and mitigation*. Retrieved from www.fao.org/climatechange/29765-08b92911662298c69c61a88337c72a578.pdf

The State of Food and Agriculture 2020. (2020). *In the State of Food and Agriculture 2020*. FAO. doi:10.4060/cb1447en

Taylor, M. (2018). Climate-smart agriculture: What is it good for? *Journal of Peasant Studies*, *45*(1), 89–107. doi:10.1080/03066150.2017.1312355

Vermeulen, S. J., Challinor, A. J., Thornton, P. K., Campbell, B. M., Eriyagama, N., Vervoort, J. M., Kinyangi, J., Jarvis, A., Läderach, P., Ramirez-Villegas, J., Nicklin, K. J., Hawkins, E., & Smith, D. R. (2013). Addressing uncertainty in adaptation planning for agriculture. *Proceedings of the National Academy of Sciences of the United States of America*, *110*(21), 8357–8362. doi:10.1073/pnas.1219441110

Westermann, O., Förch, W., Thornton, P., Körner, J., Cramer, L., & Campbell, B. (2018). Scaling up agricultural interventions: Case studies of climate-smart agriculture. *Agricultural Systems*, *165*, 283–293. doi:10.1016/j.agsy.2018.07.007

Yuan, N., Gong, Y., Fang, S., Liu, Y., Duan, B., Yang, K., Wu, X., & Zhu, R. (2021). Uav remote sensing estimation of rice yield based on adaptive spectral endmembers and bilinear mixing model. *Remote Sensing*, *13*(11). doi:10.3390/rs13112190

Zhang, Y., Li, H., Sun, Y., Zhang, Q., Liu, P., Wang, R., & Li, J. (2022). Temporal stability analysis evaluates soil water sustainability of different cropping systems in a dryland agricultural ecosystem. *Agricultural Water Management*, *272*. doi:10.1016/j.agwat.2022.107834

Zhao, J., Guo, J., Xu, Y., & Mu, J. (2015). Effects of climate change on cultivation patterns of spring maize and its climatic suitability in Northeast China. *Agriculture, Ecosystems and Environment*, *202*, 178–187. doi:10.1016/j.agee.2015.01.013

# 2 Plant Receptor Interaction with Herbivore-Associated Elicitors
## *Mechanisms and Implications for Insect Herbivory-Induced Defense*

*Saif ul Malook, Saiqa Maqbool, and Habib Ali*

## 2.1 INTRODUCTION

Plants, being immobile organisms, are incapable of fleeing from arthropod herbivores, thus making them significantly vulnerable to insect herbivory. However, living together with insects, plants have evolved complex defense mechanisms to resist being consumed by them (Wu and Baldwin, 2010). Plants rely on their ability to detect chemical cues, known as herbivore-associated elicitors (HAEs), to recognize herbivore attacks. During insect infestation, an insect's herbivory generates chemical signatures, and these signatures are recognized by surface receptors (receptor kinases; RKs) present in the surface of plants. Through this mechanism, plants are able to distinguish between insect feeding and mechanical wounding with the help of detection of specific conserved molecular signatures present in insect oral secretions (OSs) (Figure 2.1) (Arimura, 2021; Bonaventure *et al.*, 2011; Jones and Dangl, 2006; Reymond, 2021). There are a limited number of reports in the literature that indicate the constituents of OSs are influenced by the feeding behavior of insects on their host plants (Kallure *et al.*, 2022).

When insect herbivores feed on plant tissues, they deposit OSs on the wounds and thereby manipulate plant metabolism and gene expression resulting in changes in plant responses (Bonaventure *et al.*, 2011; Felton and Tumlinson, 2008). In OSs, HAEs exhibit structural diversity and can be found in the form of various molecules, including enzymes like glucose oxidase (GOX) and ß-glucosidase; lipids such as cell wall fragments like pectin, volicitin, and caeliferins; oligogalacturonides; and plant peptides such as inceptin, which are proteolytic fragments of the chloroplastic ATP synthase subunit. However, none of these molecules were found to impact the induced defenses of tomato (Alborn *et al.*, 1997; Bonaventure *et al.*, 2011; Mattiacci *et al.*, 1995; Schmelz *et al.*, 2006). The HAEs are not general elicitors in all plants, and the response of plants to insect herbivores is limited to particular associations between plants and insects, which are dependent on the specific feeding style of the insects (Acevedo *et al.*, 2015; Felton and Tumlinson, 2008; Schmelz, 2015). The unique interactions between plants and insects are a result of their shared evolutionary history and coexistence in nature (Bonaventure *et al.*, 2011). Insect herbivore-induced systemic responses are mediated by chemical signatures of insect OSs and are utilized to sustain insect resilience in response to insect attack (Chen and Mao, 2020; Erb and Reymond, 2019; Schmelz, 2015). Although it is important to have a comprehensive understanding of induced responses, the interactions between plant receptors and their HAEs still need to be explored further in the field of plant-insect interaction. As such, it remains an emerging research topic.

Plants respond to insect elicitors by triggering a defense mechanism that involves the stimulation of plasma membrane depolarization (Vm), calcium ion influx ($Ca^{+2}$), and NADPH oxidase, and the production of reactive oxygen species (ROS) and mitogen-activated protein kinases (MAPKs). Additionally, nitrogen species (NO) are activated as part of this defense response (Zebelo and

 DOI: 10.1201/9781003382089-2

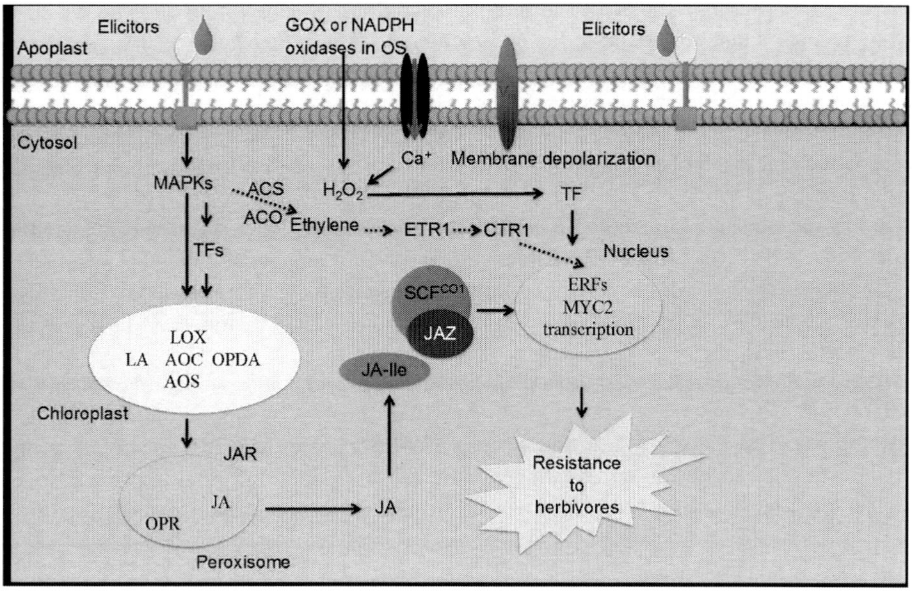


**FIGURE 2.1** The molecular signaling signature model of plant response to insect herbivory. When herbivores consume plant tissues, they deposit oral secretions (OSs) into wounds. The OSs contain herbivore-associated elicitors that are recognized by plant receptors located in the plasma membrane. The interaction triggers the activation of several signaling molecules such as reactive oxygen species, calcium ions ($Ca^{2+}$), mitogen-activated protein kinase (MAPK), and membrane depolarization (Vm) within a few minutes. This leads to the production of JA-Ile, which binds with SCFCOI1, resulting in the degradation of JAZ and the activation of downstream plant defenses.

Maffei, 2015). Upon the detection of insect elicitors, a sequence of molecular signaling pathways is activated. This results in the synthesis of defensive phytohormones including jasmonic acid (JA), salicylic acid (SA), and ethylene, as well as the factors influencing the gene transcription (transcription factors). JA, in particular, has a significant role in regulating downstream defense metabolites including glucosinolates, cyanogenic glucosides, benzoxazinoids (Bxs), phenolics, alkaloids, and proteinase inhibitors in both damaged and systemic leaves, as illustrated in Figure 2.1 (Erb and Reymond, 2019). Several valuable reviews have been published regarding the discourse on induced plant responses against arthropods (Divekar *et al.*, 2022; Erb and Kliebenstein, 2020; Erb and Reymond, 2019; Qi *et al.*, 2018). Developing pesticide-free crops that are resistant to insect pests through molecular breeding is essential to meet the food demands of a growing population. However, to achieve this goal, a deep understanding of the mechanisms underlying plant perception of herbivore attacks and the utilization of plant receptors to trigger defense responses is necessary. This chapter concentrates on presenting the latest research progress in plant receptor interactions with HAEs and the perception of herbivorous attack. The aim is to shed light on the signal transduction pathways that induce defense responses upon activation.

## 2.2 PLANT RECEPTORS PERCEIVE INSECT HERBIVORY

Although plants and insects engage in an ongoing struggle for survival, the development in evolving plant defenses is a key factor in determining the ultimate outcome of a plant's life. While researchers in plant chemical ecology have focused on studying insect elicitors and their interactions with plant receptors over the past decades, however, information on the interactions between HAEs and plant receptor proteins is at an evolving stage. The initial step in the recognition of HAEs involves detecting these bioactive molecules in the oral secretions of herbivorous insects, which is mediated by receptors that are bound to the cell membrane (Howe and Jander, 2008; Xu *et al.*, 2015). However, while much research has focused on investigating HAEs, information on their interactions with plant receptors proteins (PRRs) is still emerging. The list of some known plant receptors and their respective HAEs is shown in Table 2.1. PRRs can be classified into two categories: receptor-like kinases (RLKs) and proteins that have both ecto- and cytosolic domains located in the plasma membrane, where they can bind to specific ligands. Plasma membrane–localized leucine-rich repeat receptor-like kinases (LRR-RLKs), which are a type of RLKs, are involved in the recognition of insect herbivory when OS elicitors bind with the plasma membrane in plants. For instance, active volicitin, such as [3H]-L-volicitin, binds with the plasma membrane in maize, and its binding increases when foliar application of methyl jasmonate (MeJA) is applied (Poretsky *et al.*, 2020; Prince *et al.*, 2014).

When *Spodoptera exigua* feeds on a plant, it can enhance the plasma membrane's ability to bind with volicitin. This indicates that the binding process involves receptors that are specific to fatty acid–amino acid conjugates (FACs) and are responsible for detecting MeJA and insect herbivory. As a result, genes that encode protein binding with the plasma membrane are transcribed. Nonetheless, the receptors and signaling pathways involved in plant defense mechanisms activated by JA and

**TABLE 2.1**
**Herbivore-Associated Elicitors and Their Known Receptors against Insect Herbivory.**

| Elicitors | Receptors | Source of Elicitors | Host Plant | References |
|---|---|---|---|---|
| DNA | n.d. | These elicitors are of plant source | Bean, maize | (Barbero *et al.*, 2016) |
| Pep | *Pep receptor (PEPR)* | | Maize | (Huffaker *et al.*, 2013; Yamaguchi *et al.*, 2010) |
| ATP | ATP receptors (*DORN1/P2K1*) | | *Arabidopsis* | (Choi *et al.*, 2014) |
| Systemin | *Systemin receptor (SYR1)* | | Tomato | (Wang *et al.*, 2018) |
| FACs (volicitin) | Unknown membrane proteins | *Spodoptera exigua* | Maize | (Alborn *et al.*, 1997) |
| β-Glucosidase | n.d. | *Pieris brassicae* | Maize | (Hopke *et al.*, 1994) |
| Caeliferins | n.d. | *Schistocerca americana* | Maize | (Alborn *et al.*, 2007) |
| Inceptin | *Inceptin receptor (INR)* | *Spodoptera frugiperda* | Maize | (Schmelz *et al.*, 2006) |
| Lipase | n.d. | *Schistocerca gregaria* | Arabidopsis | (Schafer *et al.*, 2011) |
| Porin-like proteins | n.d. | *Spodoptera littoralis* | Arabidopsis | (Guo *et al.*, 2013) |
| β-Galactofuranose polysaccharide | *HAK/PBL27* | *Spodoptera* spp. | Arabidopsis | (Bricchi *et al.*, 2013) |
| Bruchins | n.d. | *Bruchus pisorum*, *Nilaparvata lugens* | Cowpea, pea | (Doss *et al.*, 2000) |
| Glucose oxidase | n.d | *Helicoverpa zea*, *Spodoptera exigua*, *Helicoverpa armigera* | Nicotiana | (Diezel *et al.*, 2009; Musser *et al.*, 2002a) |
| Mucin-like protein | n.d. | *Callosobruchus maculatus* | Rice | (Shangguan *et al.*, 2018) |
| Oligouronides | n.d. | Produced by breakdown of plant cell walls by insect feeding | Tomato | (Doares *et al.*, 1995) |

*Note:* n.d. = not detected.

MeJA are still not fully comprehended (Truitt *et al.*, 2004). Further investigations are required to gain a better understanding of the perception of MeJA and volicitin binding with the plasma membrane. In rice, several pathogen-associated molecular pattern recognition receptors have been linked to herbivore resistance. For instance, Bph3, a cluster of three genes in rice that encodes G-type lectin receptor kinases (OsLecRK1-OsLecRK3), has been shown to increase brown planthopper resistance in susceptible cultivars (Liu *et al.*, 2015). Similarly, leucine-rich repeat receptor kinase (LRR-RK) has been linked to *Chilo suppressalis* resistance in rice by inducing plant responses to the elicitor OS of *S. frugiperda* (Hu *et al.*, 2018). The native tobacco plant of western North America, *Nicotiana attenuata*, has developed defense mechanisms to protect itself from insect herbivory. Lectin receptor kinase (LecRK1) in tobacco plays a crucial role in perceiving the elicitors in larval OSs. Silencing of the LecRK1 gene in tobacco plants resulted in increased susceptibility to *Manduca sexta*, as larvae grew better and gained higher weight on ir-LecRK1 compared to wild-type plants, indicating that LecRK1 is essential in regulating plant responses to *M. sexta* (Gilardoni *et al.*, 2011).

When eggs are laid on plants, they cause damage to the surrounding tissue, leading to the formation of a necrotic zone at the egg-laying site. This, in turn, activates leucine-rich repeat receptor-like kinases (LRR-RLKs), which produce a signal that alerts the plant to initiate defense responses. Recent research indicates that the Lecrk-I.8 gene plays a crucial role in this process. Mutant plants with a T-DNA insertion in the coding sequence of this gene have been shown to respond to egg extract by inducing plant responses. Additionally, treating plants with egg extracts leads to a partial increase in PR-1 expression. Notably, Col-0 plants exhibit a significantly higher level of PR-1 expression, indicating that Lecrk-I.8 is the primary receptor responsible for detecting egg-derived elicitors in *Arabidopsis* (Gouhier-Darimont *et al.*, 2013). An intriguing experiment to conduct would be to explore the capability of Lecrk-I.8 to identify other elicitors. The function of a nucleotide-binding leucine-rich repeat protein (NB-LRR) known as Mi-1 is to provide protection against caterpillar larvae, including the potato aphid, *Macrosiphum euphorbiae*. This function relies on the presence of the plant receptor-like kinase somatic embryogenesis receptor kinase 1 (SERK1), which is a unique member of the SERK family. When the expression of SlSERK1 was suppressed in mutant tomato plants, potato aphids exhibited a higher survival rate compared to wild-type plants.

This suggested the significance of SlSERK1 in detecting potato aphid attack, as the aphids thrived while feeding on the mutant plants (Mantelin *et al.*, 2011). With mechanical wounding or under insect attack, it releases endogenous peptides into the apoplast. These peptides are known as secondary danger signals. One such peptide is systemin, which is 18 amino acids long and is produced when prosystemin cleaves it. Systemin spreads throughout the plant and activates proteinase inhibitors that can impede the growth of caterpillars (Wang *et al.*, 2018). The signaling events associated with plant resistance initially observed in tomato plants, such as those triggered by systemin, are now known to be present across various plant species. In *Arabidopsis*, the detection of Pep peptide elicitors involves two leucine-rich repeat receptors, AtPEPR1 and AtPEPR2, that are closely related to each other (Gust *et al.*, 2017; Krol *et al.*, 2010; Yamaguchi *et al.*, 2010). In tomatoes, SYR1, a high-affinity receptor belonging to the LRR-RLK family, binds to systemin. Researchers examined the resistance of tomatoes to *S. littoralis* by utilizing introgression lines that did not contain the SYR1 transgene. The findings revealed that *S. littoralis* larvae exhibited greater growth and weight gain on introgression line IL3-3, which lacked SYR1, compared to wild-type tomato plants. Interestingly, the expression of the PIN1 proteinase inhibitor gene remained unaffected in both transgenic lines' local and systemic tissues, indicating that SYR1 may trigger alternative defense mechanisms against *S. littoralis* larvae. This revealed the significance of comprehending the intricate interactions between plants and pests in the development of effective pest management strategies (Wang *et al.*, 2018). Treatment with exogenous ATP in *Arabidopsis* exhibited a greater defense response, including higher levels of cytosolic $Ca^{+}$. The DORN1 mutant, which is insensitive to ATP, has a defect in Lecrk-1.9, a lectin receptor kinase that binds with ATP and is critical for ATP-induced responses. Increasing the expression of DORN1 in plants enhanced their reaction to injuries, suggesting that DORN1 plays a role in recognizing extracellular ATP (Choi *et al.*, 2014).

The Schmelz research team recently employed a forward genetic mapping technique to prove that leucine-rich receptors in cowpea and common bean can recognize and stimulate defense responses to proteolytic fragments of chloroplastic ATP synthase, known as inceptin, in herbivores. They found that the leucine-rich repeat receptor controls the inceptin receptor (INR), which plays a crucial role in regulating plant defenses and providing resistance against *S. exigua*. These findings were demonstrated through heterologous expression in tobacco, indicating the potential for broad application in various plant species (Snoeck *et al.*, 2022; Steinbrenner *et al.*, 2020). In contrast to pathogens, the receptors that plants use to sense insect herbivory have not been extensively studied and require further investigation.

## 2.3 RESPONSES TO HERBIVORY, OVIPOSITION, AND FRASS DEPOSITION

Plants employ herbivorous feeding signatures to differentiate between actual insect feeding and mechanical wounding, which is in most cases done by recognition of components of insect elicitors in OSs (Schmelz *et al.*, 2009). For instance, by applying representative elicitors to the wounds generated by mechanical device in plant leaves, it was observed that plants respond differently to insect-derived elicitors. FACs are the primary constituent in the OSs of *M. sexta*. Application of FACs in the wounds exhibited faster metabolism of *N*-linolenoyle-glutamic acid (18:3-Gln) in *N. attenuata* leaves. In the same way, when OSs from *M. sexta* were applied in *Nicotiana attenuata* leaves, metabolism of *N*-linolenoyle-glutamic acid (18:3-Gln) was reduced and transformed into a modified form that corresponds to 13-hydroxy-18:3-Glu, 13-hydroperoxy-18:3-Glu, and 13-oxo-13:2-Glu. *N. attenuata* mutant JA deficient plants (*lipoxygenase 2* and *lipoxygenase 3*) showed a significant suppression in the modified form of 18:3-Gln, indicating that modified forms of FACs are responsible for triggering plant responses in *N. attenuata* (VanDoorn *et al.*, 2010).

Egg laying on the plant surface is also a significant threat to plant survival. In response to egg deposition, plants have developed countermechanisms that inflict necrosis in the egg-laying area. Necrosis in the egg-laying region is associated with a high mortality of neonates and reduction in the larval hatching rate (Hilker and Fatouros, 2015). The difference was observed between egg deposition by sucking and chewing herbivores. The gene expression of hundreds of transcripts was increased after three days of egg deposition by *Pieris brassicae*. The transcript accumulation was remarkably different in *Arabidopsis* leaves by egg deposition done by chewing caterpillars. However, egg deposition shared similarities with a biotrophic pathogen, both of which caused stress that increased defense-related transcription changes in the host plant (Little *et al.*, 2007). Hence, egg deposition shared not only transcript accumulation with the pathogen but also the impact on phytochemicals. Egg laying increased SA levels in the region of egg deposition, and a similarly high level of phytohormones was reported when pathogen-associated molecular patterns (PAMPs) were applied on *Arabidopsis* leaves. The egg extract from *P. brassicae* showed increased accumulation of PAMP-responsive transcripts indicating common evolution of insect elicitors and PAMP in *Arabidopsis* (Gouhier-Darimont *et al.*, 2013).

Frass deposition by larvae in plant whorls can suppress plant defenses. Insect herbivores deposit a significant amount of frass in the maize whorls during infestation. The extract of fall armyworm's frass was applied into the wounds of maize leaves generated by mechanical device. Afterward, treatment increased the performance of fall army worm (FAW) on maize leaves compare with nontreated plants. Moreover, treatment with frass extract suppressed the transcript accumulation of defense-related genes and JA levels in maize leaves indicating that frass contains proteins that are involved in the modulation of plant defenses (Ray *et al.*, 2015). Later, it was discovered that FAW feeding increased levels of chitinases Pr4 and endochitinase A in maize leaves. After FAW feeding, the chitinase proteins deposit in frass and cause a reduction in proteinase inhibitors in maize tissues which is toxic to FAW; thereby, FAW performs much better on plants treated with frass (Ray *et al.*, 2016a, 2016b). FAW specifically influences plant responses; for example, plants emit a blend of volatile organic compounds (VOCs) upon insect feeding, and FAW becomes able to suppress VOCs in maize in the same way as suppression of direct defenses. It was observed that FAW-induced VOCs were much weaker than the VOCs emitted by *Helicoverpa armigera*, *S. littoralis*, and *S. exigua* in maize. However, FAW feeding could not suppress an indirect defense in cotton (De Lange *et al.*, 2020).

## 2.4 PLANT AND INSECT ORIGIN ELICITORS

Plants encounter various kinds of biotic stresses such as pathogens and insect attacks. To defend against insect herbivory, plants have developed recognition systems and responses that rely on specific HAEs. HAEs play a role in signaling pathways and can trigger the plant's defense reaction system. These HAEs can be found in the components of OSs, as well as in bacteria, caterpillar frass, oviposition fluid, and some insect pheromones. The presence of these molecules can either disrupt or induce plant defenses. Some of these molecules include enzymes like glucose oxidase and ß-glycosidase, peptides such as inception, and fatty acid conjugates like volicitin (Bonaventure *et al.*, 2011; Kallure *et al.*, 2022; Wu and Baldwin, 2009). Over time, some plants have adapted to recognize these molecules from insects and have overcome the inhibition of their defense response. One of the first types of elicitors identified was FACs or fatty acid amides, found in the saliva of insects (Alborn *et al.*, 1997). It was exhibited that VOC emissions increased plant defense by *M. sexta* deposition of OS constituents' FACs in plants and increased predation on insects (Gaquerel *et al.*, 2009). The discoveries of different elicitors are ongoing since the FACs were discovered in 1997 by Alborn and his colleagues (Schmelz *et al.*, 2009). Two molecular elicitors known as inceptions and caeliferins are found in OSs which are involved in activation of defensive pathways in insects (Rodriguez and Bos, 2013). Furthermore, previous studies have reported the induction of defense signaling in response to the presence of glucose oxidase (GOX) in insect saliva. For example, the salivary component of *Ostrinia nubilalis* produces proteinase inhibitor 2 (PIN2), which induces a response in maize and tomato (Louis and Shah, 2013; Tian *et al.*, 2012). However, OSs also contain proteins that inhibit the plant's defense pathway. For example, OSs from *S. littoralis* and *P. brassicae* at different larval stages inhibited defensive pathways and promoted larval growth (Consales *et al.*, 2011).

## 2.5 ELICITORS OF PLANT'S INTRACELLULAR PRODUCTS

The most basic type of HAEs involves intracellular products that are released as a result of leaf tissue damage caused by insect feeding (Jeter *et al.*, 2004). When the intracellular liquid reaches the apoplast, it is recognized by DORN1/P2K1 in neighboring undamaged cells, which triggers ATP-induced $Ca^{+}$ defenses, as depicted in Figure 2.2 and Table 2.1 (Choi *et al.*, 2014). In tomato plants, *H. zea* OS assay revealed the degradation activity of adenosine-5-triphosphate (ATP) in tomato leaf fluid. However, when *H. zea's* salivary glands secrete apyrase and ATP-hydrolyzing enzymes, the ATP signaling is disrupted, and the defense-related genes in tomatoes are suppressed (Wu *et al.*, 2012). Insect feeding on plants causes regurgitation on leaves, providing signaling cues that are recognized by plants. *Arabidopsis* receptors recognize lignocellulose deposited in the herbivore gut during feeding and digested products that function as elicitors. However, it is unclear whether these compounds are gut-derived or plant cell wall degradation products (Arimura, 2021; Vadassery *et al.*, 2012). In maize and lima bean, insect feeding results in the production of extracellular self-DNA (esDNA), which triggers plant responses upon exposure. Plants exposed to esDNA and extracellular heterologous DNA exhibited an increase in plasma membrane potential (Vm) and calcium flux ($Ca^{+}$), which confirmed that esDNA triggers plant responses (Barbero *et al.*, 2016). An interesting research question to answer is whether the perception of esDNA requires specific receptors different from those involved in insect herbivory.

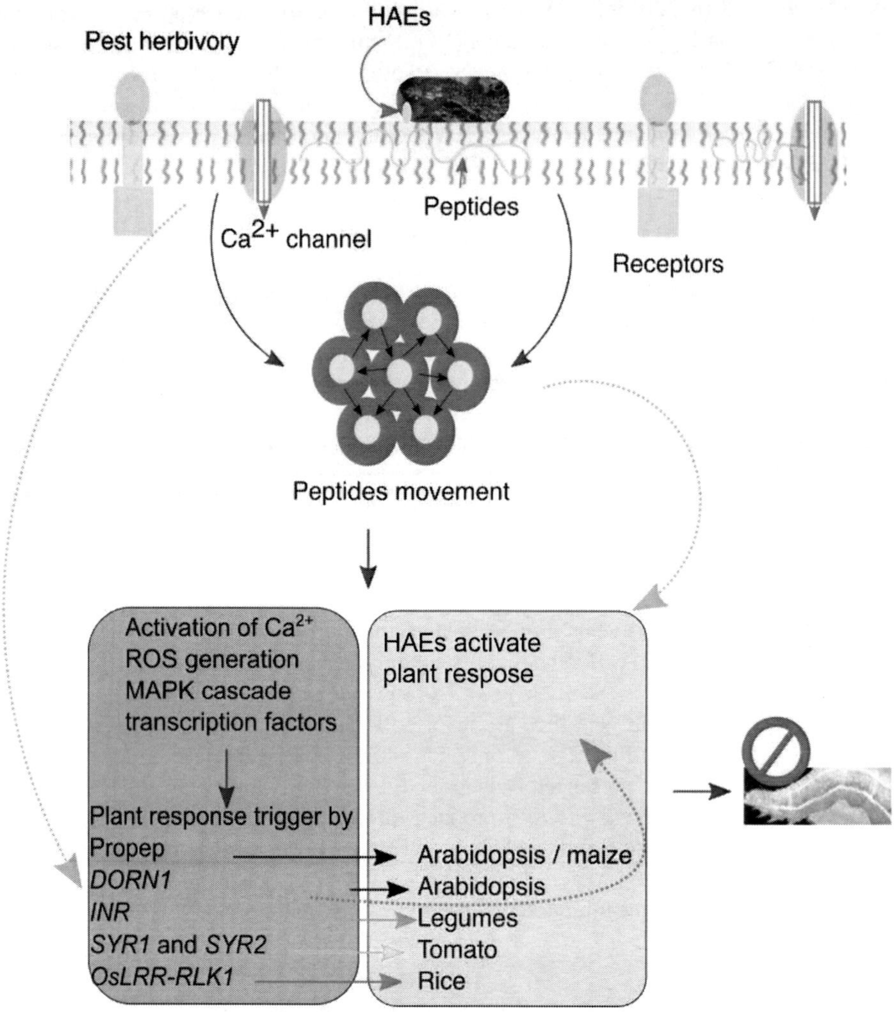


**FIGURE 2.2** How plant pattern recognition receptors recognize herbivore-associated elicitors (HAEs) on insect feeding, leading to plant responses against herbivory. Intracellular molecules released from damaged cells move into the apoplast and to undamaged neighboring cells, inducing plant responses. Insect feeding also induces endogenous peptides that bind with HAEs to elicit downstream plant defenses.

## 2.6 PEPTIDE ELICITORS

Endogenous peptides are small molecules that are secreted by plant cells and found exclusively in the Solanaceae family. They function by inducing plant defenses against herbivores, as depicted in Figure 2.2. In tomatoes, for instance, prosystemin is transported to the apoplast and processed, leading to the activation of plant defenses through its interaction with membrane-localized leucine-rich receptors (LRR) known as PEPRs (Maffei *et al.*, 2007). Similarly, in Arabidopsis, the endogenous peptide signals of AtPepl are amplified by the PEP receptor (PEPR1 and PEPR2), leading to enhanced defense responses (Yamaguchi *et al.*, 2006). Systemin, which is a larger precursor protein of about 200 amino acids and a plant peptide hormone of 18 amino acids derived from prosystemin, activates JA-dependent defenses and negatively affects the growth of chewing herbivores (Ryan, 2000). Furthermore, Spodoptera litura-induced a higher level of transcript accumulation of PEPR1, PEPR2, and PROSPEP3 in *Arabidopsis*, suggesting that these receptors play a crucial role in the perception of insect herbivory (Klauser *et al.*, 2015). Genetic evidence has also shown that larvae grow much bigger on pepr1 and pepr2 mutant plants than on wild-type plants.

In addition to direct defenses, systemic signals trigger the emission of green leaf volatiles (GLVs) to attract natural enemies of caterpillars as part of the role in tritrophic interactions. They also increase the accumulation of ROS and phytohormones, further enhancing plant defense (Degenhardt *et al.*, 2010; Degenhardt, 2009). Pep elicitors, which are similar to systemin, are recognized in *Arabidopsis* (AtPep1–8) and maize (ZmPep1,3). In maize, for example, ZmPep3 induced increased levels of JA, ethylene, proteinase inhibitors, volatiles, benzoxazinoids, and defense-related genes, similar to those elicited by *S. exigua* (Huffaker *et al.*, 2013). Similarly, in rice, brown planthopper infestation led to a significant increase in transcript levels of both OsPep receptors and Pep precursors. Knockout mutant plants that were impaired in the function of OsPEPRs showed increased susceptibility to brown planthopper feeding, while the exogenous application of OsPep3 improved resistance levels in rice seedlings against brown planthopper infestation as well as against the fungal pathogen *Magnaporthe oryzae* and the bacterial pathogen *Xanthamonas oryzae* pv. *oryzae* (Shen *et al.*, 2022). These findings demonstrate the essential role of OsPEPR signaling in plants' defense against insect herbivores.

## 2.7 ELICITORS IN OSS OF INSECTS

Insect herbivore elicitors are the entities involved in triggering plant response against herbivores. Among the insect elicitors, FACs are a well-studied group of elicitors (Alborn *et al.*, 1997; Bonaventure *et al.*, 2011; Schmelz *et al.*, 2009). FACs are the plant and insect derivative chemical compounds present in lepidopteran's OSs and have been reported to elevate the plant response in many plant species (Alborn *et al.*, 1997; Bonaventure *et al.*, 2011; Schmelz *et al.*, 2009). Volicitin (a hydroxyl FAC [*N*-(17-hydroxylinolenoyl)-L-glutamine]) derived from *S. exigua* is one of the best studied FACs among the reported elicitors of the same class (Alborn *et al.*, 1997). Applying volicitin into the wounds of maize leaves elevated emission of VOCs that inflicted parasitic wasps (natural enemies of *S. exigua*) (Alborn *et al.*, 1997). Literature has shown that many lepidopteran species–derived FACs have been reported, and their biological functional studies have been conducted in *N. attenuata*. Insect feeding by *M. sexta* larvae increased the level of JA, MAPKs, as well as accumulation of ethylene biosynthesis transcripts in systemic leaves. Some of the published studies also reported metabolite changes and reprograming in transcriptome in infested leaves (Giri *et al.*, 2006; Halitschke, 2001; Meldau *et al.*, 2009; Pohnert *et al.*, 1999; Skibbe *et al.*, 2008; Wu *et al.*, 2007). In addition, another study has shown that applying FACs into the *N. attenuata* leaves exhibited increased accumulation of WRKY6. However, mechanical wounding only elicited the transcript level of WRKY3 that is required to induce the WRKY6 level (Skibbe *et al.*, 2008).

GLVs also play a role in plant defense against herbivores. In a study by Meldau *et al.* (2009), it was shown that decreased emission of GLVs in transgenic plants increases the performance of *M. sexta*. The addition of synthetic GLVs in transgenic plants restored the performance phenotype of *M. sexta* in transgenic plants (Meldau *et al.*, 2009). *Medicago trancatula* (alfalfa) produces saponins and terpenoids in response to larval herbivory. However, the transcript accumulation of genes involved in saponin biosynthesis and terpenoid production is suppressed by *S. exigua* (beet armyworm) during feeding. To explain this, researchers proposed that GOX might play a role in inhibiting gene expression after insect herbivory. To validate this hypothesis, additional experiments were conducted. By comparing the effects of wounding, the addition of GOX into wounds, and insect feeding, the researchers confirmed that GOX indeed functions in suppressing defense responses during herbivore attacks (Bede *et al.*, 2006).

The analysis of the OSs in *S. exigua* revealed that FACs consist of plant-derived fatty acids, specifically linoleic acid (LA) and linolenic acid, and insect-derived amino acids, such as glutamine and glutamic acid (Pare *et al.*, 1998). The midgut of insects undergoes hydroxylation and conjugation, which is crucial for the biological activity of FACs that release plant volatiles to attract natural enemies of herbivores. Radiolabeled feeding experiments with *S. litura* caterpillars suggested that FACs play a vital role in nitrogen assimilation and function as glutamine storage in the insect, where glutamine is the primary component in insect nitrogen metabolism (Yoshinaga *et al.*, 2008). The production of FACs in caterpillars is not controllable, even when feeding on plants. Although plants recognize caterpillar feeding in the presence of FACs, they continue to produce them. Apart from FACs, other elicitors are present in insect OSs. Inceptin, a proteolytic fragment of the chloroplastic ATP synthase γ-subunit, is one such elicitor isolated in the OSs of *S. frugiperda* that detects insect herbivory and triggers the production of ethylene, phenylpropanoid accumulation, VOCs, and protease inhibitor in *Vigna unguiculata* (cowpea) (Schmelz *et al.*, 2006).

## 2.8 ELICITORS IN OSs OF NON-LEPIDOPTERAN INSECTS

A new class of elicitors, known as "caeliferins," has been discovered in the OSs of *Schistocerca americana*, a grasshopper species. Unlike previously known elicitors found in lepidopteran species, FACs have been found in non-lepidopteran species such as crickets and fruit flies, with a similar composition of *N*-linolenoyl-L-glutamic acid and *N*-linoleoyl-L-glutamic acid (Yoshinaga *et al.*, 2007). When maize seedlings were treated with 5 μl OS of grasshopper, a blend of volatiles similar to that emitted by herbivores was produced, comparable to that produced by seedlings treated with 100 pmol of volicitin (Alborn *et al.*, 2007). Plants have evolved a sophisticated system to detect herbivorous feeding through insect molecular signatures derived from insect feeding, saliva, OSs, eggs, volatiles, and microbes, as part of a coevolutionary process (Erb and Reymond, 2019; Hilker and Fatouros, 2015; Hilker and Meiners, 2010). In rice, a mucin-like protein secreted by the brown planthopper (*Nilaparvata lugens*), called NlMLP, alters the transcriptome of defense-related genes and increases the level of callose deposition in rice tissues during feeding (Shangguan *et al.*, 2018). The egg deposition by pea weevil (*Bruchus pisorum*) induced tumor growth or neoplasms in the egg-laying area. This egg deposition hinders the entry of insects in the pods in *Pisum sativum* (Doss *et al.*, 2000).

## 2.9 PLANT-INDUCED RESPONSES

Plants respond by releasing a cascade of mobile signaling signatures and producing defense-related phytochemicals when they are attacked by insect herbivores (Figure 2.2) (Erb and Reymond, 2019). Membrane depolarization and calcium influx occur in response to feeding by *S. littoralis* within a few minutes in *Phaseolus lunatus* (lima bean). The membrane depolarization is highest at the site of feeding and decreases as the distance from the wound site increases (Maffei *et al.*, 2004). In addition, research has shown that herbivore infestation and mechanical wounding can induce membrane depolarization in systemic tissues from damaged tissues, which is dependent on glutamate receptor-like channels (Maffei *et al.*, 2004). Electric signals and ROS signaling play important roles in the plant response to chewing insects. For example, wound-induced signaling systems require increased production of ROS by membrane-bound RBOHD protein (Toyota *et al.*, 2018). Aphid feeding induces a strong accumulation of ROS 3 hours after feeding on a resistant near-isogenic line but not in a susceptible line (Zhang *et al.*, 2020). Moreover, insect feeding can activate the MAPK signaling cascade, which is a highly conserved signaling mechanism among all eukaryotes (Wu *et al.*, 2007). In rice, OsMPK3 has been found to positively regulate the defense response against striped rice stemborer (*C. suppressalis*) by modulating JA biosynthesis (Wang and Wu, 2013).

The accumulation of phytohormones like JA and Ile conjugate is initiated by the signaling events. These phytohormones then bind to a receptor complex, which includes JASMONATE-ZIM DOMAIN (JAZ) and CORONATINE-INSENSITIVE1 (COI1). The interaction between the receptor complex and phytohormones leads to the degradation of JAZ and, consequently, the release of transcription factors that were inhibited by JAZ. Subsequently, downstream defense genes against insect herbivores are activated (Erb and Reymond, 2019). As a result, effective early signaling mechanisms are critical for coordinating the spatial, temporal, and quantitative aspects of plant defense responses. These mechanisms trigger downstream defense responses, such as the production of secondary metabolites and toxic proteins, which act to deter herbivores. While this process is well understood, further research is necessary to identify and characterize the receptors responsible for detecting chemical compounds that initiate these defense pathways in plants.

## 2.10 PLANT DEFENSE AGAINST GALL-INDUCING INSECTS

Galls refer to abnormal developments of plant organs caused by insects and other organisms. The diverse and distinct morphologies of insect-induced galls have garnered the attention of researchers (Murakami *et al.*, 2021; Price *et al.*, 1987). In fact, gall-inducing insects are estimated to range from 21,000 to 211,000, and they induce galls on different parts of the plant exhibiting different shapes (Espírito-Santo, 2009; Stone and Schönrogge, 2003). These insect galls can be induced on different parts of the plant, including leaves, stems, floral buds, flowers, fruits, or roots, and they exhibit unique shapes (Takeda *et al.*, 2021; Zagorchev *et al.*, 2021). With the advent of high-throughput technology, researchers have identified effector proteins from ovaries, venom glands in wasps, and salivary gland of grape phylloxera. However, no direct evidence supports gall formation in the host plant (Cambier *et al.*, 2019).

Host plants produce tannins to protect themselves against galls and gall-inducing insects (Chen *et al.*, 2018). In *Rosa chinensis*, aphid galls accumulate gallotannin, and genes related to gallotannin (Wang *et al.*, 2017) and gallic acid synthesis, as well as lignin biosynthesis, have been identified (Hirano *et al.*, 2020). On the other hand, gall development by Cheshunt gall wasp altered the transcriptome and increased the transcript level of genes involved in metabolic pathways in infested leaves (Zhu *et al.*, 2019). Similarly, the expression of genes encoding lignocellulose synthase was observed in galls induced by a gall-inducing aphid, *Tetraneura akinire*, on elm leaves, suggesting cell wall reinforcement to improve resistance to aphid damage (Lu *et al.*, 2019).

## 2.11 REGULATION OF PLANT RESPONSES AT PRIMED STAGE

Plant defense responses can be enhanced through priming, a phenomenon that has been well-documented in plant-pathogen interactions (Galis *et al.*, 2009). Over the past decade, studies have shown that priming can be triggered by various biotic factors including insect herbivory, egg deposition, herbivore-induced plant volatiles (HIPVs), β-aminobutyric acid, systemin, and cytokinin (Engelberth *et al.*, 2004; Frost *et al.*, 2008). For instance, when neighboring plants perceive the indole signal, they are primed to release HIPVs and activate early defense signaling genes (Erb *et al.*, 2015; Ye *et al.*, 2019). Moreover, HIPVs released upon infestation by *Mythimna separata* prime maize resistance to insect herbivory, as demonstrated by an increase in the relative transcript levels of Bowman-Birk–type trypsin inhibitor (TI) for five days after infestation, as well as in a set of methylation sites that were found demethylated in the promoter region of TI (Ali *et al.*, 2013). Similarly, *M. separata* herbivory on maize leads to the elevated accumulation of benzoxazinoids, JA/JA-Ile, and defense-related transcripts to *M. separata* feeding in systemic leaves (Malook *et al.*, 2021). According to Bandoly *et al.* (2015), when plants are subjected to oviposition, their defense mechanisms are triggered, leading to increased resistance against insect herbivores. For example, *N. attenuata* plants that have experienced oviposition by the generalist insect *S. exigua* exhibit reduced mortality, retarded development, and less feeding damage compared to those that have not undergone oviposition. Oviposited plants also show higher levels of caffeoylputrescine (CP) and trypsin protease inhibitors. In addition, a cross-resistance experiment indicated that *S. exigua* larvae exhibited decreased performance on *N. attenuata* plants that had been oviposited by *M. sexta*, similar to the reduced performance on *S. exigua*–oviposited plants (Bandoly *et al.*, 2015). However, the mechanism of priming and the role of epigenetic regulation are still in the early stages of study, and further research is needed to determine how epigenetic modifications affect priming responses in plants (Zhu *et al.*, 2016). A theoretical framework outlining how lepidopteran insects may trigger priming effects in maize has been put forward in Figure 2.3. However, there is still a need to conduct further research into the process of priming induced by insect herbivory, both locally and systemically in undamaged leaves, as it remains unexplored.

When caterpillars feed on plant leaves, they secrete OSs that contain elicitors. These elicitors are specifically perceived by plant pattern recognition receptors on the surface of the plasma membrane, which triggers the activation of the cellular signal amplifier. The cellular signal amplifier enhances the plant's defense response to herbivory by increasing the production of compounds such as benzoxazinoids and regulating gene expression and DNA methylation. This priming effect enables the plant to respond faster and more strongly to subsequent herbivory. The mechanisms of cellular signal amplifier activation and the downstream effects on plant defenses are an active area of research (Malook *et al.*, 2022).

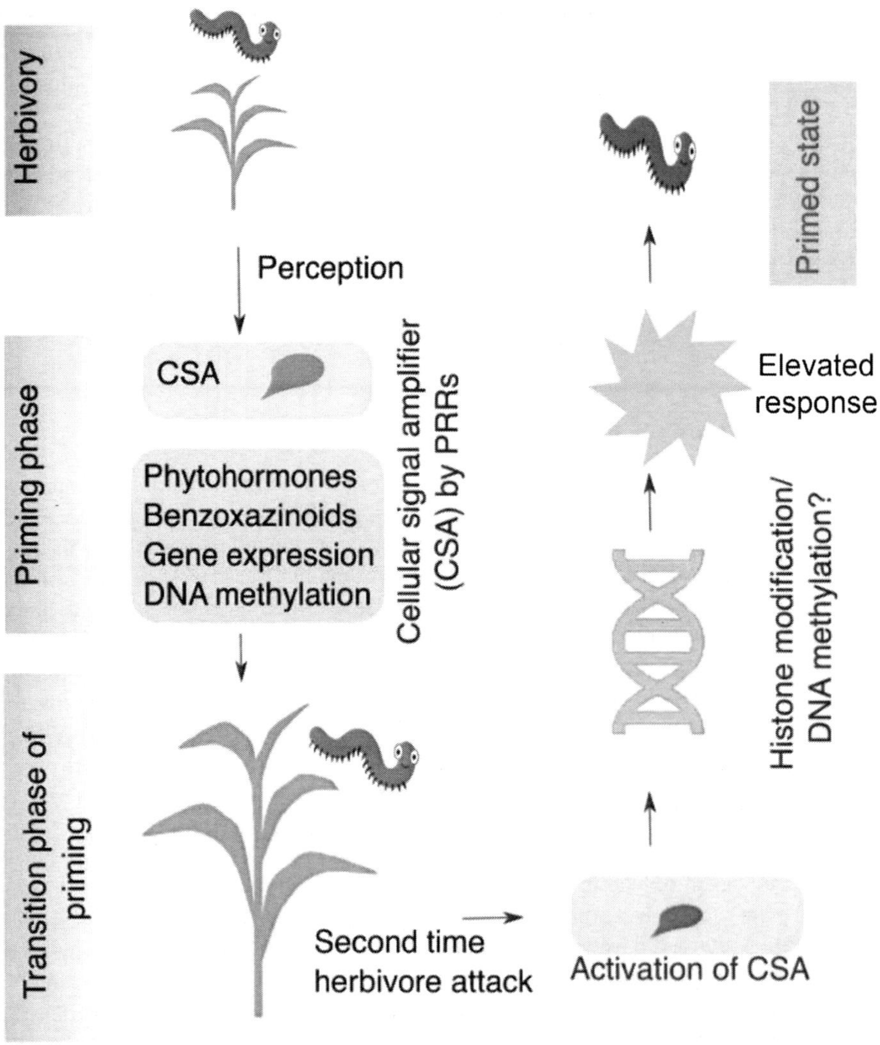


**FIGURE 2.3** The cellular signal amplifier magnifies priming defense in response to herbivory.

## 2.12 CONCLUSIONS

### 2.12.1 Concluding Remarks

To ensure effective plant protection, it is crucial to understand the mechanism of induced responses. While recent research has focused on how plants activate downstream-induced defenses by perceiving HAEs, further research is necessary to explore the plant surface receptors that detect insect attack. The plant surface receptors detect danger signatures from both endogenous and exogenous sources to trigger induced defenses. When plants detect herbivory, they can employ sophisticated defense strategies that are more effective against caterpillars if the initial perception is stronger. Plants can distinguish between mechanical damage and specific types of insect herbivory, indicating their ability to perceive the chemical cues present in attacking herbivores' OSs and specific host plants. Plant responses to insect herbivory depend largely on the nature of the elicitors perceived and are highly specific to HAEs. Repeated herbivore attacks lead to stronger and faster plant responses, and the extent to which plant receptors induce long-term responses remains to be determined.

### 2.12.2 Outstanding Questions

- The mechanisms of molecular signaling and biosynthesis for various HAEs like FACs, caeliferins, egg deposition, and frass that trigger plant defense responses have not received extensive research. Functional characterization of HAEs in plants and insects through techniques such as genome editing, and comparative transcriptomics could aid in investigating their molecular signal transduction mechanisms and biosynthesis pathways.
- Indirect defenses have a significant role in repelling herbivores, and identifying the receptors responsible for perceiving insect herbivory is crucial to determine whether they could emit volatiles to attract natural enemies for herbivore defense. Genetically and functionally characterizing these receptors is essential.
- Insect infestations can trigger short- and long-term plant responses, but defense mechanisms come at a cost. Investigating a trade-off mechanism hypothesis through studies can shed light on how plants sustain long-term defenses.
- The molecular mechanisms involved in defense priming induction are not yet fully understood. Identifying HAEs that elicit plant responses and prime them for enhanced resistance requires exploring the specificity of plant responses to insect herbivory.

## REFERENCES

Acevedo FE, Rivera-Vega LJ, Chung SH, Ray S, Felton GW. 2015. Cues from chewing insects—the intersection of DAMPs, HAMPs, MAMPs and effectors. Current Opinion in Plant Biology **26**, 80–86.

Alborn HT, Hansen TV, Jones TH, Bennett DC, Tumlinson JH, Schmelz EA, Teal PEA. 2007. Disulfooxy fatty acids from the American bird grasshopper *Schistocerca americana*, elicitors of plant volatiles. Proceedings of the National Academy of Sciences of the United States of America **104**, 12976–12981.

Alborn HT, Turlings TCJ, Jones TH, Stenhagen G, Loughrin JH, Tumlinson JH. 1997. An elicitor of plant volatiles from beet armyworm oral secretion. Science **276**, 945–949.

Ali M, Sugimoto K, Ramadan A, Arimura G. 2013. Memory of plant communications for priming anti-herbivore responses. Scientific Reports **3**, 1872–1876.

Arimura GI. 2021. Making sense of the way plants sense herbivores. Trends in Plant Science **26**, 288–298.

Bandoly M, Hilker M, Steppuhn A. 2015. Oviposition by *Spodoptera exigua* on *Nicotiana attenuata* primes induced plant defence against larval herbivory. Plant Journal **83**, 661–672.

Barbero F, Guglielmotto M, Capuzzo A, Maffei ME. 2016. Extracellular self-DNA (esDNA), but not heterologous plant or insect DNA (etDNA), induces plasma membrane depolarization and calcium signaling in Lima Bean (*Phaseolus lunatus*) and Maize (*Zea mays*). International Journal of Molecular Sciences **17**.

Bede JC, Musser RO, Felton GW, Korth KL. 2006. Caterpillar herbivory and salivary enzymes decrease transcript levels of *Medicago truncatula* genes encoding early enzymes in terpenoid biosynthesis. Plant Molecular Biology **60**, 519–531.

Bonaventure G, VanDoorn A, Baldwin IT. 2011. Herbivore-associated elicitors: FAC signaling and metabolism. Trends in Plant Science **16**, 294–299.
Bricchi I, Occhipinti A, Bertea CM, Zebelo SA, Brillada C, Verrillo F, De Castro C, Molinaro A, Faulkner C, Maule AJ, Maffei ME. 2013. Separation of early and late responses to herbivory in *Arabidopsis* by changing plasmodesmal function. Plant Journal **73**, 14–25.
Cambier S, Ginis O, Moreau SJM, Gayral P, Hearn J, Stone GN, Giron D, Huguet E, Drezen J-M. 2019. Gall wasp transcriptomes unravel potential effectors involved in molecular dialogues with oak and rose. Frontiers in Physiology **10**, 926–926.
Chen C-Y, Mao Y-B. 2020. Research advances in plant-insect molecular interaction. F1000Research **9**, F1000.
Chen H, Liu J, Cui K, Lu Q, Wang C, Wu H, Yang Z, Ding W, Shao S, Wang H, Ling X, King-Jones K, Chen X. 2018. Molecular mechanisms of tannin accumulation in *Rhus* galls and genes involved in plant-insect interactions. Scientific Reports **8**, 9841–9841.
Choi J, Tanaka K, Cao Y, Qi Y, Qiu J, Liang Y, Yeo S, Lee GS. 2014. Identification of a plant receptor for extracellular ATP. Science **343**, 290–294.
Consales F, Schweizer F, Erb M, Gouhier-Darimont C, Bodenhausen N, Bruessow F, Sobhy I, Reymond P. 2011. Insect oral secretions suppress wound-induced responses in *Arabidopsis*. Journal of Experimental Botany **63**, 727–737.
Degenhardt DC, Refi-Hind S, Stratmann JW, Lincoln DE. 2010. Systemin and jasmonic acid regulate constitutive and herbivore-induced systemic volatile emissions in tomato, *Solanum lycopersicum*. Phytochemistry **71**, 2024–2037.
Degenhardt J. 2009. Indirect defense responses to herbivory in grasses. Plant Physiology **149**, 96–102.
De Lange ES, Laplanche D, Guo HJ, Xu W, Vlimant M, Erb M, Ton J, Turlings TCJ. 2020. *Spodoptera frugiperda* caterpillars suppress herbivore-induced volatile emissions in maize. Journal of Chemical Ecology **46**, 344–360.
Diezel C, von Dahl CC, Gaquerel E, Baldwin IT. 2009. Different lepidopteran elicitors account for cross-talk in herbivory-induced phytohormone signaling. Plant Physiology **150**, 1576–1586.
Divekar PA, Narayana S, Divekar BA, Kumar R, Gadratagi BG, Ray A, Singh AK, Rani V, Singh V, Singh AK, Kumar A, Singh RP, Meena RS, Behera TK. 2022. Plant secondary metabolites as defense tools against herbivores for sustainable crop protection. International Journal of Molecular Sciences **23**, 2690.
Doares SH, Syrovets T, Weiler EW, Ryan CA. 1995. Oligogalacturonides and chitosan activate plant defensive genes through the octadecanoid pathway. Proceedings of the National Academy of Sciences of the United States of America **92**, 4095–4098.
Doss RP, Oliver JE, Proebsting WM, Potter SW, Kuy SR, Clement SL, Williamson RT, Carney JR, DeVilbiss ED. 2000. Bruchins: Insect-derived plant regulators that stimulate neoplasm formation. Proceedings of the National Academy of Sciences of the United States of America **97**, 6218–6223.
Engelberth J, Alborn HT, Schmelz EA, Tumlinson JH. 2004. Airborne signals prime plants against insect herbivore attack. Proceedings of the National Academy of Sciences of the United States of America **101**, 1781–1785.
Erb M, Kliebenstein DJ. 2020. Plant secondary metabolites as defenses, regulators, and primary metabolites: The blurred functional trichotomy. Plant Physiology **184**, 39–52.
Erb M, Reymond P. 2019. Molecular interactions between plants and insect herbivores. Annual Review Plant of Biology **70**, 4.1–4.31.
Erb M, Veyrat N, Robert CA, Xu H, Frey M, Ton J, Turlings TC. 2015. Indole is an essential herbivore-induced volatile priming signal in maize. Nature Communications **6**, 6273–6283.
Espírito-Santo M. 2009. How many species of gall-inducing insects are there on earth, and where are they? Annals of the Entomological Society of America **100**, 95–99.
Felton GW, Tumlinson JH. 2008. Plant-insect dialogs: Complex interactions at the plant-insect interface. Current Opinion in Plant Biology **11**, 457–463.
Frost CJ, Mescher MC, Carlson JE, De Moraes CM. 2008. Plant defense priming against herbivores: Getting ready for a different battle. Plant Physiology **146**, 818–824.
Galis I, Gaquerel E, Pandey SP, Baldwin IT. 2009. Molecular mechanisms underlying plant memory in JA-mediated defence responses. Plant Cell and Environment **32**, 617–627.
Gaquerel E, Weinhold A, Baldwin IT. 2009. Molecular interactions between the specialist herbivore *Manduca sexta* (Lepidoptera, Sphigidae) and its natural host *Nicotiana attenuata*. VIII. An unbiased GCxGC-ToFMS analysis of the plant's elicited volatile emissions. Plant Physiology **149**, 1408–1423.
Gilardoni PA, Hettenhausen C, Baldwin IT, Bonaventure G. 2011. *Nicotiana attenuata* LECTIN RECEPTOR KINASE1 suppresses the insect-mediated inhibition of induced defense responses during *Manduca sexta* herbivory. Plant Cell **23**, 3512–3532.

Giri AP, Wunsche H, Mitra S, Zavala JA, Muck A, Svatos A, Baldwin IT. 2006. Molecular interactions between the specialist herbivore *Manduca sexta* (Lepidoptera, Sphingidae) and its natural host *Nicotiana attenuata*. VII. Changes in the plant's proteome. Plant Physiology **142**, 1621–1641.

Gouhier-Darimont C, Schmiesing A, Bonnet C, Lassueur S, Reymond P. 2013. Signalling of *Arabidopsis thaliana* response to *Pieris brassicae* eggs shares similarities with PAMP-triggered immunity. Journal of Experimental Botany **64**, 665–674.

Guo HJ, Wielsch N, Hafke JB, Svatos A, Mithofer A, Boland W. 2013. A porin-like protein from oral secretions of *Spodoptera littoralis* larvae induces defense-related early events in plant leaves. Insect Biochemistry and Molecular Biology **43**, 849–858.

Gust AA, Pruitt R, Nurnberger T. 2017. Sensing danger: Key to activating plant immunity. Trends in Plant Science **22**, 779–791.

Halitschke R. 2001. Molecular interactions between the specialist herbivore *Manduca sexta* (Lepidoptera, Sphingidae) and its natural host *Nicotiana attenuata*. III. Fatty acid-amino acid conjugates in herbivore oral secretions are necessary and sufficient for herbivore-specific plant responses. Plant Physiology **125**, 711–717.

Hilker M, Fatouros NE. 2015. Plant responses to insect egg deposition. Annual Review of Entomology **60**, 493–515.

Hilker M, Meiners T. 2010. How do plants "notice" attack by herbivorous arthropods? Biological Reviews **85**, 267–280.

Hirano T, Kimura S, Sakamoto T, Okamoto A, Nakayama T, Matsuura T, Ikeda Y, Takeda S, Suzuki Y, Ohshima I, Sato MH. 2020. Reprogramming of the developmental program of *Rhus javanica* during initial stage of gall induction by *Schlechtendalia chinensis*. Frontiers in Plant Science **11**.

Hopke J, Donatha J, Blechertb S, Boland W. 1994. Herbivore-induced volatiles—the emission of acyclic homoterpenes from leaves of *Phaseolus lunatus* and *Zea mays* can be triggered by a beta-glucosidase and jasmonic acid. FEBS Letters **352**, 146–150.

Howe GA, Jander G. 2008. Plant immunity to insect herbivores. Annual Review of Plant Biology **59**, 41–66.

Hu LF, Ye M, Kuai P, Ye MF, Erb M, Lou YG. 2018. OsLRR-RLK1, an early responsive leucine-rich repeat receptor-like kinase, initiates rice defense responses against a chewing herbivore. New Phytologist **219**, 1097–1111.

Huffaker A, Pearce G, Veyrat N, Erb M, Turlings TCJ, Sartor R, Shen ZX, Briggs SP, Vaughan MM, Alborn HT, Teal PEA, Schmelz EA. 2013. Plant elicitor peptides are conserved signals regulating direct and indirect antiherbivore defense. Proceedings of the National Academy of Sciences of the United States of America **110**, 5707–5712.

Jeter CR, Tang WQ, Henaff E, Butterfield T, Roux SJ. 2004. Evidence of a novel cell signaling role for extracellular adenosine triphosphates and diphosphates in *Arabidopsis*. Plant Cell **16**, 2652–2664.

Jones JD, Dangl JL. 2006. The plant immune system. Nature **444**, 323–329.

Kallure GS, Kumari A, Shinde BA, Giri AP. 2022. Characterized constituents of insect herbivore oral secretions and their influence on the regulation of plant defenses. Phytochemistry **193**, 113008.

Klauser D, Desurmont GA, Glauser G, Vallat A, Flury P, Boller T, Turlings TCJ, Bartels S. 2015. The *Arabidopsis* Pep-PEPR system is induced by herbivore feeding and contributes to JA-mediated plant defence against herbivory. Journal of Experimental Botany **66**, 5327–5336.

Krol E, Mentzel T, Chinchilla D, Boller T, Felix G, Kemmerling B, Postel S, Arents M, Jeworutzki E, Al-Rasheid KA, Becker D, Hedrich R. 2010. Perception of the *Arabidopsis* danger signal peptide 1 involves the pattern recognition receptor AtPEPR1 and its close homologue AtPEPR2. Journal of Biological Chemistry **285**, 13471–13479.

Little D, Gouhier-Darimont C, Bruessow F, Reymond P. 2007. Oviposition by pierid butterflies triggers defense responses in *Arabidopsis*. Plant Physiology **143**, 784–800.

Liu YQ, Wu H, Chen H, Liu YL, He J, Kang HY, Sun ZG, Pan G, Wang Q, Hu JL, Zhou F, Zhou KN, Zheng XM, Ren YL, Chen LM, Wang YH, Zhao ZG, Lin QB, Wu FQ, Zhang X, Guo XP, Cheng XI, Jiang L, Wu CY, Wang HY, Wan JM. 2015. A gene cluster encoding lectin receptor kinases confers broad-spectrum and durable insect resistance in rice. Nature Biotechnology **33**, 301.

Louis J, Shah J. 2013. *Arabidopsis thaliana–Myzus persicae* interaction: Shaping the understanding of plant defense against phloem-feeding aphids. Frontiers in Plant Science **4**.

Lu HB, Jin LP, Wei D, Huang ZH. 2019. Study on the differential gene expression of elm leaves fed on by *Tetraneura akinire* Sasaki. Genes Genomics **41**, 1505–1516.

Maffei M, Bossi S, Spiteller D, Mithofer A, Boland W. 2004. Effects of feeding *Spodoptera littoralis* on lima bean leaves. I. Membrane potentials, intracellular calcium variations, oral secretions, and regurgitate components. Plant Physiology **134**, 1752–1762.

Maffei ME, Mithofer A, Boland W. 2007. Before gene expression: Early events in plant-insect interaction. Trends in Plant Science **12**, 310–316.
Malook SU, Maqbool S, Hafeez M, Karunarathna SC, Suwannarach N. 2022. Molecular and biochemical mechanisms of elicitors in pest resistance. Life **12**, 844.
Malook SU, Xu Y, Qi J, Li J, Wang L, Wu J. 2021. *Mythimna separata* herbivory primes maize resistance in systemic leaves. Journal of Experimental Botany **72**(10), 3792–3805.
Mantelin S, Peng HC, Li BB, Atamian HS, Takken FLW, Kaloshian I. 2011. The receptor-like kinase SlSERK1 is required for Mi-1-mediated resistance to potato aphids in tomato. Plant Journal **67**, 459–471.
Mattiacci L, Dicke M, Maartin AP. 1995. β-Glucosidase—an elicitor of herbivore-induced plant odor that attracts host-searching parasitic wasps. Proceedings of the National Academy of Sciences of the United States of America **92**, 2036–2040.
Meldau S, Wu J, Baldwin IT. 2009. Silencing two herbivory-activated MAP kinases, SIPK and WIPK, does not increase *Nicotiana attenuata*'s susceptibility to herbivores in the glasshouse and in nature. New Phytologist **181**, 161–173.
Murakami R, Ushima R, Sugimoto R, Tamaoki D, Karahara I, Hanba Y, Wakasugi T, Tsuchida T. 2021. A new galling insect model enhances photosynthetic activity in an obligate holoparasitic plant. Scientific Reports **11**, 13013.
Musser RO, Hum-Musser SM, Eichenseer H, Ervin G, Brad Murphy J, Felton GW. 2002a. Caterpillar saliva beats plant defences. Nature **416**, 599.
Pare PW, Alborn HT, Tumlinson JH. 1998. Concerted biosynthesis of an insect elicitor of plant volatiles. Proceedings of the National Academy of Sciences of the United States of America **95**, 13971–13975.
Pohnert G, Jung V, Haukioja E, Lempa K, Boland W. 1999. New fatty acid amides from regurgitant of lepidopteran (Noctuidae, Geometridae) caterpillars. Tetrahedron **55**, 11275–11280.
Poretsky E, Dressano K, Weckwerth P, Ruiz M, Char SN, Shi D, Abagyan R, Yang B, Huffaker A. 2020. Differential activities of maize plant elicitor peptides as mediators of immune signaling and herbivore resistance. Plant Journal **104**, 1582–1602.
Price PW, Fernandes GW, Waring GL. 1987. Adaptive nature of insect galls. Environmental Entomology **16**, 15–24.
Prince DC, Drurey C, Zipfel C, Hogenhout SA. 2014. The leucine-rich repeat receptor-like kinase BRASSINOSTEROID INSENSITIVE1-ASSOCIATED KINASE1 and the cytochrome P450 PHYTOALEXIN DEFICIENT3 contribute to innate immunity to aphids in Arabidopsis. Plant Physiology **164**, 2207–2219.
Qi J, Su M, Shen G, Gao L, Zhang C, Li J, Zhang J, Wang L, Wu J. 2018. Current understanding of maize and rice defense against insect herbivores. Plant Diversity **40**, 189–195.
Ray S, Alves PC, Ahmad I, Gaffoor I, Acevedo FE, Peiffer M, Jin S, Han Y, Shakeel S, Felton GW, Luthe DS. 2016a. Turnabout is fair play: Herbivory-induced plant Chitinases excreted in fall armyworm frass suppress herbivore defenses in maize. Plant Physiology **171**, 694–706.
Ray S, Basu S, Rivera-Vega LJ, Acevedo FE, Louis J, Felton GW, Luthe DS. 2016b. Lessons from the far end: Caterpillar FRASS-induced defenses in maize, rice, cabbage, and tomato. Journal of Chemical Ecology **42**, 1130–1141.
Ray S, Gaffor I, Acevedo FE, Helms A, Chuang WP, Tooker J, Felton GW, Luthe DS. 2015. Maize plants recognize herbivore-associated cues from caterpillar frass. Journal of Chemical Ecology **41**, 781–792.
Reymond P. 2021. Receptor kinases in plant responses to herbivory. Current Opinion in Biotechnology **70**, 143–150.
Rodriguez PA, Bos JI. 2013. Toward understanding the role of aphid effectors in plant infestation. Molecular Plant-Microbe Interactions **26**, 25–30.
Ryan CA. 2000. The systemin signaling pathway: Differential activation of plant defensive genes. Biochimica Et Biophysica Acta-Protein Structure and Molecular Enzymology **1477**, 112–121.
Schafer M, Fischer C, Meldau S, Seebald E, Oelmuller R, Baldwin IT. 2011. Lipase activity in insect oral secretions mediates defense responses in *Arabidopsis*. Plant Physiology **156**, 1520–1534.
Schmelz EA. 2015. Impacts of insect oral secretions on defoliation-induced plant defense. Current Opinion in Insect Science **9**, 7–15.
Schmelz EA, Carroll MJ, LeClere S, Phipps SM, Meredith J, Chourey PS, Alborn HT, Teal PE. 2006. Fragments of ATP synthase mediate plant perception of insect attack. Proceedings of the National Academy of Sciences of the United States of America **103**, 8894–8899.
Schmelz EA, Engelberth J, Alborn HT, Tumlinson JH, Teal PEA. 2009. Phytohormone-based activity mapping of insect herbivore-produced elicitors. Proceedings of the National Academy of Sciences of the United States of America **106**, 653–657.

Shangguan XX, Zhang J, Liu BF, Zhao Y, Wang HY, Wang ZZ, Guo JP, Rao WW, Jing SL, Guan W, Ma YH, Wu Y, Hu L, Chen RZ, Du B, Zhu LL, Yu DZ, He GC. 2018. A mucin-like protein of planthopper is required for feeding and induces immunity response in plants. Plant Physiology **176**, 552–565.

Shen W, Zhang X, Liu J, Tao K, Li C, Xiao S, Zhang W, Li J-F. 2022. Plant elicitor peptide signalling confers rice resistance to piercing-sucking insect herbivores and pathogens. Plant Biotechnology Journal **20**, 991–1005.

Skibbe M, Qu N, Galis I, Baldwin IT. 2008. Induced plant defenses in the natural environment: *Nicotiana attenuata* WRKY3 and WRKY6 coordinate responses to herbivory. Plant Cell **20**, 1984–2000.

Snoeck S, Guayazán-Palacios N, Steinbrenner AD. 2022. Molecular tug-of-war: Plant immune recognition of herbivory. Plant Cell **34**, 1497–1513.

Steinbrenner AD, Munoz-Amatriain M, Chaparro AF, Aguilar-Venegas JM, Lo S, Okuda S, Glauser G, Dongiovanni J, Shi D, Hall M, Crubaugh D, Holton N, Zipfel C, Abagyan R, Turlings TCJ, Close TJ, Huffaker A, Schmelz EA. 2020. A receptor-like protein mediates plant immune responses to herbivore-associated molecular patterns. Proceedings of the National Academy of Sciences of the United States of America **117**(49), 31510–31518.

Stone GN, Schönrogge K. 2003. The adaptive significance of insect gall morphology. Trends in Ecology & Evolution **18**, 512–522.

Takeda S, Hirano T, Ohshima I, Sato MH. 2021. Recent progress regarding the molecular aspects of insect gall formation. International Journal of Molecular Sciences **22**.

Tian D, Peiffer M, Shoemaker E, Tooker J, Haubruge E, Francis F, Luthe DS, Felton GW. 2012. Salivary glucose oxidase from caterpillars mediates the induction of rapid and delayed-induced defenses in the tomato plant. PLOS ONE **7**, e36168.

Toyota M, Spencer D, Sawai-Toyota S, Jiaqi W, Zhang T, Koo AJ, Howe GA, Gilroy S. 2018. Glutamate-triggers-long-distance-calcium-based-plant-defense-signaling. Science Advances **361**, 1112–1115.

Truitt CL, Wei HX, Pare PW. 2004. A plasma membrane protein from *Zea mays* binds with the herbivore elicitor volicitin. Plant Cell **16**, 523–532.

Vadassery J, Reichelt M, Mithofer A. 2012. Direct proof of ingested food regurgitation by *Spodoptera littoralis* caterpillars during feeding on *Arabidopsis*. Journal of Chemical Ecology **38**, 865–872.

VanDoorn A, Kallenbach M, Borquez AA, Baldwin IT, Bonaventure G. 2010. Rapid modification of the insect elicitor *N*-linolenoyl-glutamate via a lipoxygenase-mediated mechanism on *Nicotiana attenuata* leaves. BMC Plant Biology **10**, 164.

Wang H, Cui K, Shao S, Liu J, Chen H, Wang C, Wu H, Yang Z, Lu Q, King-Jones K, Chen X. 2017. Molecular response of gall induction by aphid *Schlechtendalia chinensis* (Bell) attack on *Rhus chinensis* Mill. Journal of Plant Interactions **12**, 465–479.

Wang L, Einig E, Almeida-Trapp M, Albert M, Fliegmann J, Mithofer A, Kalbacher H, Felix G. 2018. The systemin receptor SYR1 enhances resistance of tomato against herbivorous insects. Nature Plants **4**, 152–156.

Wang L, Wu J. 2013. The essential role of jasmonic acid in plant-herbivore interactions—using the wild tobacco *Nicotiana attenuata* as a model. Journal of Genetics and Genomics **40**, 597–606.

Wu J, Baldwin IT. 2009. Herbivory-induced signalling in plants: Perception and action. Plant Cell and Environment **32**, 1161–1174.

Wu J, Hettenhausen C, Meldau S, Baldwin IT. 2007. Herbivory rapidly activates MAPK signaling in attacked and unattacked leaf regions but not between leaves of *Nicotiana attenuata*. Plant Cell **19**, 1096–1122.

Wu JQ, Baldwin IT. 2010. New insights into plant responses to the attack from insect herbivores. Annual Review of Genetics **44**, 1–24.

Wu S, Peiffer M, Luthe DS, Felton GW. 2012. ATP hydrolyzing salivary enzymes of caterpillars suppress plant defenses. PLOS ONE **7**.

Xu SQ, Zhou WW, Pottinger S, Baldwin IT. 2015. Herbivore associated elicitor-induced defences are highly specific among closely related *Nicotiana* species. BMC Plant Biology **15**.

Yamaguchi Y, Huffaker A, Bryan AC, Tax FE, Ryan CA. 2010. PEPR2 is a second receptor for the Pep1 and Pep2 peptides and contributes to defense responses in *Arabidopsis*. Plant Cell **22**, 508–522.

Yamaguchi Y, Pearce G, Ryan CA. 2006. The cell surface leucine-rich repeat receptor for AtPep1, an endogenous peptide elicitor in *Arabidopsis*, is functional in transgenic tobacco cells. Proceedings of the National Academy of Sciences of the United States of America **103**, 10104–10109.

Ye M, Glauser G, Lou Y, Erb M, Hu L. 2019. Molecular dissection of early defense signaling underlying volatile-mediated defense regulation and herbivore resistance in rice. Plant Cell **31**, 687–698.

Yoshinaga N, Aboshi T, Abe H, Nishida R, Alborn HT, Tumlinson JH, Mori N. 2008. Active role of fatty acid amino acid conjugates in nitrogen metabolism in *Spodoptera litura* larvae. Proceedings of the National Academy of Sciences of the United States of America **105**, 18058–18063.

Yoshinaga N, Aboshi T, Ishikawa C, Fukui M, Shimoda M, Nishida R, Lait CG, Tumlinson JH, Mori N. 2007. Fatty acid amides, previously identified in caterpillars, found in the cricket *Teleogryllus taiwanemma* and fruit fly *Drosophila melanogaster* larvae. Journal of Chemical Ecology **33**, 1376–1381.

Zagorchev L, Atanasova A, Albanova I, Traianova A, Mladenov P, Kouzmanova M, Goltsev V, Kalaji HM, Teofanova D. 2021. Functional characterization of the photosynthetic machinery in *Smicronix* galls on the parasitic plant *Cuscuta campestris* by JIP-test. Cells **10**.

Zebelo SA, Maffei ME. 2015. Role of early signalling events in plant-insect interactions. Journal of Experimental Botany **66**, 435–448.

Zhang Y, Fu Y, Wang Q, Liu X, Li Q, Chen J. 2020. Transcriptome analysis reveals rapid defence responses in wheat induced by phytotoxic aphid *Schizaphis graminum* feeding. BMC Genomics **21**, 339.

Zhu C, Shi F, Chen Y, Wang M, Zhao Y, Geng G. 2019. Transcriptome analysis of Chinese chestnut (*Castanea mollissima* Blume) in response to *Dryocosmus kuriphilus* Yasumatsu infestation. International Journal of Molecular Sciences **20**.

Zhu QH, Shan WX, Ayliffe MA, Wang MB. 2016. Epigenetic mechanisms: An emerging player in plant-microbe interactions. Molecular Plant-Microbe Interactions **29**, 187–196.

# 3 Insect Biodiversity Informatics
## *Conservation and Decline*

*Maqsood Razzaq, Adnan Khalil, Chunhao Liang, and Taswar Ahsan*

## 3.1 INTRODUCTION

Insects comprise more than half of all described species in the animal kingdom and account for a considerable proportion of all biodiversity on the planet (Tihelka et al., 2021). The considerable diversity observed among insect species can be attributed to their unique genetic, morphological, and functional adaptations that enable them to effectively adapt to the intricate and ever-changing environments they inhabit (Deng et al., 2019). Insects are commonly recognized as pests or vectors of disease that have a significant impact on agriculture, human health, and natural resources. Numerous species of these organisms provide advantages to human beings, including but not limited to plant pollination, production of valuable substances, regulation of pest populations, participation in waste disposal, and consumption by other animals. Additionally, there is potential for these organisms to serve as a food source for humans in the foreseeable future (Stork, 2018). The global distribution and abundance of species are currently experiencing swift alterations. The statement emphasizes the importance of dependable and cohesive data in directing and evaluating measures and strategies aimed at administering and preserving the diverse roles and advantages of various species (Kazdin, 2021). The scientific investigation of insects has traditionally relied on a simple experimental approach that employs a restricted range of tools (such as a microscope, a net, etc.) to facilitate observation, hypothesis formulation, experimentation, and forecasting. The advent of digital technologies such as genomics libraries, automated taxonomic tools, and distributed databases, along with the widespread use of computers and the internet, has obviated the need for laborious examination of source material (Dong et al., 2023).

## 3.2 DATABASES: THE FOUNDATION OF INFORMATICS

The database serves as the foundational element of biodiversity informatics. There exists a diverse array of software options that are commercially available or open-source, which span a spectrum of complexity from rudimentary to intricate relational databases. The search ability of data or information is contingent upon its structure, which ultimately influences its compatibility with various forms of data storage and retrieval. The forms of information can vary in their organization, which may include unstructured and un-atomized information such as a word processor document, a table of atomized information like a spreadsheet, and atomized and normalized content in a relational database such as Microsoft Access, File Maker Pro, and others. In essence, a relational database is designed to store data in the form of one or more two-dimensional arrays, which are commonly referred to as relations or tables (Woodin, 2011). The records contained within each table signify individual instances, while the fields within those records denote the specific characteristics or attributes associated with said instances. Relational databases often consist of numerous interconnected tables that can be utilized by other tables to retrieve data and generate novel tables.

DOI: 10.1201/9781003382089-3

**TABLE 3.1**
**Databases, Indexes, and Applications**

| Database | Indexes and Applications |
|---|---|
| Taxonomic database | Index to Organism Names, International Plant Name Index, uBio project, ZooBank, The BioSystematic Database of World Diptera |
| Genetic sequences database | National Center for Biotechnology Information, European Molecular Biology Laboratory, DNA Data Bank of Japan, FlyBase |
| Phylogeny databases | TreeBASE |
| Digitized text | Nomenclator Zoologicus and Biologia Centrali-Americana, Nomenclator Zoologicus, electronic Biologia Centrali-Americana |
| Images database | Morphbank, Web Services and Life Science Identifiers |
| Specimen-level database | Label data, geo-referencing, loan information, taxonomy, nomenclature, bibliography, specimen images |
| Encyclopedia of Life, Tree of Life Web Project (TOL) | Web-based species data collection, Wikispecies, a multilingual, web-based, directory |

Normalization is a crucial aspect of maintaining data integrity, especially as relational databases grow in size. Its primary function is to minimize information duplication and prevent logic conflicts that may arise from non-coordinated modifications of record duplicates. The structure of databases exhibits a high degree of idiosyncrasy, particularly within the fields of entomology and systematics, which encompasses dipterous. Often, such structures reflect the localized requirements of the compiler, such as collection, rather than the necessities of the ultimate user. The efficacy and practicality of biodiversity informatics lies in the amalgamation of taxonomic data from various origins, achieved through the integration of information from divergent databases. A distributed database refers to a database that is physically stored across two or more computer systems. A distributed database system is capable of managing and controlling the entire database as a unified collection of data, despite being geographically dispersed. The coordination of communication across various locations is facilitated through the utilization of a standardized set of fields and language. This process is overseen by a centralized management system and a singular user interface. The process of integrating and sharing information across distinct databases necessitates the establishment of standardization protocols to ensure that the information is consistent and relevant across all databases. Additionally, a common language must be developed to facilitate comprehension by both data providers and users, as noted by Tang et al. (2019). Diverse protocols are being formulated to facilitate the transmission of data, contingent on the nature of the data. The principal objective of Biodiversity Information Standards (TDWG) [previously known as the Taxonomic Database Working Group] is to facilitate the creation and dissemination of a uniform collection of data pertaining to biodiversity information. The aforementioned collective is responsible for formulating criteria and procedures pertaining to the dissemination of biodiversity information, predominantly utilizing XML schema, including but not limited to the Structure for Descriptive Data as outlined by Walls et al. (2014). The objectives of the Global Biodiversity Information Facility (GBIF) encompass various aspects such as enabling the extraction of data through metadata analyses, converting collections data into digital format, creating an electronic compendium of names, developing species information pages, conducting outreach activities, and digitizing literature. Various databases exist that are founded on distinct tools, as shown in Table 3.1.

### 3.2.1 Digital Taxonomy

The taxonomic description's highly structured nature renders it well suited for matrix presentation and subsequent digitization. A matrix can be utilized to assign scores to taxa based on the presence or absence of a variety of characters and character states that can be enumerated. While binary scoring is a common practice for characters, it is important to note that the potential number of character states is infinite. The aforementioned matrix has the potential to serve as a fundamental component of a character repository. It could be employed for the purpose of producing interactive diagnostic instruments (as indicated later), or the resulting data could be utilized to generate meticulously structured and uniform depictions of language in its natural form. The Descriptive Language for Taxonomy (DELTA) format, which was commonly utilized in the past, has been predominantly employed for creating taxonomic descriptions in natural language (Liu et al., 2022) and for producing interactive keys. Although it has a high learning curve for novice users, DELTA is no longer actively developed or supported. Subsequent iterations of Lucid, specifically those from version 3.1 onward, possess the capability to produce intricate natural language explanations through the utilization of XML and XML transformations.

The taxonomic impediment poses chronic and extensive consequences, as resources and expertise for taxonomy are declining while biological diversity remains largely undescribed. Despite the potential for this tool to enhance the efficiency of species description, its utilization has not been widespread. The process of species description is a meticulous and time-intensive endeavor that necessitates specialized expertise and a comprehensive understanding of a particular group of organisms. Certain stages of this procedure can be optimized for enhanced efficacy in data management by evaluating characters in morphological matrices and utilizing this information in electronic form to construct taxonomic accounts of novel species without intermediaries, as well as producing interactive identification tools based on the same data. The digitization of character data allows taxonomists to shift from manual and laborious creation of species descriptions using word processors, to a process that involves verifying relevant character states and subsequently converting them into species descriptions in XML format (Bevilacqua et al., 2021). Shih et al. (2020) have incorporated online resources into their taxonomic analysis of a novel genus of Evaniidae (Hymenoptera). This includes the inclusion of image links from Morphbank, species pages on a taxon-specific website and the Tree of Life Web Project, interactive keys, and registration of taxon names in ZooBank, with the allocation of LSIDs (Life Sciences Identifiers) for each name.

The manuscript has been annotated using TaxonX, an XML schema designed to encode taxonomic literature in a structured manner, thereby enabling the extraction of textual elements such as scientific names, localities, morphological characters, etc., by external web applications and their utilization in distributed web resources. The comprehension of intricate adult fly anatomy and morphology has been facilitated by the Anatomical Atlas of Flies, an online resource published by Kõljalg et al. (2020). The present interactive atlas has been created utilizing high-resolution images as a supplement to interactive keys for Diptera, as demonstrated in Winterton (2010), or as a pedagogical tool for teaching Diptera morphology. Table 3.2 lists commonly used websites in the field of insect informatics.

**TABLE 3.2**
**General Websites and Taxon Related to the Insect Biodiversity Informatics**

| Resource | Websites |
|---|---|
| NCBI (Genbank) | www.ncbi.nlm.nih.gov |
| ZooKeys | https://pensoftonline.net/zookeys |
| ION | www.organismnames.com |
| GBIF | www.gbif.org |
| UBio | https://github.com/ubio |
| TDWG | www.tdwg.org |
| Biodiversity Heritage Library | www.biodiversitylibrary.org |
| Wikispecies | http://species.wikimedia.org |
| Zootaxa | www.mapress.com/zootaxa |
| MorphoBank | http://morphobank.org |

### 3.2.2 Character Data and Matrices: Interactive Identification Tools

Conventional techniques for identification are based on the utilization of dichotomous keys, which are typically included in a taxonomic analysis of a specific group (such as a genus revision) and are presented as a tangible document in a journal publication. Dichotomous keys are a taxonomic tool that utilizes a series of contrasting statements, typically consisting of two characteristics, referred to as a couplet. Subsequently, these characteristics give rise to either a taxonomic group embodying those traits or an alternative pair of opposing descriptive statements. The individual seeking to classify a specimen utilizes a series of couplets to trace a path toward a taxonomic classification. Dichotomous keys are commonly characterized by the presence of taxonomic terminology and a limited number of visual aids. They have been subject to criticism for their perceived usability solely by taxonomists who possess prior knowledge of the group (Egloff et al., 2016).

Despite variations in interface and format, dichotomous keys are effectively utilized on the internet, with the fundamental principle of a pathway key remaining consistent. Dichotomous keys are commonly accessible on the World Wide Web in the form of uncomplicated text-based keys that are integrated into web pages, resembling their presentation in scholarly journal publications. The enhancement of the interface of the "flat" key can be achieved by utilizing HTML links and anchors. This involves the representation of a couplet as an individual web page, with pointers to subsequent couplets being established through the use of links to other pages or anchors to couplets on the same web page. This allows the writer to include additional visual and textual content in the key, which may not be accessible through the printed version of the publication. Lucid Phoenix represents the most advanced form of the multimedia dichotomous key. The software package has been developed with the purpose of creating new or importing preexisting digitized dichotomous keys, which can be obtained through optical character recognition. The keys are then enriched with multimedia content and compiled into an interface that resembles a matrix-based interactive key, as illustrated later.

The aforementioned software application holds significant value in terms of rendering historically significant dichotomous keys accessible on the World Wide Web, supplemented with multimedia content, as per the research conducted by Choraś et al. (2021). One of the limitations of dichotomous keys, also known as pathway keys, is the presence of an unsolvable couplet. This phenomenon arises when the entities featured in a given exposition are not readily discernible due to various factors such as impaired specimens, distinct life stages, or incongruent gender, or when the reader encounters difficulty in comprehending the author's intended meaning. The prompt consequence is the discontinuation of crucial functionality at said pair of lines, or potentially more severe, the individual speculates, resulting in erroneous recognition (i.e., a positive result that is actually negative). Matrix-based keys offer users the flexibility to select characters in any sequence, thereby earning the monikers of multi-access or random-access keys. Matrix-based keys consist of a database that contains character or feature states that are scored for taxa in the key. Additionally, a program is utilized to query the database, and a user interface is provided, which typically comprises a set of windows that display character state selection and taxon remaining or eliminated. This is illustrated in Figure 14.3 and has been discussed in previous literature (Howell et al., 2019).

The selection of character states by the user results in the program discarding taxa that do not exhibit the specific state. This process of elimination leads to the identification of the taxon. The issue pertaining to the aforementioned conundrum of the unresolvable couplet can be circumvented by providing the user with the option to opt for an alternative character in the event that a particular character is indiscernible or perplexing. To optimize the identification process, certain programs may incorporate features such as "Best" (which typically favors characters that result in an equal parsing of remaining taxa) or "Bingo" (a compilation of distinctive characters that lead to the derivation of a single remaining taxon). Interactive keys based on matrices for different insect groups, such as Diptera (Table 14.1), can be accessed through either CD or the internet, where the interface is typically operated through a web browser. There exists a plethora of interactive key software products that are available for use, either free of charge or for a fee. Among these, the most frequently employed ones include 20q (Discover Life), Intkey, Lucid (Lucidentral), and Linnaeus II (ETI BioInformatics) (Winterton, 2010).

### 3.2.3 Data Analyses

The utilization of online specimen-based data can facilitate diverse metadata analyses in the field of Diptera biodiversity research. In their study, Meier and Dikow (2004) utilized nonparametric species richness estimators to analyze specimen data from a revision of a genus of robber flies (Asilidae: Leptogasterinae: *Euscelidia* Westwood). Their objective was to estimate the global species diversity for the aforementioned genus and to evaluate whether the established biodiversity hotspots based on plant data also correspond to significant biodiversity richness for invertebrates. Several studies have been conducted on species richness in various fly families: Novotny et al. (2005) examined Tephritidae, McCreadie et al. (2005) investigated Simuliidae, Roque et al. (2010) studied Chironomidae, Foley et al. (2014) explored Culicidae, and Petersen et al. (2007) focused on Danish Diptera.

Distribution maps are a frequently employed method for illustrating the geographical distribution of Diptera. The optimal approach for presenting this information would be to utilize dot points on the map to depict the collection or observation occurrences of a specific taxonomic group. The representation of distributions on a map is a common practice in research, wherein shaded areas are used to depict the predicted range of a taxon based on a prior set of distribution points. However, it is important to note that such predicted distributions are fraught with assumptions regarding the homogeneity of occurrence of the taxon across the landscape, which can have significant and potentially hazardous implications. The data points utilized for plotting on a distribution map are obtained either from label data present on specimens or from field observations, and they are subsequently recorded as coordinates in terms of latitude and longitude. Contemporary field collection practices typically incorporate the utilization of a global positioning system (GPS) to enable the inclusion of GPS coordinates within collection labels. These coordinates can be directly input into geographic information system (GIS) software (Banerjee et al., 2014).

Geo-referencing older specimens lacking GPS coordinates can be a laborious and time-consuming task that incurs significant costs for manual data entry. According to Hair et al. (2019), the process of identifying the collection sites mentioned on labels can be facilitated by utilizing online or networked gazetteers and geographic name servers. Google Earth is a significant advancement in the field of GIS on the internet. It is an online virtual globe that portrays the earth through a sequence of superimposed images acquired from high-resolution satellite and aerial photography. Despite the potential benefits of this powerful technology, dipterists have been slow to adopt its use. However, other areas of entomology have already begun to leverage it for the purpose of creating distribution maps of ant species on a digital globe, as evidenced by Saunders (2022). Software programs like Bioclim (Hijmans and Elith, 2019) can be utilized to estimate the prediction of distributions based on specimen label data. Ecological niche modeling (ENM), also known as bioclimatic modeling, is a method that involves the use of geo-referenced collection data from specimen labels, along with environmental parameters on distribution maps, to develop models of the ecological requirements of individual species. The digital representation of species distribution is achieved by mapping the conditions that correspond to the predefined characteristics of the species. This mapping is done using tools such as Climex (Creative Research Systems), which provides a range of probabilities. The potential distribution of the species is then indicated on the digital maps. This methodology has been discussed in the works of Neyns and Canters (2022). Instances of employment in Diptera are Lutzomyia Franca (de Aguiar et al., 2014) and Ceratopogonidae (Venter, 2015). The aforementioned modeling technique holds significant utility in the context of pathway analysis pertaining to the incursion of exotic insects, as well as in the assessment of the impact of climate change on distribution patterns. Lifemapper is an automated software package that can be utilized for analogous ecological niche modeling analyses, serving as a counterpart to DesktopGARP. The acquisition of species distribution data by Lifemapper is facilitated through the retrieval of specimen collection data from dispersed databases, which are accessed through the employment of DiGIR.

### 3.2.4 Insect Success from a Conservation Perspective

Insects are the most diverse organisms on the earth, making up 70% of all organisms. Except for the coldest and saltiest situations, they rule them all. They live in swampy ponds, tropical woods, and roaring waterfalls, in addition to deserts (Samways, 2018). They are the majority, hiding in plain sight so that very few of us notice. Even mosses and liverworts, which are land and freshwater plants, have relationships with insects. The majority of plants contain flowers, and pollination by insects is necessary for them to reproduce. Almost all frogs and lizards depend on insects for food. Most fish, birds, and small mammals depend on insects for sustenance. A third of insects eat other insects as a result (Buchmann, 2015). Insects are, in a nutshell, the basic warp and weft of all ecosystems related to land. Additionally, we cannot survive without them since a third of our food—especially the healthiest parts, like fruit and nuts—depends entirely or mostly on insect pollination (Gaffney, 2016).

The insect's three tagmata (head, thorax, and abdomen), extremely adaptable skeletal structure, and wide range of physiology have all contributed to its success. There is an enormous variety of sensory equipment in the cranium. The thorax contains exceptionally strong legs for a variety of settings and, most significantly, wings for dispersing and locating supplies and mates quickly

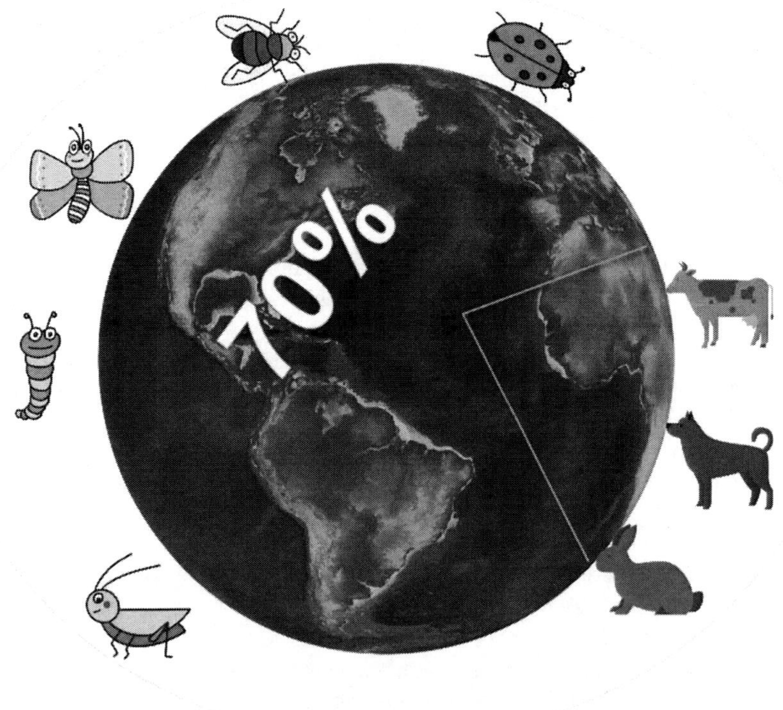


**FIGURE 3.1** In spite of the fact that land constitutes only one-third of the planet, insects comprise 70% of the species on the earth. Insects are considered to be the most ecologically interconnected organisms in both terrestrial and freshwater systems.

(McGavin and Davranoglou, 2023). Insects have different mouth parts made from the strong chitin of their skeletons that allow them to chew, rasp, suck, and dig through all kinds of plant, animal, and fungal tissue, both living and dead. Also, they can make a huge number of chemicals that help them digest food, hide, attract mates, and attack (McGavin and Davranoglou, 2023). In addition, certain species of moths have an amazing sensitivity to specific compounds, which allows them to seek a mate from a large distance. Other species, such as certain parasitoids, have the ability to discover prey that is buried deep inside plant tissue (Samways, 2018).

Insects are interactors as well as items. They are among the creatures with the strongest ecological connections. One thousand species in a straightforward biotope can interact in half a million different ways. Insects interact with almost every component in the terrestrial and freshwater realms, according to the over 1 million documented insect species and the estimate that there are over 5 million species overall, as shown in Figure 3.1 (based on the concept presented by Anderegg et al., 2015).

### 3.2.5 Insects in Today's World

The fact that insects seemed to have mostly survived the last mass extinction, which happened at the end of the Cretaceous period 66 million years ago, is shocking. They moved around to reestablish themselves in thermal optimum conditions and, to a certain extent, were independent of plants, which helped them survive the numerous glacial maxima and minima (Wagner et al., 2021). It is not yet possible to move on such a vast scale. Numerous obstacles to unrestricted mobility have been produced by the patchwork of man-made, unique ecosystems that humans have caused. The future of insect variety hinges on three possibilities: (1) adapt locally, (2) breach barriers induced by humans, or (3) become extinct. This is because global temperature change and landscape fragmentation are a "deadly anthropogenic cocktail" (Benton et al., 2022). Despite their significance, insects have only recently been integrated into biodiversity conservation efforts. Major choices are made at the size of country states through National Biodiversity and Strategy Action Plans. This is being done at several operational levels from the species level up to the landscape level of conservation (Heinrich et al., 2020). Furthermore, it emphasizes the proverb "think global, act local" given the current global initiatives for insect conservation. These operational levels—species, landscape, national, and global—are not antagonistic to one another but rather cooperative (Heinrich et al., 2020).

Insect conservation in the 21st century can be seen against six interrelated themes: (1) philosophy (establishing the ethical foundation), (2) research (the finding out), (3) policy (the framework for action), (4) psychology (understanding how humans engage in insect conservation action), (5) practice (implementation of action), and (6) validation (establishing how well we are doing at conserving insects), as shown in Figure 3.2 (based on the concept presented by Antunes and Franco, 2016).

### 3.2.6 Philosophy for Insect Conservation

To refute this perspective, we must take into account two factors. First, insects have inherent value and should be preserved for their own purpose, especially because they showcase the enormous complexity of life (Speijer, 2023). We are sentient and, ideally, a smart and caring species—hence the name "Sapiens," which we have given ourselves. This is the case for intrinsic worth. Simply said, a staggering array of living forms and bug forms coexist with us on this isolated world. Are we so callous that we merely consign them to obscurity? Second, they are simply and selfishly valuable as tools, that is, they are valuable only to us (Eiben and Smith, 2015).

We cannot continue as we have been doing things since the world's ecosystems are seriously declining, and humanity has received its "Second Warning." We need a profound shift in how we think and behave. For directing practice, we need a solid intellectual foundation. In essence, it is regarded by many as the sole tangible form of insect conservation currency (Washington et al., 2020). The binary distinction between intrinsic and instrumental value has drawbacks since it concentrates on things like bug species or landscapes rather than on how we interact with nature. There

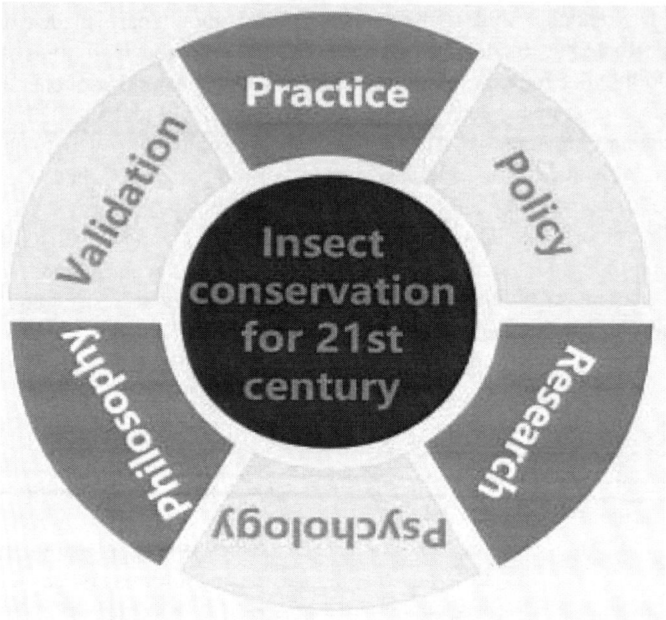


**FIGURE 3.2** A potential approach to examining insect conservation in the current era involves emphasizing six interconnected themes: philosophy, research, policy, psychology, practice, and validation.

is currently a shift away from this dichotomous perspective toward one that focuses on individual and societal welfare, with this well-being achieved through our relationship with and appreciation of the environment (Jenny et al., 2020).

Our dependence on nature and the idea that humans share a common destiny with biodiversity as a whole underpin the emphasis on relational value. Insects, for the most part, are essential to our health and happiness, since in the absence of them, the planet would be unsafe and impoverished as resources become scarcer. Simply said, if we take care of insects, they will take care of us. If there is any hope for the future of our grandkids, we must stop disregarding this fact, as shown in Figure 3.3 (based on the concept presented by Admiraal et al., 2017).

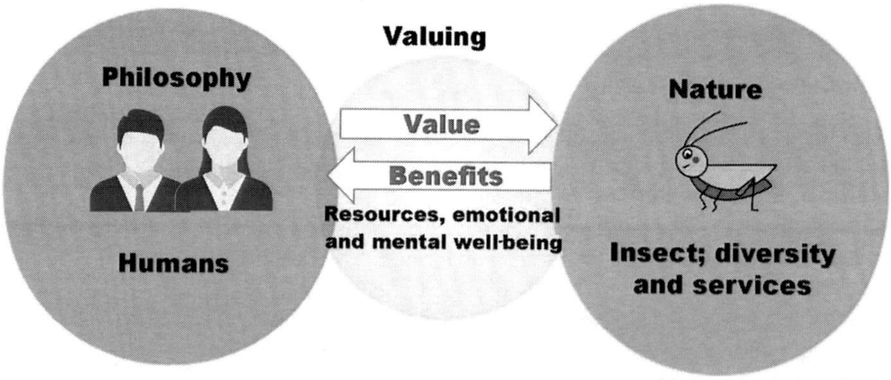


**FIGURE 3.3** The act of justifying and subsequently engaging in insect conservation necessitates a philosophical perspective that is grounded in the appreciation of the intrinsic worth of the natural world. Valuing the human relationship with nature yields benefits in physical resources, such as food, as well as emotional and mental well-being.

## 3.3 RESEARCH NEEDED FOR 21ST-CENTURY INSECT CONSERVATION

### 3.3.1 Operational Levels of Insect Conservation Research

The goal of research is to find new knowledge. The goal of this research is to develop fresh, efficient strategies for preserving insect species, populations, and variety. We study the ideal environmental circumstances that enable insect survival because we need to comprehend the ecological fabric around them in order to offer practical insect conservation solutions (Admiraal et al., 2017). Abiotic environmental factors include things like temperature patterns, fire frequency and intensity, rainfall frequency and intensity, insolation, elevation, rockiness, water, pH, and dissolved oxygen, as well as toxins, pollutants, pesticides, and a host of other factors. Furthermore, biotic environmental factors include vegetation form and composition, pollen and seed availability, fungal presence or absence, host availability (including other insects and vertebrates), mutualist presence, dung availability, mimic models, and others (Habel et al., 2019).

Many insects have developmental polymorphism, where the larva is considerably different from the adult in terms of morphology and function. As a result, they may need a variety of abiotic and biotic circumstances as well as resources to survive in their habitat (Tougeron et al., 2020). While a butterfly's adult needs specific nectar supplies, mate-meeting locations like hilltops, ovipositional sites, sunny conditions for flying, as well as enemy- and disease-free space, the butterfly's larva need a specific host plant or plants as well as enemy- and disease-free space (Stahl et al., 2023).

Although an insect's home is part of an ecosystem, some insects need to be supported by multiple ecosystems. Conditions in their habitat are altered by anthropogenic landscape modification as well as those surrounding it. The terrain matrix around a habitat may not only lack essential elements and supplies, making it impossible for generations to survive there, but it also has an impact on the habitat itself, frequently negatively (Demurtas et al., 2023). In terms of research, in addition to examining the habitat itself, we need to identify the context and contrast of the terrain. One of the main study directions in the 21st century is how to make this matrix more favorable for insects. Here is where the demands of humans and insects must be balanced. Although progress is being made, in the upcoming years we must speed up that process (Karban, 2021).

### 3.3.2 Species-Level Insect Conservation

We can think of "a species" from the standpoint of conservation as populations made up of a collection of individuals. In reality, we preserve specific members of the species rather than "a species." In a population as a whole, there may or may not be occasional allele exchange (Fišer et al., 2018). Importantly, a population's ability to adapt to environmental change, whether for the better or worse, whether in response to local (e.g., pollution) or global (e.g., climate change) impacts, or both, depends on the range of genetic variation in populations and how it is distributed among individuals (Frankham et al., 2017).

The movement of genes that are very valuable for adjusting to the environment affects the viability of metapopulations. These variables are currently evolving quickly and frequently have a negative synergistic effect (Figure 3.4). For instance, fragmentation and insecticidal impact together present a higher problem than either factor acting alone. Metapopulation dynamics can be disturbed without gene flow, creating an environment where adaptation is the only option. Because of varying or stochastic environmental occurrences, genetic impoverishment, and/or isolation, there may not be sufficient genetic variation within the isolated group of individuals for long-term survival. Landscape fragmentation can isolate populations, and this drive to adapt to the new environment may result in anthropovicariance, or human-induced evolution (Hilty et al., 2019).

### 3.3.3 Landscape-Level Insect Conservation

Insect variety conservation employing landscapes is probably a luxury above species conservation, at least from a global standpoint. Although exclusive, protecting insect species is a morally right

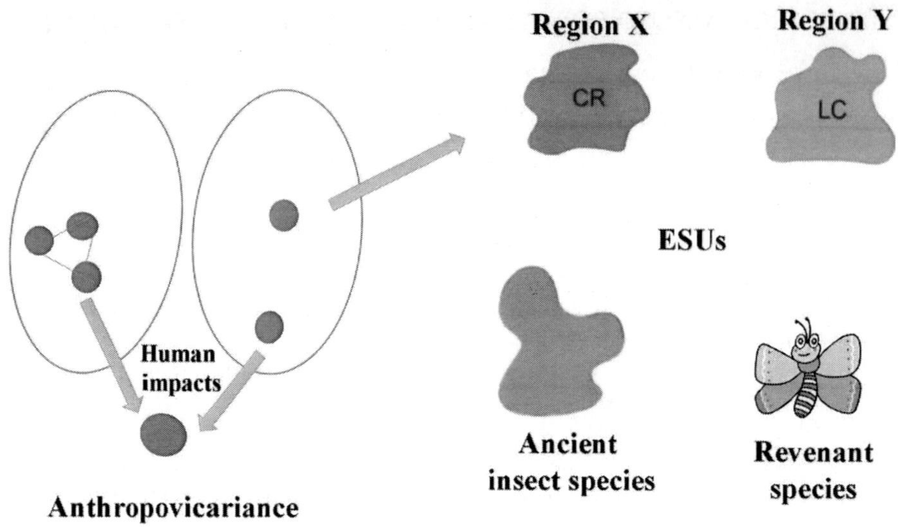


**FIGURE 3.4** Given a comprehensive understanding of the genetic, behavioral, and ecological aspects that define a particular species, there exist numerous opportunities for the conservation of insect species at the species level. The diagram depicts two distinct species, denoted as A and B. Species A is characterized by the presence of metapopulations that facilitate genetic exchange, while species B comprises two subpopulations that do not engage in such exchange. Anthropogenic influences may lead to genetic alterations in either of these, a phenomenon referred to as anthropovicariance (lower left corner). The existence of two evolutionarily significant units (ESUs) in geographically distant locations (X and Y) comprising two subpopulations may exhibit genetic and physical variations from each other (upper right as Complement receptor (CR) and Langerhans cell (LC)).

thing to do, and there is not much time left to do it. There is a good incentive to practice excellent insect conservation in regions of the world that can afford to do so, i.e., in those with large gross domestic products (GDPs) and where there is a lot of public interest and involvement. It results in the exploratory approaches and procedures that the world will need in the future. Additionally, sustaining high habitat heterogeneity in good-quality landscapes will benefit both economically developed countries and those with lower GDPs, as it will help numerous insect species. Figure 3.5 represents the concept (drawn based on the concept of van de Water et al., 2022).

### 3.3.4 Policy for Insect Conservation

Regarding policy, there are two ways to look at insect conservation. In addition to contributing to biodiversity, insects also play a crucial role in maintaining the health of the ecosystem. These viewpoints can be taken into account at a range of spatial scales, from the global to the national (keeping in mind that nation-states are a crucial unit for conservation action), and finally the small (Akhtar et al., 2021). Since insects work at multiple trophic levels and interact with a variety of other creatures, including plants, fungi, animals, and protozoa, from the standpoint of biodiversity, insect conservation is essential to biodiversity conservation. The Convention on Biological Diversity (CBD), which is ratified by the majority of nation-states, serves as the global framework for the preservation of biodiversity.

Both the species level and insect diversity are being conserved for all of the world's insects. The International Union for the Conservation of Nature (IUCN)/Species Survival Commission (SSC), which creates the Red List of Threatened Species (Sheppard et al., 2019).

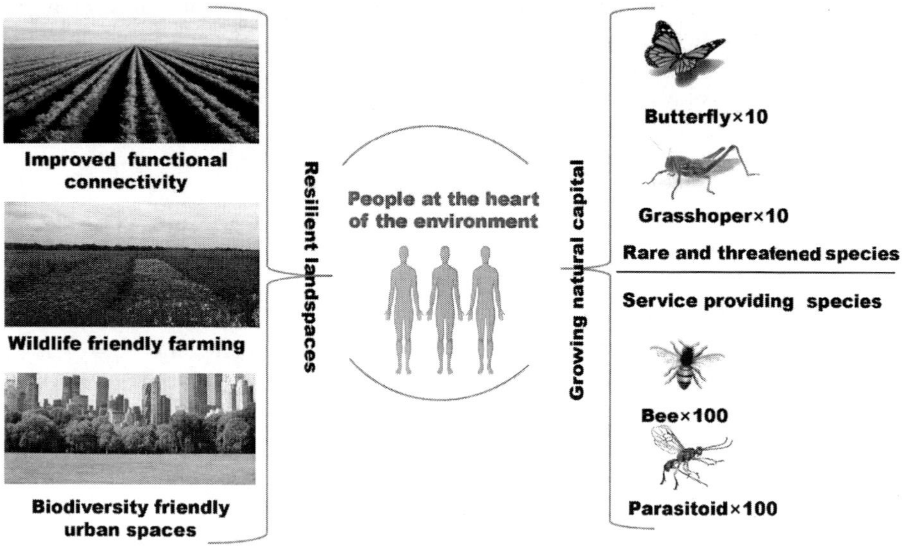


**FIGURE 3.5** The implementation of landscape-level insect conservation will be guided by three fundamental concepts, namely, the construction of resilient landscapes, prioritization of human involvement in environmental matters, and expansion of natural capital as defined by Natural England. Resilient landscapes exhibit greater functional connectivity, as an example. The implementation of grassland conservation corridors amid exotic pine plantation tree blocks, adoption of wildlife-friendly farming practices such as organic methods and preservation of natural habitat, and creation of biodiversity-friendly areas within urban settings are some of the measures being taken to promote conservation. Augmenting the natural capital involves not only amplifying the variety of indigenous species but also elevating the number of the intended species.

To stop the extinction of insects, the following conservation methods could be used (Taylor et al., 2018):

- To stop further decrease, especially in places where intensive agriculture is practiced, habitat restoration along with reduced agrochemical use and agricultural "redesign" is likely to be the most effective alternative.
- To enhance the status of threatened species, more efforts must be made to quantify insect threat levels and develop biodiversity regulations.
- The most endangered insect species require the development and implementation of conservation strategies.
- Current farming methods should be rethought and awareness raised of the benefits of ecologically sound methods and prudent pesticide use.
- The amount of dangerous chemicals, especially pesticides, that flow off into waterways and leak into the waterways must be reduced.
- More funding is required for conservation initiatives centered on insect species that are identified as vulnerable on the IUCN Red List.
- Every ten years or as soon as new data become available, the IUCN Red List of Insects should be updated.
- To stop further deforestation, forests should be considered as a priceless natural resource that needs to be immediately protected.
- National and regional governments must create regulations for the protection and restoration of natural habitats, as well as take drastic action to reduce greenhouse gas emissions and stop the harmful impacts of overusing many taxa.

- The loss of many agriculturally significant insects, particularly in annual crops, represents a serious issue due to the agricultural ecosystem's extreme dynamics and instability. In order to reduce this and preserve these insects, public-private partners must create entomophage parks (Taylor et al., 2018).

## 3.4 INSECT BIODIVERSITY DECLINES

Several insect species' populations are dwindling (Didham et al., 2020). The status and trends are at least as alarming as they are for vertebrate species for the vanishingly small part of them for whom conservation risk has been appraised (Cowie et al., 2022). Entomologists would naturally not be surprised by this since they have been reading about declining insect numbers in periodicals like *Insect Conservation and Diversity* for at least ten years. But for the general public, it has come as a startling revelation, brought to light by numerous recent studies that attracted international media attention (Didham et al., 2020). Some of these studies have had rather poor quality, either as a result of incorrect data interpretation or overly optimistic claims. This has led to a quick rise in critical reevaluation of the findings, which is typical of science as a self-correcting activity (Shapiro and Willig, 2019).

The response from the international media has been much varied, but it has successfully raised the issue of insect conservation to the top of the public and policy agenda. It has never been more crucial to pause and consider carefully what constitutes evidence for a drop in the first place amid the rush to solve "global insect losses." High interannual variability is more common than unusual for most insects, but it complicates efforts to establish what the historical abundance "reference state" should be and inevitably lengthens the time series needed to distinguish signal from noise (Willig, 2019).

Without a clear trajectory of population size drop through time, there could be any number of errors in the data collection, processing, or interpretation of the results that appear to alter anything from being present to being absent or from being abundant to being scarce between two time points (Laroche et al., 2017). To accurately determine long-term population decline rates, it is essential to establish a dependable quantitative evaluation of historical population levels, as an inaccurate baseline can significantly distort the results. Unfortunately, these figures are rarely available because the great majority of "decline" studies are often only initiated in response to perceived changes in population size; a similar situation is found with pest insects, where studies are frequently initiated in response to outbreaks (Glister, 2020).

At first glance, this might imply that current estimates of decline should, on average, be underestimates of the longer-term trends. For example, saproxylic insects, like *Rhysodes sulcatus* Fabricius 1787, are believed to have been declining throughout Europe for the past 3,000 years due to the gradual loss of old-growth forests. We anticipate that measured baselines may typically understate genuine historical baselines, e.g., Wiser et al. (2021), but speculative backcasting from decreases like those documented by Didham et al. (2020) and others (e.g., 75% decline in insect biomass since the 1980s) may be problematic from the start. In the UK, for instance, Eggleton (2020) discovered that the 1980s to the present were indeed a period of declining moth biomass, but data from an even earlier "baseline" period (1967–1982) revealed that moth biomass was previously much lower than at present and had actually increased to a peak in the 1980s (for unknown reasons) prior to the more recent decline. Several studies employed a haphazard collection of historical data as proxies for the missing baseline in the absence of long time series of decline, but even if this is frequently all we have, it is unusual that they are likely to be appropriate. For example, previous quantitative surveys may have been created to answer an unrelated ecological question and (inadvertently) deviate from the statistical presumptions required to obtain an accurate estimate of historical population abundance.

Fournier et al. (2019) demonstrate how nonrandom site-selection bias in historical baseline conditions measurement could considerably enhance the risk of inferring a false drop, even in the absence of a long-term trend in the data. This is because it is much more likely that researchers will choose locations where the study species are known to occur and/or are sufficiently numerous to sample. It is plausible that a "false baseline effect"—an above-average beginning point in a time series comparison—could cause an apparent drop over time as a straightforward statistical artifact of regression to the mean, particularly when abundances vary greatly from year to year (Fournier et al., 2019).

In actuality, mistakes of baseline estimation almost certainly include both shifting baseline effects (past abundance is presumed to be accurately assessed, but there is no knowledge of trends prior to that point in time) and false baseline effects (the appropriate reference window is known, but historical abundance is inaccurately estimated). The corollary of site-selection bias inflating baseline average local abundance is known as the "missing zero effect," and it states that empty sites (i.e., actual absences, not detection errors) would be underrepresented in baseline local population estimates (Glister, 2020; Laroche et al., 2017).

At local scales, the missing zero impact wastes the chance to measure future recolonization of uninhabited places, whether through extra-limital processes such as shifting geographic ranges brought on by climate change and species invasion (Yackel Adams et al., 2019) or natural processes linked to spatially structured population dynamics (Drake et al., 2022). The relationship between local abundance measurements and regional occupancy changes at the regional scale is further mediated by missing zeros (Didham et al., 2020).

## REFERENCES

Admiraal, J. F., Van den Born, R. J., Beringer, A., Bonaiuto, F., Cicero, L., Hiedanpää, J., . . . Musters, C. J. (2017). Motivations for committed nature conservation action in Europe. *Environmental Conservation, 44*(2), 148–157.

Akhtar, M. N., Shaikh, A. J., Khan, A., Awais, H., Bakar, E. A., & Othman, A. R. (2021). Smart sensing with edge computing in precision agriculture for soil assessment and heavy metal monitoring: A review. *Agriculture, 11*(6), 475.

Anderegg, W. R., Hicke, J. A., Fisher, R. A., Allen, C. D., Aukema, J., Bentz, B., . . . McDowell, N. (2015). Tree mortality from drought, insects, and their interactions in a changing climate. *New Phytologist, 208*(3), 674–683.

Antunes, A., & Franco, M. (2016). How people in organizations make sense of responsible leadership practices: Multiple case studies. *Leadership & Organization Development Journal, 37*(1), 126–152.

Banerjee, S., Carlin, B. P., & Gelfand, A. E. (2014). *Hierarchical modeling and analysis for spatial data.* CRC Press.

Benton, M. J., Wilf, P., & Sauquet, H. (2022). The angiosperm terrestrial revolution and the origins of modern biodiversity. *New Phytologist, 233*(5), 2017–2035.

Bevilacqua, S., Anderson, M. J., Ugland, K. I., Somerfield, P. J., & Terlizzi, A. (2021). The use of taxonomic relationships among species in applied ecological research: Baseline, steps forward and future challenges. *Austral Ecology, 46*(6), 950–964.

Buchmann, S. (2015). *The reason for flowers: Their history, culture, biology, and how they change our lives.* Simon and Schuster.

Choraś, M., Demestichas, K., Giełczyk, A., Herrero, Á., Ksieniewicz, P., Remoundou, K., . . . Woźniak, M. (2021). Advanced Machine Learning techniques for fake news (online disinformation) detection: A systematic mapping study. *Applied Soft Computing, 101*, 107050.

Cowie, R. H., Bouchet, P., & Fontaine, B. (2022). The sixth mass extinction: Fact, fiction or speculation? *Biological Reviews, 97*(2), 640–663.

de Aguiar, G. M., de Azevedo, A. C. R., de Medeiros, W. M., Alves, J. R. C., & Rendeiro, V. (2014). Aspects of the ecology of phlebotomines (Diptera: Psychodidae: Phlebotominae) in an area of cutaneous leishmaniasis occurrence, municipality of Angra dos Reis, coast of Rio de Janeiro State, Brazil. *Revista do Instituto de Medicina Tropical de São Paulo, 56*, 143–149.

Demurtas, O. C., Nicolia, A., & Diretto, G. (2023). Terpenoid transport in plants: How far from the final picture? *Plants, 12*(3), 634.

Deng, J., Guo, Y., Cheng, Z., Lu, C., & Huang, X. (2019). The prevalence of single-specimen/locality species in insect taxonomy: An empirical analysis. *Diversity, 11*(7), 106.

Didham, R. K., Basset, Y., Collins, C. M., Leather, S. R., Littlewood, N. A., Menz, M. H., . . . Schönrogge, K. (2020). Interpreting insect declines: Seven challenges and a way forward. *Insect Conservation and Diversity, 13*(2), 103–114.

Dong, H. T., Chaijarasphong, T., Barnes, A. C., Delamare-Deboutteville, J., Lee, P. A., Senapin, S., . . . Bondad-Reantaso, M. G. (2023). From the basics to emerging diagnostic technologies: What is on the horizon for tilapia disease diagnostics? *Reviews in Aquaculture*, *15*, 186–212.

Drake, J., Lambin, X., & Sutherland, C. (2022). Spatiotemporal connectivity dynamics in spatially structured populations. *Journal of Animal Ecology*, *91*(10), 2050–2060.

Eggleton, P. (2020). The state of the world's insects. *Annual Review of Environment and Resources*, *45*, 61–82.

Egloff, W., Agosti, D., Kishor, P., Patterson, D., & Miller, J. A. (2016). Copyright and the use of images as biodiversity data. *bioRxiv*, 087015.

Eiben, A. E., & Smith, J. E. (2015). *Introduction to evolutionary computing*. Springer.

Fišer, C., Robinson, C. T., & Malard, F. (2018). Cryptic species as a window into the paradigm shift of the species concept. *Molecular Ecology*, *27*(3), 613–635.

Foley, D. H., Linton, Y. M., Ruiz-Lopez, J. F., Conn, J. E., Sallum, M. A. M., Póvoa, M. M., . . . Wilkerson, R. C. (2014). Geographic distribution, evolution, and disease importance of species within the Neotropical *Anopheles albitarsis* Group (Diptera, Culicidae). *Journal of Vector Ecology*, *39*(1), 168–181.

Fournier, A. M., White, E. R., & Heard, S. B. (2019). Site-selection bias and apparent population declines in long-term studies. *Conservation Biology*, *33*(6), 1370–1379.

Frankham, R., Ballou, J. D., Ralls, K., Eldridge, M., Dudash, M. R., Fenster, C. B., . . . Sunnucks, P. (2017). *Genetic management of fragmented animal and plant populations*. Oxford University Press.

Gaffney, M. (2016). Nature, economy, and equity: Sacred water, profane markets. *American Journal of Economics and Sociology*, *75*(5), 1064–1231.

Glister, P. (2020). A 20th anniversary review of Ward and Brownlee's 'rare earth'. *Centauri Dreams*, June, p. 26.

Habel, J. C., Samways, M. J., & Schmitt, T. (2019). Mitigating the precipitous decline of terrestrial European insects: Requirements for a new strategy. *Biodiversity and Conservation*, *28*, 1343–1360.

Hair, J. F., Page, M., & Brunsveld, N. (2019). *Essentials of business research methods*. Routledge.

Heinrich, M., Scotti, F., Andrade-Cetto, A., Berger-Gonzalez, M., Echeverría, J., Friso, F., . . . Maake, C. (2020). Access and benefit sharing under the Nagoya protocol—Quo Vadis? Six Latin American case studies assessing opportunities and risk. *Frontiers in Pharmacology*, *11*, 765.

Hijmans, R. J., & Elith, J. (2019). Spatial distribution models. b978-008045405. https://rspatial.org/sdm/SDM.pdf.

Hilty, J. A., Keeley, A. T., Merenlender, A. M., & Lidicker Jr, W. Z. (2019). *Corridor ecology: Linking landscapes for biodiversity conservation and climate adaptation*. Island Press.

Howell, K. L., Davies, J. S., Allcock, A. L., Braga-Henriques, A., Buhl-Mortensen, P., Carreiro-Silva, M., . . . Game, C. A. (2019). A framework for the development of a global standardised marine taxon reference image database (SMarTaR-ID) to support image-based analyses. *PLOS ONE*, *14*(12), e0218904.

Jenny, J.-P., Anneville, O., Arnaud, F., Baulaz, Y., Bouffard, D., Domaizon, I., . . . Dorioz, J.-M. (2020). Scientists' warning to humanity: Rapid degradation of the world's large lakes. *Journal of Great Lakes Research*, *46*(4), 686–702.

Karban, R. (2021). Plant communication. *Annual Review of Ecology, Evolution, and Systematics*, *52*, 1–24.

Kazdin, A. E. (2021). *Research design in clinical psychology*. Cambridge University Press.

Kõljalg, U., Nilsson, H. R., Schigel, D., Tedersoo, L., Larsson, K.-H., May, T. W., . . . Lindahl, B. D. (2020). The taxon hypothesis paradigm—On the unambiguous detection and communication of taxa. *Microorganisms*, *8*(12), 1910.

Laroche, O., Wood, S. A., Tremblay, L. A., Lear, G., Ellis, J. I., & Pochon, X. (2017). Metabarcoding monitoring analysis: The pros and cons of using co-extracted environmental DNA and RNA data to assess offshore oil production impacts on benthic communities. *PeerJ*, *5*, e3347.

Liu, Y., Guo, L., Gao, H., You, Z., Ye, Y., & Zhang, B. (2022). Machine vision based condition monitoring and fault diagnosis of machine tools using information from machined surface texture: A review. *Mechanical Systems and Signal Processing*, *164*, 108068.

McCreadie, J. W., Adler, P. H., & Hamada, N. (2005). Patterns of species richness for blackflies (Diptera: Simuliidae) in the Nearctic and Neotropical regions. *Ecological Entomology*, *30*(2), 201–209.

McGavin, G. C., & Davranoglou, L.-R. (2023). *Essential entomology*. Oxford University Press.

Meier, R., & Dikow, T. (2004). Significance of specimen databases from taxonomic revisions for estimating and mapping the global species diversity of invertebrates and repatriating reliable specimen data. *Conservation Biology*, *18*(2), 478–488.

Neyns, R., & Canters, F. (2022). Mapping of urban vegetation with high-resolution remote sensing: A review. *Remote Sensing*, *14*(4), 1031.

Novotny, V., Clarke, A. R., Drew, R. A., Balagawi, S., & Clifford, B. (2005). Host specialization and species richness of fruit flies (Diptera: Tephritidae) in a New Guinea rain forest. *Journal of Tropical Ecology*, *21*(1), 67–77.

Petersen, F., Damgaard, J., & Meier, R. (2007). DNA taxonomy: How many DNA sequences are needed for solving a taxonomic problem? The case of two parapatric species of louse flies (Diptera: Hippoboscidae: *Ornithomya* Latreille, 1802). *Arthropod Systematics & Phylogeny*, *65*, 119–125.

Roque, F. O., Siqueira, T., Bini, L. M., Ribeiro, M. C., Tambosi, L. R., Ciocheti, G., & Trivinho-Strixino, S. (2010). Untangling associations between chironomid taxa in neotropical streams using local and landscape filters. *Freshwater Biology*, *55*(4), 847–865.

Samways, M. J. (2018). Insect conservation for the twenty-first century. In *Insect science-diversity, conservation and nutrition*. IntechOpen.

Saunders, E. (2022). *Wild bees, wasps and ants and other stinging insects*. DigiCat.

Shapiro, C., & Willig, R. D. (2019). Economic rationales for the scope of privatization. In *The political economy of public sector reform and privatization* (pp. 55–87). Routledge.

Sheppard, A., Paynter, Q., Mason, P., Murphy, S., Stoett, P., Cowan, P., . . . Shaw, R. (2019). *IUCN SSC Invasive Species Specialist Group*. Paper presented at the the Application of Biological Control for the Management of Established Invasive Alien Species Causing Environmental Impacts. The Secretariat of the Convention on Biological Diversity Technical Series.

Shih, P. J., Li, L., Li, D., & Ren, D. (2020). Application of geometric morphometric analyses to confirm three new wasps of Evaniidae (Hymenoptera: Evanioidea) from mid-Cretaceous Myanmar amber. *Cretaceous Research*, *109*, 104249.

Speijer, D. (2023). How mitochondria showcase evolutionary mechanisms and the importance of oxygen. *BioEssays*, 2300013.

Stahl, E., Maier, L.-P., & Reymond, P. (2023). Insect egg-induced innate immunity: Who benefits? *PLoS Pathogens*, *19*(1), e1011072.

Stork, N. E. (2018). How many species of insects and other terrestrial arthropods are there on Earth? *Annual Review of Entomology*, *63*, 31–45.

Tang, S., Shelden, D. R., Eastman, C. M., Pishdad-Bozorgi, P., & Gao, X. (2019). A review of building information modeling (BIM) and the internet of things (IoT) devices integration: Present status and future trends. *Automation in Construction*, *101*, 127–139.

Taylor, G. S., Braby, M. F., Moir, M. L., Harvey, M. S., Sands, D. P., New, T. R., . . . Glatz, R. V. (2018). Strategic national approach for improving the conservation management of insects and allied invertebrates in Australia. *Austral Entomology*, *57*(2), 124–149.

Tihelka, E., Cai, C., Giacomelli, M., Lozano-Fernandez, J., Rota-Stabelli, O., Huang, D., . . . Pisani, D. (2021). The evolution of insect biodiversity. *Current Biology*, *31*(19), R1299–R1311.

Tougeron, K., Brodeur, J., Le Lann, C., & van Baaren, J. (2020). How climate change affects the seasonal ecology of insect parasitoids. *Ecological Entomology*, *45*(2), 167–181.

van de Water, A., Henley, M., Bates, L., & Slotow, R. (2022). The value of elephants: A pluralist approach. *Ecosystem Services*, *58*, 101488.

Venter, G. J. (2015). *Culicoides* spp. (Diptera: Ceratopogonidae) as vectors of bluetongue virus in South Africa—A review. *Vet Ital*, *51*, 325–333. doi:10.12834/VetIt.505.2436.2

Wagner, D. L., Grames, E. M., Forister, M. L., Berenbaum, M. R., & Stopak, D. (2021). Insect decline in the Anthropocene: Death by a thousand cuts. *Proceedings of the National Academy of Sciences*, *118*(2), e2023989118.

Walls, R. L., Deck, J., Guralnick, R., Baskauf, S., Beaman, R., Blum, S., . . . Endresen, D. (2014). Semantics in support of biodiversity knowledge discovery: An introduction to the biological collections ontology and related ontologies. *PLOS ONE*, *9*(3), e89606.

Washington, H., Lowe, I., & Kopnina, H. (2020). Why do society and academia ignore the 'Scientists Warning to Humanity' on population. *Journal of Futures Studies*, *25*(1), 93–106.

Willig, C. (2019). What can qualitative psychology contribute to psychological knowledge? *Psychological Methods*, *24*(6), 796.

Winterton, S. L. (2010). Chapter fourteen. Diptera biodiversity informatics. In *Diptera diversity: Status, challenges and tools* (pp. 381–406). Brill.

Wiser, R., Rand, J., Seel, J., Beiter, P., Baker, E., Lantz, E., & Gilman, P. (2021). Expert elicitation survey predicts 37% to 49% declines in wind energy costs by 2050. *Nature Energy*, *6*(5), 555–565.

Woodin, E. M. (2011). A two-dimensional approach to relationship conflict: Meta-analytic findings. *Journal of Family Psychology*, *25*(3), 325.

Yackel Adams, A., Nafus, M., Klug, P., Lardner, B., Mazurek, M., Savidge, J., & Reed, R. (2019). Contact rates with nesting birds before and after invasive snake removal: Estimating the effects of trap-based control. *NeoBiota*, *49*, 1–17.

# 4 Understanding Anthropogenic Climate Change, Its Consequences on Insect Pests, and Techniques in Forecasting and Monitoring Pest Dynamics

## *A Contemporary Scenario*

*Deepak Kumar Mahanta, Ipsita Samal, J. Komal, Tanmaya Kumar Bhoi, Prasanta Kumar Majhi, and Mohammad Abbas Ahmad*

## 4.1 INTRODUCTION

Scientists are concerned that a loss of biodiversity will eventually lead to the breakdown of ecological communities, which would have severe consequences for many aspects of ecosystem functioning (Steffen et al., 2015). Human actions such as habitat alteration (Samal et al., 2022), agrochemical pollution, the introduction of invasive species, and changes in global temperature are the primary causes of biodiversity loss in the biosphere (Venter et al., 2016). Since the beginning of industrialisation, the biosphere has warmed by 1.1°C; without significant reductions in greenhouse gas (GHG) emissions, this warming is projected to increase to between 2°C and 5°C by the year 2100 (IPCC, 2021). It is possible that some species will go extinct, and climate change will affect the ranges, populations, communities, and interactions of others (Schleuning et al., 2020; Singh et al., 2023). Additionally, it may have a multiplicative influence on the consequences of other variables already known to threaten biodiversity. Decreases in insect populations have been observed globally, across several regions and taxonomic or functional groupings (Warren et al., 2021). The effects of population declines on crucial species in agriculture, forestry, and human settlements are becoming increasingly apparent (Wagner et al., 2021). Insects contribute to ecosystem services, above- and belowground biodiversity, and the proper functioning of ecosystems (Dangles and Casas, 2019). Just a few of the numerous crucial functions performed by insects are pollination, pest control, and nutrition cycling (Schowalter et al., 2018). Food webs that include insects and insect products play an essential role in human diets (Schowalter, 2013; Mahanta et al., 2023). Aphids and mosquitoes, for example, can spread diseases to crops and, in turn, to humans (Schowalter et al., 2018; Bhoi et al., 2022a). Deforestation, habitat fragmentation, and agricultural development threaten the delicate equilibrium between food web linkages and ecological variables that regulate disease and vector populations (Burkett-Cadena and Vittor, 2018; Mahanta et al., 2022). Depending on the specific pathogen, the incubation period and geographic spread of infectious diseases may lengthen as the world heats (Woodward et al., 2014). Warmer temperatures may encourage a rise in pest populations by making it simpler for insects to reproduce, expand their range, and invade new areas (Skendžić et al., 2021). Because of their small size and the fact that most insect species are

 DOI: 10.1201/9781003382089-4

ectothermic, insects are particularly sensitive to variations in temperature and humidity (Halsch et al., 2021). Climate change may alter the reproductive cycles of insects, as well as the lifespans of certain species (Wagner, 2020). Insects can either get smaller or bigger as temperatures rise, and ectothermic species, which prefer cooler habitats, would flourish rapidly and then disappear if the world suddenly got much hotter (Verberk et al., 2021). It has been found that size disparities among tropical insect populations are not the product of body size plasticity but are instead inherited, and that body size losses are not a universal consequence of global warming (García-Robledo et al., 2020). As the climate changes, it may affect many insect orders and families. Many insects are able to flourish in areas with higher average temperatures or longer growth seasons than would normally be conceivable due to their unique evolutionary histories, behavioural and ecophysiological adaptations, habitats, and climatic conditions. Due to global warming, mosquitoes and other thermophilic pests and disease vectors are spreading into new areas (Jactel et al., 2019). It is unclear how climate change will affect agriculture and public health. Insect populations and other functional groups are adapting to climate change in both natural and human-dominated habitats, including woods, wetlands, agricultural landscapes, and cities.

A recent study shows that climate change and intensive agriculture are damaging insect diversity (Outhwaite et al., 2022). This study assesses what is known so far regarding insect reactions to climate change, what is not known yet, and what research needs to be done to fill in the gaps and better inform policymakers and the general public. The climate can change slowly, become more variable, or change suddenly and dramatically, and global warming can affect all three. These changes have implications for the management and protection of insect populations (Cardoso et al., 2020). Numerous studies have examined how human ecophysiology affects the longevity of insect species. Insect ecosystem processes and activities will be disrupted by temperature changes, weakening insect-dependent food webs and mitigating these consequences (Tylianakis and Morris, 2017). Extinction scepticism is on the upswing, much like climate change scepticism (Lees et al., 2020). Lacking knowledge of the significance of insects in natural and managed terrestrial and freshwater ecosystems prevents us from taking measures to reduce the negative impacts of climate change and other human-caused stressors on insects and the ecosystems in which they reside (Losey and Vaughan, 2006). Protecting the planet requires an understanding of the effects climate change is having on insects and the implementation of measures to lessen those effects. This chapter discusses anthropogenic climate change and its resultant effects on insects' spatial distribution, interspecies interactions, and phenological changes, along with attempts taken to understand the effects of climate change on plant pests, like forecasting and monitoring the insect population, that have future implications in pest management.

## 4.2 THE EVER-SHIFTING CLIMATE PATTERNS

Climate affects diverse ecosystems including terrestrial ecosystems, forestry, and agriculture (Rosenzweig et al., 2007). The phenomenon involves long-term variations in temperature, humidity, and precipitation. Rising temperatures, climate extremes, $CO_2$ and other GHGs, and shifting precipitation patterns threaten world food supplies (Shrestha, 2019). Both air temperature and sea level rise have hit record highs (Field et al., 2014). According to the World Meteorological Organization (WMO), human activity is responsible for a 10°C increase in the average world temperature. According to the Intergovernmental Panel on Climate Change (IPCC), the 2000s were the warmest decade on record. Warming of 1.40°C–5.80°C is forecast over the next century by a wide range of global climate models and development scenarios (Pachauari and Reisinger, 2007). Rising GHG levels like $CO_2$, methane, and nitrous oxide ($N_2O$) are common atmospheric gases that are major drivers of global warming. Human activities like using fossil fuels and changing farming practises also contribute to emission of GHGs (Yoro and Daramola, 2020). Compared to pre-industrial levels, GHG concentrations have increased since the Industrial Revolution 200 years ago (Rogelj et al., 2018). $CO_2$ is a major GHG, and the progressive rise in $CO_2$ levels is a well-documented global trend over the last 50 years (Prentice et al., 2001). By 2100, the amount of carbon dioxide in the air is expected to double from its current level of 416 ppm to 880 ppm. Since the middle of the 20th century, extreme weather and climate have become more common and stronger. Many of these changes are caused by human actions, such as a decrease in extreme cold temperatures and an increase in extreme hot temperatures, a faster rise in sea level, and more heavy rain in many places. According to predictions about climate change, heat waves will get worse and happen more often, and there will be more rain than usual. The equatorial Pacific and higher latitudes are facing more yearly precipitation, while average precipitation will decrease in drier mid-latitudes and subtropics while it will increase in wetter mid-latitudes. Extreme precipitation is expected to rise in mid-latitudes and moist tropics (Field et al., 2014). The United Nations and IPCC have taken many viable options to reduce GHGs, help poorer countries restore their economies, and improve their ability to adapt to changing environmental conditions.

## 4.3 ANTHROPOGENIC CLIMATE CHANGE AND ITS RESULTANT EFFECTS ON INSECTS

Insect populations do not typically decrease precipitously due to long-term, continuous warming. Changes occur in the dynamics, dispersion, phenology, or abundance of insect populations after certain thresholds of reproduction, survival, and other performance are met (Harvey et al., 2020). It is becoming warmer in the temperate, boreal, and arctic regions at higher latitudes than it is in the tropics and subtropics (Post et al., 2018). Ecosystems are shifting upslope as the tropical highlands warm by 0.017°C every year (Feeley et al., 2013). Nights are getting warmer and warmer as winter progresses (IPCC, 2014). Ectothermic organisms have a greater risk of resource depletion as their metabolic rates increase in response to higher temperatures (Williams et al., 2012). Warming

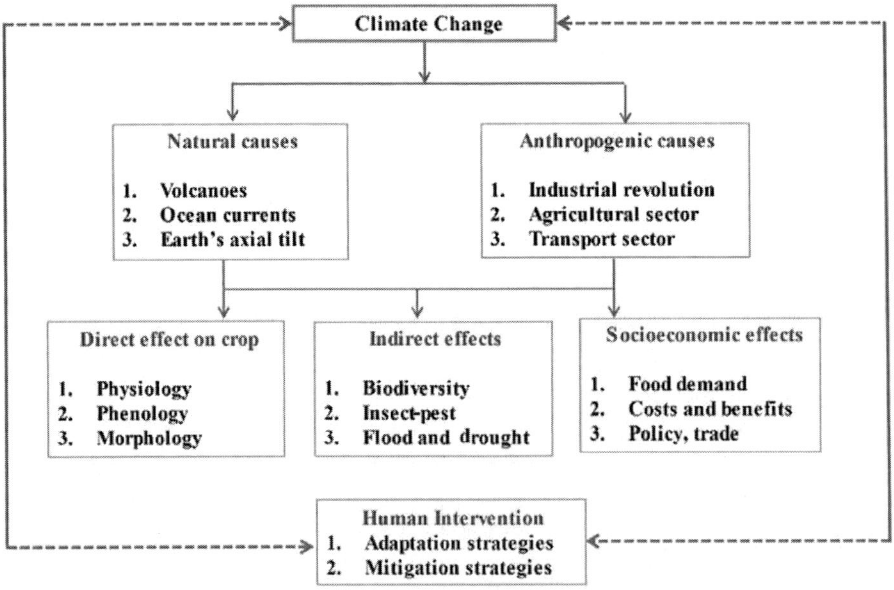


**FIGURE 4.1** The natural and anthropogenic causes of climate change and their impact on crops (direct, indirect, and socioeconomic effects). Direct effects include impact on crop physiology, phenology, and morphology. Indirect impacts include changes in biodiversity, insect pest status, interaction, and altered climatic extremes. Further, the socioeconomic effects of climate change include rising food demand, altered costs and benefits ratio, and impact on policy and trade. These altogether can be managed by diverse human interventions including adaptation and mitigation strategies.

promotes growth, development, and reproduction at optimal temperatures but causes developmental failure, lower fecundity, diminished dispersal capability, decreased fitness, and higher mortality above this point due to disproportionately high metabolic costs and accumulating thermal injuries (Gilbert and Raworth, 1996). High temperatures hinder insects' capacity to grow, reproduce, and move around, which could hinder their ability to adapt to climate change (Sweeney et al., 2018). Potentially detrimental effects of climatic change on population persistence include bottlenecks, the loss of genetic diversity independent of selection, and outbreeding or inbreeding depression (Halsch et al., 2021). Due to a lack of genetic variety in thermal tolerance features, local adaptation in some insect populations lags behind current temperatures (García-Robledo and Baer, 2021). Ant studies demonstrate that, above certain temperatures, the fitness of a colony is neutralised. Different species and ecological niches dictate their heat tolerance (or lack thereof) (Diamond et al., 2013). When insects' bodies get too hot, they shed their melanin and become a different shade (Clusella-Trullas and Nielsen, 2020). It is possible that lighter insects have an easier time managing their body temperature in warm environments. Potentially mitigating the physiological effects of climate change is the capacity for phenotypic plasticity. Thus, it can be concluded that the most significant impact of climate change is the destabilisation of insect communities, which raises the risk of population declines and even extinction of entire species if geographic distributions, behavioural activity patterns, and seasonal phenologies are not adapted to the new climatic conditions (Halsch et al., 2021). Ecological communities are reorganised in some areas, which has an influence on species interactions, the health of ecosystems, and the provision of ecosystem services. Figure 4.1 depicts diverse causes of climate change and their subsequent mitigation and adaptation strategies that will be helpful for agriculture.

### 4.3.1 Influence on the Spatial Distribution of Species

Species distributions are highly dynamic and respond to climatic change, increasing or contracting (Hewitt, 2000). Insects' ability to track climate change depends on how quickly the environment is changing. Possible habitat loss due to global warming is examined using bioclimatic envelope models, insect sensitivities to severe temperatures, and simple graphical models of species elevational ranges (Colwell et al., 2008). All these hypotheses say creatures in lower latitudes or elevations must travel to cooler locations to escape extinction (Sunday et al., 2014). Insects' capacity to monitor altering thermoclines is influenced by ecological and evolutionary variables (Wellenreuther et al., 2022). Genome sequencing and modelling are likely to guide future management strategies (Wellenreuther et al., 2022). Low, moderate, and severe warming will extirpate 34,000 insect species, according to bioclimatic models (Warren et al., 2018). By 2100, temperature and range grew exponentially. Half of all insect species would lose 50% or more of their range at +3.2°C, compared to 18% at +2°C and 6% at +1.5°C (Warren et al., 2018). Insects, like many other species, are going poleward due to global warming (Grewe et al., 2013). Lower latitude or altitude range contractions have gotten less attention since population loss is slower. As climate appropriateness of trailing edge populations declines, more insect species, including butterflies, will become extinct (Devictor et al., 2012).

Only two studies (Deutsch et al., 2008; García-Robledo and Baer, 2021) have tried to figure out how global warming affects insect populations. Deutsch et al. (2008) used life table analysis and world temperature data to predict fitness in a warming environment. These fitness estimates are only for pantropical crop pests and tropical biocontrol agents that have been developed in the lab and can handle high temperatures or have adapted to the lab environment. Their findings back up the idea that tropical insect species are more likely to become extinct than their northern counterparts (Deutsch et al., 2008). The leaf beetle *Cephaloleia belti* exhibits mitochondrial haplotypes suited to high and low elevations. At or near the equator, haplotypes may interbreed. Long-term temperature statistics coupled to demographic models show that even a 2°C increase in average global temperature will induce inhabitants to recede from the trailing edge and seek refuge from rising temperatures (García-Robledo and Baer, 2021). Many species may withstand climate change if range expansions offset population decreases at lower latitudes or elevations (Deutsch et al., 2008). Active or passive displacement, population trends, genetic variation, thermal-safety margins, habitat accessibility, interactions between and among species, and co-shifting food supplies all affect a species' ability to change its range. Different insects respond differently to changes in climate, so we should be careful when judging changes in species ranges. Higher rates of local extinction were linked to "hot" years, which suggests that more than half of insect species may not be able to move fast enough to avoid going extinct (Román-Palacios and Wiens, 2020). Adaptation to warming near range boundaries and genetic alternation for thermal tolerance imply some species can adjust quickly to shifting thermoclines (Dudaniec et al., 2018). Insects are threatened by habitat loss, fragmentation, and isolation caused by human land-use changes, especially in uneven terrain with altitude gradients (Yadav et al., 2021). Through niche overlapping, if old and new niches overlap, natural competitors could go extinct due to climate change (or vice versa).

### 4.3.2 Resultant Impact on Interspecies Interactions

Insects, like most other creatures, exist in intricate social networks where interactions with members of other species impact their life. Predation and other multitrophic interactions are frequently the product of long-term coevolutionary processes that are unique to a particular ecosystem and occur in relatively stable climatic conditions. Temperature fluctuations, on the other hand, may have varied effects on the biology of the diverse species that comprise a system. Research on multitrophic systems with plants, herbivores, parasitoids, predators, and diseases, as well as hyperparasitoids, backs this up (Bhoi et al., 2022b; Pardikes et al., 2022). Research shows that the effects on survival, growth, and reproduction are mostly negative. Herbivore fitness and related biocontrol services may suffer if plants acclimated to hotter, more carbon monoxide–rich, and/or drier circumstances offer them with suboptimal nutritional resources (Han et al., 2022). Warmer temperatures combined with an increase in herbivorous insect populations could be bad for host plants (Wang et al., 2021), leading to increased feeding period and subsequently lengthening the damaging instar. Phytophagous insects, for example, may shift their feeding choices from alien plants to native ones as temperatures rise (Liu et al., 2021). It is commonly assumed that when the climate changes, plant and insect pollination interactions will decrease (Tylianakis et al., 2008). The effects are expected to be significantly more pronounced at higher trophic levels because lower trophic levels are less abundant and their affects depend on host specificity and their capacity to adapt to these changes (Monticelli et al., 2021). The effects could be more severe in less diverse communities, such as those found on islands or in the increasingly homogenous, chemically changed agro-landscapes (Raven and Wagner, 2021). It is expected that a fall in insect populations in a warmer world will have far-reaching consequences for larger ecological systems, as many creatures (including small mammals, birds, reptiles, and amphibians) rely on insects for nourishment.

Most of the food we eat depends on plants that have been pollinated by insects, hence the health of natural ecosystems depends on these interactions. Components of Global Environmental Change (GEC), such as greater temperatures, increased $CO_2$ levels, and different patterns of precipitation, can directly affect the relationships between plants and pollinators by modifying the phenology, morphology, and distribution of plants and insects. Plant-pollinator interactions can also be affected by GEC variables by modifying the phytochemicals (nectar and volatile chemistry) required to attract pollinators. Components of climate change can alter floral biogenic volatile organic compounds (BVOCs), which play an important role in attracting pollinators and facilitating plant-pollinator mutualisms. The majority of the temperature impacts on floral BVOCs have been demonstrated, with a consistent positive influence on global warming BVOC emissions. *Petunia hybrida* flowers actively transport BVOCs via a protein through the plasma membrane and then expel them. There is a positive correlation between temperature and protein behaviour. But man-made air pollutants like ozone and diesel exhaust can neutralise flower VOCs after they have been emitted, prolonging the time it takes pollinators to go out and find food. It was observed that the right amount of ozone damaged the BVOCs in the flowers of *Brassica nigra* in a compound-specific way, changing the ratio of bouquet compounds that greatly reduced the attraction of the generalist bumblebee pollinator, *Bombus terrestris*. Clearly, pollinators' attraction to flowers is drastically impacted by airborne pollutants, and in ways that are not always predicted (e.g. by changing BVOC ratios).

The risk of geographical and temporal incompatibilities among interacting species is one of the most visible consequences of shifting insect distribution and phenology. Interacting species may exhibit generational differences due to changes in reactivity at different trophic levels, for example (Gérard et al., 2020). It is probable that specialised species, those restricted by their hosts' or prey's activities, or those that already occupy specific ecological niches, are more vulnerable to this (Damien and Tougeron, 2019). The discrepancy between insects' food availability in the spring and long-distance migratory insects' establishment success is a well-studied example (Both et al., 2006). These mismatches can make important interactions like those between plants and pollinators or herbivores and their natural enemies less stable and less effective (Peralta et al., 2020). Mismatches can be fixed by putting strong selection pressure on the phenology, range, or physiology of interacting species (Schleuning et al., 2020). A warmer climate may also result in the reunification of species that have been separated for different factors (Shah et al., 2020). When species respond differently to warming and subsequent adaptation (including the value of those species as interaction partners, such as nutritional quality), when species respond differently to changes in space or time, or when species lose their interaction partners and must find new ones, novel interactions may emerge (Tylianakis and Morris, 2017). Mismatches in thermal tolerance can arise even within a species, as in the instance of whiteflies and damselflies, which are closely related but have different sensitivities to high temperatures and hence can have sexual conflicts with one another (Abarca and Spahn, 2021). The thermal response curves for pathogen development and host defence reveal that nonlinear responses to even mild temperature changes are possible (Thomas and Blanford, 2003). These "thermal mismatch" effects may alter the stability of the system, possibly leading to the extinction of some components, and may also disrupt the normal movement of mass and energy via food webs (Thakur, 2020). Research using simulated warming to mimic real conditions has shown that increasing temperatures can decrease the interaction intensity and the fitness of each interacting species (Sentis et al., 2013). Due to species-specific variations in thermal performance, temperature may influence insect populations even when no mediating biotic interactions are present (Thierry et al., 2021).

Insects have evolved highly sophisticated symbiotic connections with the microbial communities that live in their cells, cuticle, digestive tracts, and hemolymph. For instance, mutualists in these microbiomes can boost resistance to disease, aid in the elimination of toxic plant chemicals, or make digestion easier (Feldhaar, 2011). Antagonists that bacteria may harbour include, for instance, pathogens that are detrimental to the health of insects (Gurung et al., 2019). It is expected that the interactions between hosts, symbionts, and illnesses would shift as a result of the effects of climate change on insect immunity. Examples include the increased immunity observed in insects that are exposed to higher and more variable temperature regimes compared to those that are not (Van Dievel et al., 2017). Their significance at the level of ecological communities is still little understood, despite the fact that mutualistic bacteria have been demonstrated to have favourable ecological consequences such as greater resilience to environmental stress, pathogen infections, and natural enemies (Ferrari and Vavre, 2011). As a result of their shortened genomes through coevolution with their hosts, mutualistic endosymbionts are the weak link in the insect's defences against a warming planet (Renoz et al., 2019). Insect hosts have been shown to interact with both friendly and hostile microbes, and evidence suggests that temperature plays a role in this dynamic. For instance, studies have linked shifts in cold tolerance to shifts in the composition of gut bacteria (Ferguson et al., 2018). As temperatures rise, the benefits and drawbacks that symbionts provide may shift, as certain symbionts reside in very heat-sensitive organs like bacteriocytes (Tougeron and Iltis, 2022). Further study is required to ascertain how shifts in seasonal strategies brought on by climate change genuinely influence microbial communities, despite the fact that diapause in arthropods may entirely reorganise the microbiome (Mushegian and Tougeron, 2019).

Aquatic insects are less likely to adapt to climate change in the same way as terrestrial insects because they face different ecophysiological challenges. In the same way as rising temperatures and changed patterns of temperature variation threaten terrestrial insects, they also threaten aquatic insect populations (Birrell et al., 2020). Since water has a far higher heat capacity than air, local temperature variance is restricted in lotic habitats, while individuals' behavioural thermoregulation options are severely constricted. This raises a new set of problems. As a result, it is possible that thermal performance curves for aquatic insects are more limited than for terrestrial insects (Birrell et al., 2020). Furthermore, the oxygen carrying capacity of water is significantly lower than that of air, adding to the difficulties of maintaining aerobic respiration in water. However, it is possible that aquatic insects' low heat tolerance is due to a lack of oxygen. Oxygen availability reduces little, but demand (from metabolism) typically increases dramatically as water temperatures rise (Frakes et al., 2021). When average temperatures rise, aquatic ectotherms like aquatic insects may be pushed beyond their top critical limits (Rubalcaba et al., 2020). As stated by Chown et al. (2015), the heat tolerance limits of aquatic insects are often lower than those of terrestrial insects. The significance of flow in delivering oxygen to the organism is reported by Birrell et al. (2020). Low flows caused by shifts in hydrological cycles or seasonality may amplify the difficulties created by high temperatures and low oxygen levels. Some insects found on land, such leaf gall former and leaf miners, are believed to have a similar high temperature–hypoxia relationship (Pincebourde and Casas, 2016). Eggs, larvae, and pupae of many insect species develop in water before emerging as flightless, land-dwelling adults. It is believed that the oxygen constraint inherent to aquatic existence makes these early life stages (eggs and pupae) more vulnerable to heat stress. Additionally, their limited mobility prevents them from taking advantage of temperature and oxygen availability gradients (Kingsolver et al., 2011). Therefore, the fate of terrestrial adults depends heavily on how aquatic adolescents respond to rising temperatures.

When considering the effects of climate change on aquatic communities, these physiological and physical aspects may be crucial. Compared to tropical mountains of the same altitude, the seasonal temperature changes experienced by aquatic insects in temperate highlands are more severe. It is possible that warmer-blooded animals have a wider thermal tolerance range and greater ease of acclimatisation to higher temperatures (Shah et al., 2017). Further, it was observed that tropical aquatic insects are especially vulnerable to rising temperatures (Tewksbury et al., 2008).

### 4.3.3 Consequence for Phenology

Shifts in phenology are another well-studied insect reaction to climate change. Some insect species can boost their reproductive production thanks to voltinism, which takes advantage of the extended growth season and faster development at higher temperatures. Seasonal or annual generation increase can occur despite slight temperature changes (Parmesan and Yohe, 2003). Furthermore, according to Altermatt (2010), warmer temperatures after 1980 increased generation sizes for 44 butterfly and moth species. Higher summer temperatures allow for more generations, although voltinism may generate smaller adults (Verberk et al., 2021). Overwintering strategy and photoperiod (onset of diapause and shorter days) affect later-season generation augmentation for many temperate-zone insects (Marshall et al., 2020). When predicting the impact of severe temperatures and climate change on insect voltinism and body size, it is important to consider these features (Verberk et al., 2021). Little is known about the ecological implications of overwintering and activity scheduling changes (Williams et al., 2015). An extended warm period that encourages development may delay winter diapause. According to the "lost generation idea," an extra generation born in the fall may not have the resources to reach adulthood (Kerr et al., 2020). Mild winters hinder diapause maintenance and lead to death of insects. Some insect species enter diapause before resuming normal development (Lehmann et al., 2017), and without a "cooling signal," insects may postpone or not emerge (Tougeron et al., 2019). Fitzgerald et al. (2021) found that arthropods evolved to cooler environments were more influenced by warming during mild winters. MacLean et al. (2017) observed that acorn ants "primed" their metabolism for early spring activity by spending the winter in warmer temperatures. Summer heat waves or winter warm periods (or both) can give insects the wrong signals (e.g., return to development in winter) (Forrest, 2016). *Lasiommata megera* breaks diapause near the end of the second generation to acclimatise to the warm autumn. This has caused the butterfly's rapid decrease throughout western Europe. This renders the development of a third generation in the winter season (van Dyck et al., 2015). Certain aphid parasitoids are year-round active, suggesting repeated mild winters may lead to a loss of winter diapause (Tougeron et al., 2017). Active insects need physiological and behavioural thermotolerance, such as quick cold hardening or brief, reversible rest periods, to survive winter (Diniz et al., 2017). Warmer winters alter honeybee colony phenology, resulting in nectar and pollen differences (Nürnberger et al., 2019). Warmer winters help honeybee colonies to develop brood, but this leads to the infestation of aggressive *Varroa destructor* mite (Vercelli et al., 2021).

Winter circumstances are more locally and geographically variable than summer ones, hence latitudinal variation may have a stronger effect on biotic activities (Bonan, 2004). Global warming is causing shorter winters, later snowfall, and earlier melting of ice in temperate regions (Fontrodona Bach et al., 2018; Zhu et al., 2019). Early spring was reported to expose insects to dangerously low temperatures (Williams et al., 2015). Without snow, diapause periods dry out faster leading to metabolic and phenological changes, and mismatches with vital supplies (Renner and Zohner, 2018). When snow cover melts, diapaused insects are more vulnerable to winter predators (Cooper, 2014). Long-term impacts of seasonal temperature and precipitation fluctuations on insect demography are understudied.

### 4.3.4 Impact of Climate Change on Plant and Insect Defensive Traits

Plants use both secondary chemicals and structural characteristics to deter herbivory, and both are affected by temperature. There could be serious consequences for insects and plants if temperatures continue to rise. Host plants undergo morphological and chemical changes as temperatures fluctuate. Catechol phenolics (chlorogenic acid and rutin) were found in much higher concentrations in tomatoes when the temperature dropped to 17°C at night compared to other temperatures (Bradfield and Stamp, 2004). Tomatoes have been shown to have low polyphenol oxidase (PPO) and peroxidase (POX) activity at 35°C, as demonstrated by Rivero et al. (2003), and to have a considerable decrease in protease inhibitor activity below 22°C, as demonstrated by other researchers (Green and Ryan, 1973). The thickness of leaf trichomes often increases at high temperatures (Bickford, 2016). Under stress, plants release 10% of recently absorbed carbon as BVOCs (Peñuelas and Llusià, 2003), and these molecules mediate antagonistic and mutualistic multitrophic interactions and determine the troposphere's oxidative capacity, which affects BVOC-driven interactions. Herbivores and diseases trigger certain BVOC mixes that are helpful to reveal plant location, quality, infestation history, and insect presence (Holopainen and Gershenzon, 2010; Bruce and Pickett, 2011; Ode, 2013). In addition, at higher temperatures, the amounts of secondary plant metabolites (sapogenins and saponins) in alfalfa (*Medicago sativa*) were reported to increase, hence inhibiting caterpillar growth (*Spodoptera exigua*). In contrast, the green-veined butterfly, *Pieris napi*, ingested significantly more plant tissue in response to warming-induced low-quality Brassicaceae leaves (Bauerfeind and Fischer, 2013). The survival and development of aphids (*Myzus persicae* and *Brevicoryne brassicae*) fed on oilseed rape plants with varying temperatures and nutrient density had no effect (Himanen et al., 2008). Furthermore, the size of tobacco hornworm, *Manduca sexta*, on tobacco (*Nicotiana tabacum*) was found to be increased in lower temperatures (Diamond and Kingsolver, 2010).

Apart from elevated temperature, decreased precipitation, and air pollutants affect herbivore-induced BVOC emissions, but little is known about their interaction consequences (Holopainen and Gershenzon, 2010). Kivimäenpää et al. (2016) found that previous-season herbivory boosted monoterpene emissions in Scots pine (*Pinus sylvestris*) seedlings at increasing temperature while reducing $O_3$-induced increases. Previous drought may condition plants for a faster herbivore BVOC response, imparting herbivore resistance (Copolovici et al., 2014). Simultaneous herbivore and drought stress can increase parasitoid attraction or repulsion. Further, drought-stressed plants may be more susceptible to herbivore oviposition but not larval performance or parasitoid host placement (Weldegergis et al., 2015). Certain plant-feeding insect species synthesise their pheromone molecules from host-plant compounds, making them more vulnerable to $CO_2$ changes (Boullis et al., 2015). Bark beetles make pheromones via the mevalonate pathway by hydroxylating secondary compounds from plants (Blomquist et al., 2010). Higher temperatures cause plants to produce more VOCs for longer periods, changing their ecological importance in insect-plant interactions. Monoterpene emissions increase thrice for every 10°C temperature increase (Constable et al., 1999). Insect defence signalling chemicals in plants, such as methyl jasmonate and methyl salicylate, are ineffective against herbivores when present in higher amounts. Fragrant environments can hinder plant reproduction and health by disorienting pollinators and seed dispersers (Arneth et al., 2010). Changing herbivore-induced plant volatiles would affect biological control and efficiency of biocontrol agents. Temperature changes impair volatiles' olfactory perception, affecting natural enemies' host finding capacity. Higher $CO_2$ concentrations would also impact the amounts of oxalic and malic acids in chickpea (Sharma et al., 2016). Figure 4.2 depicts detailed effects of climate change on aboveground and belowground herbivory, plant traits, and beneficial insect preference and subsequent interactions.

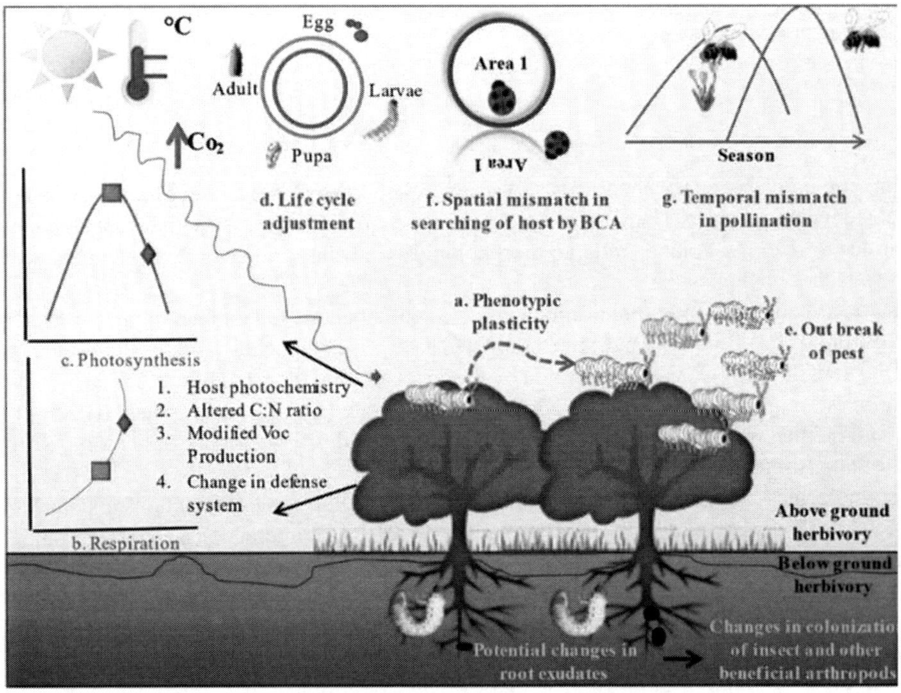


**FIGURE 4.2** Diverse impacts of climate change on host plants–insect pests–natural enemies, tritrophic interaction. Also illustrated is the impact on aboveground as well as belowground herbivory due to change in the root exudates leading to altered colonization of insect and other beneficial arthropods. (1a, b and c) Climate change impacts on plant phenotypic plasticity, photosynthesis, and respiration. (1d and e) Life-history changes and outbreak of pests as an effect of climate change. (1f and g) The impact of climatic dynamics on third trophic level and beneficial insects including spatial mismatch in searching of host by biocontrol agents and temporal mismatch leading to pollinator-plant asynchrony. Overall, it is shown that climate change can cause their geographic range to prosper, their ability to survive the winter to rise, the number of generations to rise, the synchrony between plants and pests to change, the interactions between different species to change, the risk of invasion by migratory pests to rise, the number of plant diseases spread by insects to rise, and the effectiveness of biological control, especially natural enemies, to collapse.

## 4.4 APPROACHES TO UNDERSTAND THE EFFECTS OF CLIMATE CHANGE ON PLANT PESTS

Changes in temperature, carbon dioxide levels, ozone levels, ultraviolet-B radiation, and water and humidity patterns have all been the focus of studies on the effects of climate change on insect pests over the past 30–40 years. A wide range of horticulture crops have been studied for pests, from tropical and plantation crops to forest trees and field crops like wheat, barley, rice, soybeans, and potatoes (Ghini et al., 2011). Research techniques employed in this field are summarised in Table 4.1. Several of these investigations have taken an experimental tack, watching how changing the weather influences a certain environment. Another line of research has looked at species' distributions along latitudinal or elevational gradients to infer previous climatic change. In addition to these empirical methodologies, "theoretical" approaches like meta-analysis of published results or examination of long-term data sets have been used. Lastly, studies have used the opinions of experts or made simulation models to try to predict how future changes in temperature or the composition of the air might affect the location, number, severity, and control of pests and other species.

**TABLE 4.1**
**Effect of Climate Change on Insect Pest**

| SL. No. | Response | Crop | Insect Affected | Country | Study Details | Reference |
|---|---|---|---|---|---|---|
| | | | **Effect of carbon dioxide ($CO_2$)** | | | |
| 1 | Increase in feeding rate and more damage potential | soybean | Japanese beetle (*Popilia japonica* Newman), Potato leafhopper (*Empoasca fabae* Harris), Mexican bean beetle (*Epilachna varivestis* Mulsant), and Western corn rootworm (*Diabrotica virgifera virgifera* Le Conte) | Illonis | Soybeans exhibited 57% more damage from insects at elevated atmospheric $CO_2$ on concentrations than soybean grown under ambient atmospheric conditions | Hamilton et al. (2005) |
| 2 | Reduced effectiveness of biological control agents | Wheat | English grain aphid (*Sitobion avenae* F.) | China | Study examining the efficiency of the parasitoid wasp (*Aphidius picipes* Nees) feeding on the English grain aphid (*Sitobion avenae* F.) showed that parasitism increased In correlation with elevated $CO_2$, but the same elevated $CO_2$ level resulted in lower fecundity of the wasps | Jun Chen et al. (2007) |

*(Continued)*

**TABLE 4.1 (*Continued*)**
**Effect of Climate Change on Insect Pest**

| SL. No. | Response | Crop | Insect Affected | Country | Study Details | Reference |
|---|---|---|---|---|---|---|
| | | | **Effect of temperature** | | | |
| 1 | Faster developmental rate | Coffee | Coffee Berry Borer (*Hypothenemus hampei*) | Western Kenya | Development of insect was faster as temperatures increased, with developmental time from egg to adult taking 53.7 ± 0.7 days at 20°C, in contrast to 23.3 ± 0.3 at 30°C. | Jaramillo et al. (2009) |
| 2 | Dispersal and area expansion | Fruit and vegetable crops | Brown marmorated stink bug, (*Halyomorpha halys*) | Switzerland | This pest is projected to expand into higher altitudes, produce more generations per year, and be active earlier in spring. | Stoeckli et al. (2020) |
| 3 | Range shift | NA | Olive fly (*Bactrocera oleae*) Rossi | Arizona to California and Italy | Olive fly in both Europe and North America will retreat southward and expand northward due to the effects of warmer summer temperatures and milder winters on adult flies. | Gutierrez et al. (2009) |
| 4 | Outbreak of pest | NA | Desert locust (*Schistocerca gregaria*) | India | Increasing temperatures and precipitation over desert areas, and heavy winds combined with tropical cyclones can provide a new environment for reproduction, growth, and migration of pest | Karthik et al. (2021) |
| 5 | Increased overwintering survival | Pedunculate oak | Winter moth (*Operophtera brumata* L.) | Canada | 2°C increase in temperature will dramatically alter the timing of bud burst, but the timing of larval hatching is likely to be significantly advanced, possibly leading to larval hatching before bud burst | Harrington et al. (2001) |

**TABLE 4.1 (*Continued*)**
**Effect of Climate Change on Insect Pest**

| SL. No. | Response | Crop | Insect Affected | Country | Study Details | Reference |
|---|---|---|---|---|---|---|
| | | | **Effect of precipitation** | | | |
| 1 | Increase pest population | Lowland calcareous grassland site | Wireworms, *Agriotes lineatus* | UK | Increased summer rainfall encourages a rapid rise in the soil-dwelling wireworms, *Agriotes lineatus* population and larvae of root chewing insects, *Agriotes lineatus*. | Staley et al. (2007) |
| | | | **Effect of ozone** | | | |
| 1 | Plant-pollinator interactions | Black mustard, *Brassica nigra* | Bumblebee, *Bombus terrestris* | Netherlands | Altering the ratio of bouquet compounds in flower of black mustard strongly inhibited the attraction of the generalist bumblebee. | Farré-Armengol et al. (2016) |

Few studies have been able to test how plant diseases and pests are affected by climate change (Luck et al., 2011; Pautasso et al., 2012). The elements that contribute to the spread of pests across a wide range of crops have been illuminated by research on the effects of climate change in open-topped chambers and free-air $CO_2$ enrichment (FACE) systems (Eastburn et al., 2011). Pests like weeds and insects have also been studied in such environments (Delucia et al., 2012). Increases in most insect and disease issues studied in FACE systems under high $CO_2$ settings were found in a recent assessment by Ainsworth and Long (2021).

In phytotrons, which are environmental chambers built to study the effects of different combinations of environmental factors, the effects of short-term changes in $CO_2$ and temperature on the relationships between plants and insects can be studied (Hakata et al., 2017). Such study findings can be used to develop real-world methods for dealing with anticipated future scenarios, such as assisting the plant-breeding sector (Gullino et al., 2020). Research should be conducted in various habitats throughout a latitudinal gradient, such as subtropical, temperate, and semi-arid climatic conditions, and along an elevation gradient from low- to high-elevation locations (Betz et al., 2020). The first technique has the advantage of maintaining a steady photoperiod regardless of altitude. The second technique anticipates a photoperiod latitudinal gradient. In the tropics, for example, summer days are shorter and nights are longer, whereas winter days are longer and nights are shorter. Taking photoperiod fluctuations into consideration is critical for accurate data interpretation. Nonetheless, this method is useful for identifying overarching patterns in the wild across broad environmental gradients and a variety of climatic regions, and such research can help determine whether a species is endemic to a specific climate or widespread and may invade locations that are becoming warmer (Juroszek and von Tiedemann, 2013).

Researchers did meta-analyses of data sets that had already been published to try to find broad trends in how different pests react to different environmental conditions (Vilà et al., 2021). Climate change's effects have been made clear by the warming of the last few decades, and scientists have used long-term data sets from field observations to study these effects (Huang and Hao, 2020). These long-term data are helpful for future research because scientists can tell the effects of climate change apart from those of other things (Huang and Hao, 2020). Researchers have combined data from long-term data sets, large-scale studies, and computer modelling to learn more about how global warming affects insects (Lehmann et al., 2020). For example, a meta-analysis of data from lab studies showed that higher-order species, like predators, are more likely to be affected by climate change than lower-order species, like plants or insects that eat plants (Fussmann et al., 2014). Since there is not much field data on the subject, this is very important when figuring out how natural enemies affect the life cycle and biological control of insect pests (Thomson et al., 2010).

Simulation models (Sutherst et al., 2011) can be used to predict how climate change will affect pests in the future. These predictions can be used to make decisions about how to deal with pest management (Hill and Thomson, 2015). One way to predict how pest dynamics will change in the future is to use "climate matching." This method involves looking at a place where the current climate is similar to the future climate in the area of interest and then extrapolating the results to the area of interest (Angelotti et al., 2017).

### 4.4.1 Forecasting the Climate and the Development of Models

Because forecasts of changes in average temperature and other climate parameters vary so significantly around the world, developing a priori climate change adaptation strategies for specific national or global climate change scenarios is impractical. Climate change adaptation is an integral component of every agricultural production plan. Effective pest management necessitates techniques that are adaptive enough to handle the challenges given by local climate fluctuation and change. Another alternative is to undertake sensitivity assessments for a specific region across a wide range of variable values and combine the resulting scenarios of future climate change. This type of technique could be useful for pest management professionals devising strategies for adjusting to changing environmental conditions (Sutherst et al., 2011). Accurate projections of global change can be generated when climate models are utilised in conjunction with the biological requirements of a specific pest species. Thus, modelling pest risk in conjunction with plant-host reactions to climate change allows for more precise forecasts of insect infestation outcomes. Ecologists use ecological niche models (ENMs) to predict where insect pests will establish themselves. There is a distinction to be made between mechanistic and correlative explanations for phenomena. A correlative model can forecast which places will be most suitable for the species in issue by using data on the species' past occurrence and the relationships between the values of environmental elements. MaxEnt, Bioclim, and Random Forest are three recently created and commonly used correlational models (Kumar et al., 2014). The majority of estimates of climate change impacts on biodiversity are obtained using correlative species distribution models, which are heavily used in climate change policy (Gillson et al., 2013). Correlative models could be extremely useful for predicting species distributions, assessing extinction risks, and determining conservation priorities. These models first establish a statistical relationship between a species' current geographic distribution and climatic characteristics, allowing them to predict where a species might be able to reside in a warming world (Pacifici et al., 2015). The risk of a species' introduction and subsequent establishment can be calculated by comparing the anticipated area of future climate-suitable habitats with the species' current distribution (Evans et al., 2015). Mechanistic models can be used to make predictions by integrating information on the environmental elements that influence a species with information on the species' tolerances for those factors (Kumar et al., 2014). The emphasis on the mechanisms through which environmental influences impact physiological function distinguishes correlative and mechanistic models of species dispersion. Then, to estimate future species distributions, an exclusion technique is used to exclude prospective habitats that limit physiological performance to the point that it hinders the ability to persist, develop, or reproduce (Kearney et al., 2010). Because of its ability to generalise beyond individual cases and isolate the components that affect biogeography, mechanistic models have been stated to be the most effective approach of resolving a wide range of management concerns (Cuddington et al., 2013). CLIMEX and other semi-mechanistic modelling techniques are used to forecast the best sites or habitats for a species based on physiological and behavioural features as well as climatic factor values (Kriticos et al., 2015). Developing such models and conducting in-depth assessments of meteorological and historical weather data would also make it easier to predict pest hazards. As the climate continues to change, this may take the form of novel methods of pest prevention and control (Yonow et al., 2018).

### 4.4.2 Monitoring Distribution and Abundance

Climate change may impact the population dynamics of insect pest species, hence long-term data are required to determine this (Yamamura et al., 2006). It is hard to make accurate assessments of pest population changes in response to shifting climatic regimes or forecast future population dynamics without this core information (Andrew and Hill, 2017). Long-term pest population and behaviour monitoring, particularly in climate change–vulnerable areas, may give some of the earliest signs of biological responses to climate change (Heeb et al., 2019). It is critical to track the local dynamics of vectors, illnesses, and host populations, as well as any changes in their distribution. Climate change is facilitating the spread of new alien species in many places of the world. We need strong monitoring and control techniques in place to stop invasive species before they become a financial burden in unexplored areas (Hellmann et al., 2008). This necessitates innovative biosecurity and pest control strategies. To cope with pests in the context of climate change, current pest control measures may need to be enhanced. Keeping a watch on things, making forecasts, manipulating them with chemicals, or even utilising biological agents to keep them in check are some examples (Heeb et al., 2019). Because many insect pests can traverse international boundaries freely, a worldwide strategy is required for reliable monitoring and risk assessment. Important information regarding insects, invasive alien species, and ecological conditions, including meteorological data, must be transmitted across regions, necessitating the establishment of a worldwide system. As a result, it is critical to strengthen cooperation among nations as well as among regional, national, and international organisations (Perrings et al., 2010). The U.S. Department of Agriculture's early warning and rapid response program and the European and Mediterranean Plant Protection Organization's Early Warning and Information System for invasive alien species will continue to prioritise entry point monitoring and rapid eradication (Ziska et al., 2011). Furthermore, farmers can take proactive efforts to restrict the occurrence and spread of potential pest problems by monitoring the weather and pests in tandem with climate and pest risk prediction data (Heeb et al., 2019).

## 4.5 CONCLUSIONS AND FUTURE DIRECTIONS

There may be many concerns still unresolved about climate change, but it is generally agreed that it has major effects on crop yield and the associated insect pests. Several aspects of climate change that are significant to insect pests are clouded by small-scale climatic variability, such as temperature rise, rise in $CO_2$ concentration, altered precipitation patterns, and variations in relative humidity. Due to the wide variety of insect species, their host plants, and worldwide climatic variability, it is predicted that different insect species would react differently to global warming in different regions of the globe. Although different insect species may respond to climate change in different ways, all of them will realise changes to their phenologies, populations, habitats, and sexual dimorphism as a result. Moreover, an increase in the overall number of pest outbreaks, particularly those involving a greater number and variety of insect pests, also can be expected, and this might lead to expansion of insects outside their ecological niche (especially northward). With a greater population and enhanced overwintering abilities, certain pests will multiply at an accelerated rate. An increase in insect-borne plant diseases and the potential for the introduction of invasive pest species to new areas are both consequences. Climate change may also reduce the effectiveness of biological control agents, often known as natural enemies, which could provide a serious problem for future pest management efforts. If climate change variables create conditions favourable to insect infestation and crop damage, there is a major risk of significant financial losses and a threat to human food security. It necessitates the need for a scientifically based, preventative approach to resolve this matter. Because of this, it is crucial to plan for and create adaptation and mitigation strategies, such as enhanced integrated pest management (IPM) approaches, monitoring of pests and the climate, and the use of modelling tools in pest management. The following research gaps need to be addressed in climate-change impacts on insect pests:

1. *Exploring the less explored areas:* Several areas have remained unexplored despite the fact that current research has been focused primarily on climate change–related insect abundance and management. Crop security options have yet to be investigated (Sutherst et al., 2007).
2. *Studying climate change impact on belowground and tropical herbivory:* The impact of climate change on belowground species has received less attention than that on aboveground species (Pritchard, 2011), tropical species have received less attention than species from subtropical and temperate regions (Ghini et al., 2011), and pests in unmanaged systems have received less attention than pests in managed systems (Anderson et al., 2004).
3. *Emphasizing lesser important pests:* Research has focused on a few key pest species, leaving many others unstudied or understudied (e.g. bacteria and viruses are significantly less studied than aboveground harmful fungi).
4. *Realising the need for multifactorial and multimodal research:* Many more multifactorial field research involving temperature, water, and $CO_2$ interactions is necessary (e.g. modelling future real-world situations using free-air $CO_2$ enrichment techniques) (Tenllado and Canto, 2020; Vilà et al., 2021)
5. *Widening research on impact of abiotic changes on biotic interactions:* Biotic interactions at various trophic levels, as well as species adaptation, are not well understood (Juroszek et al., 2020). It is necessary to evaluate present plant-protection measures in the context of climate change scenarios (Delcour et al., 2015). Long-term data sets are required to distinguish between potential climate change effects on pests and disease and confounding factors such as management changes (Garrett et al., 2021).
6. *Linking climate change mediated insect impacts on crop yield:* Future pest risk simulations should be linked to crop models on a more regular basis to illustrate possible yield losses; additionally, prospective adaptation and mitigation techniques should be included in model runs when practicable (Juroszek and von Tiedemann, 2015). Much more adaptation and mitigation research is needed to limit growing dangers (Hoffmann et al., 2019). Frameworks are required for adapting decision-making systems to new weather frequency distributions and, in certain cases, whole new scenarios (Garrett et al., 2021).

Thus, this review sums up the understanding of abiotic factors' effect on entomo-dynamics and, ultimately, its impact on crop yield, and the importance of developing suitable models and forecasting the yield-infestation relationship, which would be helpful for future collaborative researchers.

## Author Contribution

DKM, IS, PKM, SB, SH, JK, TKB: conceptualization, writing original draft preparation and supervision; VKD, VS, HRD, MAA, PPS: preparation of figures, and table; AN, DKM, IS, TKB, JK: conceptualization, preparation of figures, supervision, review and editing. All authors read and approved the manuscript.

## REFERENCES


Abarca, M., and Spahn, R. 2021. Direct and indirect effects of altered temperature regimes and phenological mismatches on insect populations. *Current Opinion in Insect Science* 47: 67–74. https://doi.org/10.1016/j.cois.2021.04.008

Ainsworth, E. A., and Long, S. P. 2021. 30 years of free-air carbon enrichment (FACE): What have we learned about future crop productivity and its potential for adaptation? *Global Change Biology* 27(1): 27–49. https://doi.org/10.1111/gcb.15375

Altermatt, F. 2010. Climatic warming increases voltinism in European butterflies and moths. *Proceedings. Biological Sciences* 277(1685): 1281–1287. https://doi.org/10.1098/rspb.2009.1910

Anderson, P. K., Cunningham, A. A., Patel, N. G., Morales, F. J., Epstein, P. R., and Daszak, P. 2004. Emerging infectious diseases of plants: Pathogen pollution, climate change and agrotechnology drivers. *Trends in Ecology and Evolution* 19(10): 535–544. https://doi.org/10.1016/j.tree.2004.07.021

Andrew, Nigel R., and Hill, Sarah J. 2017. Effect of climate change on insect pest management. In M. Coll and E. Wajnberg (Eds.), *Environmental pest management: Challenges for agronomists, ecologists, economists and policymakers* (pp. 197–215). John Wiley and Sons Ltd.

Angelotti, F., Hamada, E., Magalhães, E. E., Ghini, R., Garrido, L. R., and Pedro Júnior, M. J. P. 2017. Climate change and the occurrence of downy mildew in Brazilian grapevines. *Pesquisa Agropecuaria Brasileira* 52(6): 426–434. https://doi.org/10.1590/s0100-204x2017000600006

Arneth, A., Sitch, S., Bondeau, A., Butterbach-Bahl, K., Foster, P., Gedney, N., de Noblet-Ducoudré, N., Prentice, I. C., Sanderson, M., Thonicke, K., Wania, R., and Zaehle, S. 2010. From biota to chemistry and climate: Towards a comprehensive description of trace gas exchange between the biosphere and atmosphere. *Biogeosciences* 7(1): 121–149. https://doi.org/10.5194/bg-7-121-2010

Bauerfeind, S. S., and Fischer, K. 2013. Increased temperature reduces herbivore host-plant quality. *Global Change Biology* 19(11): 3272–3282. https://doi.org/10.1111/gcb.12297

Betz, O., Srisuka, W., and Puthz, V. 2020. Elevational gradients of species richness, community structure, and niche occupation of tropical roove beetles (Coleoptera: Staphylinidae: Steninae) across mountain slopes in Northern Thailand. *Evolutionary Ecology* 34(2): 193–216. https://doi.org/10.1007/s10682-020-10036-2

Bhoi, T. K., Samal, I., Mahanta, D. K., Komal, J., Jinger, D., Sahoo, M. R., Achary, G. C., Nayak, P., Sunani, S. K., Saini, V., Raghuraman, M., and Singh, S. 2022a. Understanding how silicon fertilization impacts chemical ecology and multitrophic interactions among plants, insects and beneficial arthropods. *Silicon* 1–21. https://doi.org/10.1007/s12633-022-02220-6

Bhoi, T. K., Samal, I., Majhi, P. K., Komal, J., Mahanta, D. K., Pradhan, A. K., Saini, V., Nikhil Raj, M., Ahmad, M. A., Behera, P. P., and Ashwini, M. 2022b. Insight into aphid mediated potato virus Y transmission: A molecular to bioinformatics prospective. *Frontiers in Microbiology* 13: 1001454. https://doi.org/10.3389/fmicb.2022.1001454

Bickford, C. P. 2016. Ecophysiology of leaf trichomes. *Functional Plant Biology* 43(9): 807–814. https://doi.org/10.1071/FP16095

Birrell, J. H., Shah, A. A., Hotaling, S., Giersch, J. J., Williamson, C. E., Jacobsen, D., and Woods, H. A. 2020. Insects in high-elevation streams: Life in extreme environments imperilled by climate change. *Global Change Biology* 26(12): 6667–6684. https://doi.org/10.1111/gcb.15356

Blomquist, G. J., Figueroa-Teran, R., Aw, M., Song, M., Gorzalski, A., Abbott, N. L., Chang, E., and Tittiger, C. 2010. Pheromone production in bark beetles. *Insect Biochemistry and Molecular Biology* 40(10): 699–712. https://doi.org/10.1016/j.ibmb.2010.07.013

Bonan, G. B. 2004. Biogeophysical feedbacks between land cover and climate. *Washington DC American Geophysical Union Geophysical Monograph Series* 153: 61–72.

Both, C., Bouwhuis, S., Lessells, C. M., and Visser, M. E. 2006. Climate change and population declines in a long-distance migratory bird. *Nature* 441(7089): 81–83. https://doi.org/10.1038/nature04539

Boullis, A., Francis, F., and Verheggen, F. J. 2015. Climate change and tritrophic interactions: Will modifications to greenhouse gas emissions increase the vulnerability of herbivorous insects to natural enemies? *Environmental Entomology* 44(2): 277–286. https://doi.org/10.1093/ee/nvu019

Bradfield, M., and Stamp, N. 2004. Effect of night time temperature on tomato plant defensive chemistry. *Journal of Chemical Ecology* 30(9): 1713–1721. https://doi.org/10.1023/b:joec.0000042397.42061.9f

Bruce, T. J., and Pickett, J. A. 2011. Perception of plant volatile blends by herbivorous insects—finding the right mix. *Phytochemistry* 72(13): 1605–1611. https://doi.org/10.1016/j.phytochem.2011.04.011

Burkett-Cadena, N. D., and Vittor, A. Y. 2018. Deforestation and vector-borne disease: Forest conversion favors important mosquito vectors of human pathogens. *Basic and Applied Ecology* 26: 101–110. https://doi.org/10.1016/j.baae.2017.09.012

Cardoso, P., Barton, P. S., Birkhofer, K., Chichorro, F., Deacon, C., Fartmann, T., Fukushima, C. S., Gaigher, R., Habel, J. C., Hallmann, C. A., Hill, M. J., Hochkirch, A., Kwak, M. L., Mammola, S., Ari Noriega, J., Orfinger, A. B., Pedraza, F., Pryke, J. S., Roque, F. O.,. . . Samways, M. J. 2020. Scientists' warning to humanity on insect extinctions. *Biological Conservation* 242: 108426. https://doi.org/10.1016/j.biocon.2020.108426

Chown, S. L., Duffy, G. A., and Sørensen, J. G. 2015. Upper thermal tolerance in aquatic insects. *Current Opinion in Insect Science* 11: 78–83. https://doi.org/10.1016/j.cois.2015.09.012

Clusella-Trullas, S., and Nielsen, M. 2020. The evolution of insect body coloration under changing climates. *Current Opinion in Insect Science* 41: 25–32. https://doi.org/10.1016/j.cois.2020.05.007

Colwell, R. K., Brehm, G., Cardelús, C. L., Gilman, A. C., and Longino, J. T. 2008. Global warming, elevational range shifts, and lowland biotic attrition in the wet tropics. *Science* 322(5899): 258–261. https://doi.org/10.1126/science.1162547

Constable, J. V. H., Litvak, M. E., Greenberg, J. P., and Monson, R. K. 1999. Monoterpene emission from coniferous trees in response to elevated $CO_2$ concentration and climate warming. *Global Change Biology* 5(3): 252–267. https://doi.org/10.1046/j.1365-2486.1999.00212.x

Cooper, E. J. 2014. Warmer shorter winters disrupt arctic terrestrial ecosystems. *Annual Review of Ecology, Evolution, and Systematics* 45(1): 271–295. https://doi.org/10.1146/annurev-ecolsys-120213-091620

Copolovici, L., Kännaste, A., Remmel, T., and Niinemets, Ü. 2014. Volatile organic compound emissions from *Alnus glutinosa* under interacting drought and herbivory stresses. *Environmental and Experimental Botany* 100: 55–63. https://doi.org/10.1016/j.envexpbot.2013.12.011

Cuddington, K., Fortin, M.-J., Gerber, L. R., Hastings, A., Liebhold, A., O'Connor, M., and Ray, C. 2013. Process-based models are required to manage ecological systems in a changing world. *Ecosphere* 4(2): 1–12. https://doi.org/10.1890/ES12-00178.1

Damien, M., and Tougeron, K. 2019. Prey–predator phenological mismatch under climate change. *Current Opinion in Insect Science* 35: 60–68. https://doi.org/10.1016/j.cois.2019.07.002

Dangles, O., and Casas, J. 2019. Ecosystem services provided by insects for achieving sustainable development goals. *Ecosystem Services* 35: 109–115. https://doi.org/10.1016/j.ecoser.2018.12.002

Delcour, I., Spanoghe, P., and Uyttendaele, M. 2015. Literature review: Impact of climate change on pesticide use. *Food Research International* 68: 7–15. https://doi.org/10.1016/j.foodres.2014.09.030

Delucia, E. H., Nabity, P. D., Zavala, J. A., and Berenbaum, M. R. 2012. Climate change: Resetting plant-insect interactions. *Plant Physiology* 160(4): 1677–1685. https://doi.org/10.1104/pp.112.204750

Deutsch, C. A., Tewksbury, J. J., Huey, R. B., Sheldon, K. S., Ghalambor, C. K., Haak, D. C., and Martin, P. R. 2008. Impacts of climate warming on terrestrial ectotherms across latitude. *Proceedings of the National Academy of Sciences of the United States of America* 105(18): 6668–6672. https://doi.org/10.1073/pnas.0709472105

Devictor, V., Van Swaay, C., Brereton, T., Brotons, L., Chamberlain, D., Heliölä, J., Herrando, S., Julliard, R., Kuussaari, M., Lindström, Å., Reif, J., Roy, D. B., Schweiger, O., Settele, J., Stefanescu, C., Van Strien, A., Van Turnhout, C., Vermouzek, Z., WallisDeVries, M.,. . . Jiguet, F. 2012. Differences in the climatic debts of birds and butterflies at a continental scale. *Nature Climate Change* 2(2): 121–124. https://doi.org/10.1038/nclimate1347

Diamond, S. E., and Kingsolver, J. G. 2010. Fitness consequences of host plant choice: A field experiment. *Oikos* 119(3): 542–550. https://doi.org/10.1111/j.1600-0706.2009.17242.x

Diamond, S. E., Penick, C. A., Pelini, S. L., Ellison, A. M., Gotelli, N. J., Sanders, N. J., and Dunn, R. R. 2013. Using physiology to predict the responses of ants to climatic warming. *Integrative and Comparative Biology* 53(6): 965–974. https://doi.org/10.1093/icb/ict085

Diniz, D. F. A., de Albuquerque, C. M. R., Oliva, L. O., de Melo-Santos, M. A. V., and Ayres, C. F. J. 2017. Diapause and quiescence: Dormancy mechanisms that contribute to the geographical expansion of mosquitoes and their evolutionary success. *Parasites and Vectors* 10(1): 310. https://doi.org/10.1186/s13071-017-2235-0

Dudaniec, R. Y., Yong, C. J., Lancaster, L. T., Svensson, E. I., and Hansson, B. 2018. Signatures of local adaptation along environmental gradients in a range-expanding damselfly (*Ischnura elegans*). *Molecular Ecology* 27(11): 2576–2593. https://doi.org/10.1111/mec.14709

Eastburn, D. M., McElrone, A. J., and Bilgin, D. D. 2011. Influence of atmospheric and climate change on plant–pathogen interactions. *Plant Pathology* 60(1): 54–69. https://doi.org/10.1111/j.1365-3059.2010.02402.x

Evans, T. G., Diamond, S. E., and Kelly, M. W. X. 2015. Mechanistic species distribution modelling as a link between physiology and conservation. *Conservation Physiology* 3(1): cov056. https://doi.org/10.1093/conphys/cov056

Farré-Armengol, G., Peñuelas, J., Li, T., Yli-Pirilä, P., Filella, I., Llusia, J., and Blande, J. D. 2016. Ozone degrades floral scent and reduces pollinator attraction to flowers. *New Phytologist* 209(1): 152–160. https://doi.org/10.1111/nph.13620

Feeley, K. J., Hurtado, J., Saatchi, S., Silman, M. R., and Clark, D. B. 2013. Compositional shifts in Costa Rican forests due to climate-driven species migrations. *Global Change Biology* 19(11): 3472–3480. https://doi.org/10.1111/gcb.12300

Feldhaar, H. 2011. Bacterial symbionts as mediators of ecologically important traits of insect hosts. *Ecological Entomology* 36(5): 533–543. https://doi.org/10.1111/j.1365-2311.2011.01318.x

Ferguson, L. V., Dhakal, P., Lebenzon, J. E., Heinrichs, D. E., Bucking, C., and Sinclair, B. J. 2018. Seasonal shifts in the insect gut microbiome are concurrent with changes in cold tolerance and immunity. *Functional Ecology* 32(10): 2357–2368. https://doi.org/10.1111/1365-2435.13153

Ferrari, J., and Vavre, F. 2011. Bacterial symbionts in insects or the story of communities affecting communities. *Philosophical Transactions of the Royal Society of London. Series B, Biological Sciences* 366(1569): 1389–1400. https://doi.org/10.1098/rstb.2010.0226

Field, C. B., Barros, V. R., Dokken, D. J., Mach, K. J., Mastrandrea, M. D., Bilir, T. E.2014. IPCC Summary for policymakers. In *Climatic change: Impacts, adaptation, and vulnerability. Part, A. Global and sectoral aspects; contribution of working group II to the fifth assessment report of the Intergovernmental Panel on Climate Change* (pp. 1–32). Cambridge University Press.

Fitzgerald, J. L., Stuble, K. L., Nichols, L. M., Diamond, S. E., Wentworth, T. R., Pelini, S. L., Gotelli, N. J., Sanders, N. J., Dunn, R. R., and Penick, C. A. 2021. Abundance of spring- and winter-active arthropods declines with warming. *Ecosphere* 12(4): e03473. https://doi.org/10.1002/ecs2.3473

Fontrodona Bach, A., Van der Schrier, G., Melsen, L. A., Klein Tank, A. M. G., and Teuling, A. J. 2018. Widespread and accelerated decrease of observed mean and extreme snow depth over Europe. *Geophysical Research Letters* 45(22): 12,312–12,319. https://doi.org/10.1029/2018GL079799

Forrest, J. R. K. 2016. Complex responses of insect phenology to climate change. *Current Opinion in Insect Science* 17: 49–54. https://doi.org/10.1016/j.cois.2016.07.002

Frakes, J. I., Birrell, J. H., Shah, A. A., and Woods, H. A. 2021. Flow increases tolerance of heat and hypoxia of an aquatic insect. *Biology Letters* 17(5): 20210004. https://doi.org/10.1098/rsbl.2021.0004

Fussmann, K. E., Schwarzmüller, F., Brose, U., Jousset, A., and Rall, B. C. 2014. Ecological stability in response to warming. *Nature Climate Change* 4(3): 206–210. https://doi.org/10.1038/nclimate2134

García-Robledo, C., and Baer, C. S. 2021. Positive genetic covariance and limited thermal tolerance constrain tropical insect responses to global warming. *Journal of Evolutionary Biology* 34(9): 1432–1446. https://doi.org/10.1111/jeb.13905

García-Robledo, C., Baer, C. S., Lippert, K., and Sarathy, V. 2020. Evolutionary history, not ecogeographic rules, explains size variation of tropical insects along elevational gradients. *Functional Ecology* 34(12): 2513–2523. https://doi.org/10.1111/1365-2435.13666

Garrett, K. A., Nita, M., De Wolf, E. D., Esker, P. D., Gomez-Montano, L., and Sparks, A. H. 2021. Plant pathogens as indicators of climate change. In T. M. Letcher (Ed.), *Climate change: Observed impacts on planet Earth* (3rd ed., pp. 499–513). Elsevier.

Gérard, M., Vanderplanck, M., Wood, T., and Michez, D. 2020. Global warming and plant–pollinator mismatches. *Emerging Topics in Life Sciences* 4(1): 77–86. https://doi.org/10.1042/ETLS20190139

Ghini, R., Hamada, E., and Bettiol, W. 2011. *Impactos das mudanças climáticas sobre doenças de importantes culturas no Brasil*. Embrapa. Meio ambiente.

Gilbert, N., and Raworth, D. A. 1996. Forum: Insects and temperature—a general theory. *Canadian Entomologist* 128(1): 1–13. https://doi.org/10.4039/Ent1281-1

Gillson, L., Dawson, T. P., Jack, S., and McGeoch, M. A. 2013. Accommodating climate change contingencies in conservation strategy. *Trends in Ecology and Evolution* 28(3): 135–142. https://doi.org/10.1016/j.tree.2012.10.008

Green, T. R., and Ryan, C. A. 1973. Wound-induced proteinase inhibitor in tomato leaves: Some effects of light and temperature on the wound response. *Plant Physiology* 51(1): 19–21. https://doi.org/10.1104/pp.51.1.19

Grewe, Y., Hof, C., Dehling, D. M., Brandl, R., and Brändle, M. 2013. Recent range shifts of European dragonflies provide support for an inverse relationship between habitat predictability and dispersal. *Global Ecology and Biogeography* 22(4): 403–409. https://doi.org/10.1111/geb.12004

Gullino, M. L., Tabone, G., Gilardi, G., and Garibaldi, A. 2020. Effects of elevated atmospheric $CO_2$ and temperature on the management of powdery mildew of zucchini. *Journal of Phytopathology* 168(7–8): 405–415. https://doi.org/10.1111/jph.12905

Gurung, K., Wertheim, B., and Falcao Salles, J. 2019. The microbiome of pest insects: It is not just bacteria. *Entomologia Experimentalis et Applicata* 167(3): 156–170. https://doi.org/10.1111/eea.12768

Gutierrez, A. P., Ponti, L., and Cossu, Q. A. 2009. Effects of climate warming on Olive and olive fly (*Bactrocera oleae* (Gmelin)) in California and Italy. *Climatic Change* 95(1–2): 195–217. https://doi.org/10.1007/s10584-008-9528-4

Hakata, M., Wada, H., Masumoto-Kubo, C., Tanaka, R., Sato, H., and Morita, S. 2017. Development of a new heat tolerance assay system for rice spikelet sterility. *Plant Methods* 13(1): 34. https://doi.org/10.1186/s13007-017-0185-3

Halsch, C. A., Shapiro, A. M., Fordyce, J. A., Nice, C. C., Thorne, J. H., Waetjen, D. P., and Forister, M. L. 2021. Insects and recent climate change. *Proceedings of the National Academy of Sciences of the United States of America* 118(2): e2002543117. https://doi.org/10.1073/pnas.2002543117

Hamilton, J. G., Dermody, O., Aldea, M., Zangerl, A. R., Rogers, A., Berenbaum, M. R., and Delucia, E. H. 2005. Anthropogenic changes in tropospheric composition increase susceptibility of soybean to insect herbivory. *Environmental Entomology* 34(2): 479–485. https://doi.org/10.1603/0046-225X-34.2.479

Han, P., Lavoir, A. V., Rodriguez-Saona, C., and Desneux, N. 2022. Bottom-up forces in agroecosystems and their potential impact on arthropod pest management. *Annual Review of Entomology* 67: 239–259. https://doi.org/10.1146/annurev-ento-060121-060505

Harrington, R., Fleming, R. A., and Woiwod, I. P. 2001. Climate change impacts on insect management and conservation in temperate regions: Can they be predicted? *Agricultural and Forest Entomology* 3(4): 233–240. https://doi.org/10.1046/j.1461-9555.2001.00120.x

Harvey, J. A., Heinen, R., Gols, R., and Thakur, M. P. 2020. Climate change-mediated temperature extremes and insects: From outbreaks to breakdowns. *Global Change Biology* 26(12): 6685–6701. https://doi.org/10.1111/gcb.15377

Heeb, L., Jenner, E., and Cock, M. J. W. 2019. Climate-smart pest management: Building resilience of farms and landscapes to changing pest threats. *Journal of Pest Science* 92(3): 951–969. https://doi.org/10.1007/s10340-019-01083-y

Hellmann, J. J., Byers, J. E., Bierwagen, B. G., and Dukes, J. S. 2008. Five potential consequences of climate change for invasive species. *Conservation Biology* 22(3): 534–543. https://doi.org/10.1111/j.1523-1739.2008.00951.x

Hewitt, G. 2000. The genetic legacy of the quaternary ice ages. *Nature* 405(6789): 907–913. https://doi.org/10.1038/35016000

Hill, M. P., and Thomson, L. J. 2015. Species distribution modelling in predicting response to climate change. In C. Björkman and P. Niemelä (Eds.), *Climate change and insect pests* (pp. 16–37). CABI Publishing.

Himanen, S. J., Nissinen, A., Dong, W. X., Nerg, A. M., Stewart, C. N., Poppy, G. M., and Holopainen, J. K. 2008. Interactions of elevated carbon dioxide and temperature with aphid feeding on transgenic oilseed rape: Are *Bacillus thuringiensis* (Bt) plants more susceptible to nontarget herbivores in future climate? *Global Change Biology* 14(6): 1437–1454. https://doi.org/10.1111/j.1365-2486.2008.01574.x

Hoffmann, A. A., Rymer, P. D., Byrne, M., Ruthrof, K. X., Whinam, J., McGeoch, M., Bergstrom, D. M., Guerin, G. R., Sparrow, B., Joseph, L., Hill, S. J., Andrew, N. R., Camac, J., Bell, N., Riegler, M., Gardner, J. L., and Williams, S. E. 2019. Impacts of recent climate change on terrestrial flora and fauna: Some emerging Australian examples. *Austral Ecology* 44(1): 3–27. https://doi.org/10.1111/aec.12674

Holopainen, J. K., and Gershenzon, J. 2010. Multiple stress factors and the emission of plant VOCs. *Trends in Plant Science* 15(3): 176–184. https://doi.org/10.1016/j.tplants.2010.01.006

Huang, J., and Hao, H. 2020. Effects of climate change and crop planting structure on the abundance of cotton bollworm, *Helicoverpa armigera* (Hübner) (Lepidoptera: Noctuidae). *Ecology and Evolution* 10(3): 1324–1338. https://doi.org/10.1002/ece3.5986

IPCC. 2014. Climate change 2014 the physical science basis: Final draft underlying scientific-technical assessment: Synthesis report. In T. F. Stocker et al. (Eds.), *1535. Contribution of working groups I, II and III to the fifth assessment report of the Intergovernmental Panel on Climate Change*. Cambridge University Press.

IPCC. 2021. The physical science basis. In V P. Zhai et al. (Eds.), *Contribution of working group I to the sixth assessment report of the Intergovernmental Panel on Climate Change Masson-Delmotte*. Cambridge University Press.

Jactel, H., Koricheva, J., and Castagneyrol, B. 2019. Responses of forest insect pests to climate change: Not so simple. *Current Opinion in Insect Science* 35: 103–108. https://doi.org/10.1016/j.cois.2019.07.010

Jaramillo, J., Chabi-Olaye, A., Kamonjo, C., Jaramillo, A., Vega, F. E., Poehling, H. M., and Borgemeister, C. 2009. Thermal tolerance of the coffee berry borer *Hypothenemus hampei*: Predictions of climate change impact on a tropical insect pest. *PLOS ONE* 4(8): e6487. https://doi.org/10.1371/journal.pone.0006487

Jun Chen, F., Wu, G., Parajulee, M. N., and Ge, F. (2007). Impact of elevated $CO_2$ on the third trophic level: A predator *Harmonia axyridis* and a parasitoid *Aphidius picipes*. *Biocontrol Science and Technology* 17(3): 313–324. https://doi.org/10.1080/09583150701211814

Juroszek, P., Racca, P., Link, S., Farhumand, J., and Kleinhenz, B. 2020. Overview on the review articles published during the past 30 years relating to the potential climate change effects on plant pathogens and crop disease risks. *Plant Pathology* 69(2): 179–193. https://doi.org/10.1111/ppa.13119

Juroszek, P., and von Tiedemann, A. 2013. Plant pathogens, insect pests and weeds in a changing global climate: A review of approaches, challenges, research gaps, key studies and concepts. *Journal of Agricultural Science* 151(2): 163–188. https://doi.org/10.1017/S0021859612000500

Juroszek, P., and von Tiedemann, A. 2015. Linking plant disease models to climate change scenarios to project future risks of crop diseases: A review. *Journal of Plant Diseases and Protection* 122(1): 3–15. https://doi.org/10.1007/BF03356525

Karthik, S., Reddy, M. S., and Yashaswini, G. 2021. Climate change and its potential impacts on insect-plant interactions. In *The nature, causes, effects and mitigation of climate change on the environment*. IntechOpen.

Kearney, M. R., Wintle, B. A., and Porter, W. P. 2010. Correlative and mechanistic models of species distribution provide congruent forecasts under climate change. *Conservation Letters* 3(3): 203–213. https://doi.org/10.1111/j.1755-263X.2010.00097.x

Kerr, N. Z., Wepprich, T., Grevstad, F. S., Dopman, E. B., Chew, F. S., and Crone, E. E. 2020. Developmental trap or demographic bonanza? Opposing consequences of earlier phenology in a changing climate for a multivoltine butterfly. *Global Change Biology* 26(4): 2014–2027. https://doi.org/10.1111/gcb.14959

Kingsolver, J. G., Woods, H. A., Buckley, L. B., Potter, K. A., MacLean, H. J., and Higgins, J. K. 2011. Complex life cycles and the responses of insects to climate change. *Integrative and Comparative Biology* 51(5): 719–732. https://doi.org/10.1093/icb/icr015

Kivimäenpää, M., Ghimire, R. P., Sutinen, S., Häikiö, E., Kasurinen, A., Holopainen, T., and Holopainen, J. K. 2016. Increases in volatile organic compound emissions of Scots pine in response to elevated ozone and warming are modified by herbivory and soil nitrogen availability. *European Journal of Forest Research* 135(2): 343–360. https://doi.org/10.1007/s10342-016-0939-x

Kriticos, D. J., Maywald, G. F., Yonow, T., Zurcher, E. J., Herrmann, N. I., and Sutherst, R. 2015. Exploring the effects of climate on plants, animals and diseases. *CLIMEX Version* 4: 184.

Kumar, S., Neven, L. G., and Yee, W. L. 2014. Evaluating correlative and mechanistic niche models for assessing the risk of pest establishment. *Ecosphere* 5: 1–23.

Lees, A. C., Attwood, S., Barlow, J., and Phalan, B. 2020. Biodiversity scientists must fight the creeping rise of extinction denial. *Nature Ecology and Evolution* 4(11): 1440–1443. https://doi.org/10.1038/s41559-020-01285-z

Lehmann, P., Ammunét, T., Barton, M., Battisti, A., Eigenbrode, S. D., Jepsen, J. U., Kalinkat, G., Neuvonen, S., Niemelä, P., Terblanche, J. S., Økland, B., and Björkman, C. 2020. Complex responses of global insect pests to climate warming. *Frontiers in Ecology and the Environment* 18(3): 141–150. https://doi.org/10.1002/fee.2160

Lehmann, P., Van Der Bijl, W., Nylin, S., Wheat, C. W., and Gotthard, K. 2017. Timing of diapause termination in relation to variation in winter climate. *Physiological Entomology* 42(3): 232–238. https://doi.org/10.1111/phen.12188

Liu, Z., Zhang, C., Ma, L., Zhou, X., Sun, X., and Ding, J. 2021. Elevated temperature decreases preferences of native herbivores to an invasive plant. *Entomologia Generalis* 41(2): 137–146. https://doi.org/10.1127/entomologia/2020/1050

Losey, J. E., and Vaughan, M. 2006. The economic value of ecological services provided by insects. *BioScience* 56(4): 311–323. https://doi.org/10.1641/0006-3568(2006)56[311:TEVOES]2.0.CO;2

Luck, J., Spackman, M., Freeman, A., Trebicki, P., Griffiths, W., Finlay, K., Chakraborty, S. 2011. Climate change and diseases of food crops. *Plant Pathology* 60(1): 113–121. https://doi.org/10.1111/j.1365-3059.2010.02414.x

MacLean, H. J., Penick, C. A., Dunn, R. R., and Diamond, S. E. 2017. Experimental winter warming modifies thermal performance and primes acorn ants for warm weather. *Journal of Insect Physiology* 100: 77–81. https://doi.org/10.1016/j.jinsphys.2017.05.010

Mahanta, D. K., Jangra, S., Priti, G. A., Ghosh, A., Sharma, P. K., Iquebal, M. A., Jaiswal, S., Baranwal, V. K., Kalia, V. K., and Chander, S. 2022. Groundnut bud necrosis virus modulates the expression of innate immune, endocytosis, and cuticle development-associated genes to circulate and propagate in its vector, *Thrips palmi*. *Frontiers in Microbiology* 13: 773238. https://doi.org/10.3389/fmicb.2022.773238

Mahanta, D. K., Komal, J., Samal, I., Bhoi, T. K., Dubey, V. K., Pradhan, K., Nekkanti, A., Gouda, M. N. R., Saini, V., Negi, N., Bhateja, S., Jat, H. K., and Jeengar, D. 2023. Nutritional aspects and dietary benefits of "Silkworms": Current scenario and future outlook. *Frontiers in Nutrition* 10: 1121508. https://doi.org/10.3389/fnut.2023.1121508

Marshall, K. E., Gotthard, K., and Williams, C. M. 2020. Evolutionary impacts of winter climate change on insects. *Current Opinion in Insect Science* 41: 54–62. https://doi.org/10.1016/j.cois.2020.06.003

Monticelli, L. S., Bishop, J., Desneux, N., Gurr, G. M., Jaworski, C. C., McLean, A. H. C., Thomine, E., and Vanbergen, A. J. 2021. Multiple global change impacts on parasitism and biocontrol services in future agricultural landscapes. *Advances in Ecological Research* 65: 45–304. https://doi.org/10.1016/bs.aecr.2021.10.002

Mushegian, A. A., and Tougeron, K. 2019. Host-microbe interactions in the context of diapause. *The Biological Bulletin* 237: 180–191.

Nürnberger, F., Härtel, S., and Steffan-Dewenter, I. 2019. Seasonal timing in honey bee colonies: Phenology shifts affect honey stores and varroa infestation levels. *Oecologia* 189(4): 1121–1131. https://doi.org/10.1007/s00442-019-04377-1

Ode, P. J. 2013. Plant defences and parasitoid chemical ecology. *Chemical Ecology of Insect Parasitoids* 2013: 9–36.

Outhwaite, C. L., McCann, P., and Newbold, T. 2022. Agriculture and climate change are reshaping insect biodiversity worldwide. *Nature* 605(7908): 97–102. https://doi.org/10.1038/s41586-022-04644-x

Pachauari, R. K., and Reisinger, A. 2007. *Synthesis report. Contribution of working groups I, II and III to the fourth assessment report on Intergovernmental Panel on Climate Change.* Intergovernmental Panel on Climate Change (IPCC).

Pacifici, M., Foden, W. B., Visconti, P., Watson, J. E. M., Butchart, S. H. M., Kovacs, K. M., Scheffers, B. R., Hole, D. G., Martin, T. G., Akçakaya, H. R., Corlett, R. T., Huntley, B., Bickford, D., Carr, J. A., Hoffmann, A. A., Midgley, G. F., Pearce-Kelly, P., Pearson, R. G., Williams, S. E.,. . . Rondinini, C. 2015. Assessing species vulnerability to climate change. *Nature Climate Change* 5(3): 215–224. https://doi.org/10.1038/nclimate2448

Pardikes, N. A., Revilla, T., Lue, C., Thierry, M. D., Souto-Vilarós, D., and Hrcek, J. 2022. Effects of phenological mismatch under warming are modified by community context. *Global Change Biology* 28: 4013–4026.

Parmesan, C., and Yohe, G. 2003. A globally coherent fingerprint of climate change impacts across natural systems. *Nature* 421(6918): 37–42. https://doi.org/10.1038/nature01286

Pautasso, M., Döring, T. F., Garbelotto, M., Pellis, L., and Jeger, M. J. 2012. Impacts of climate change on plant diseases—Opinions and trends. *European Journal of Plant Pathology* 133(1): 295–313. https://doi.org/10.1007/s10658-012-9936-1

Peñuelas, J., and Llusià, J. 2003. BVOCs: Plant defense against climate warming? *Trends in Plant Science* 8(3): 105–109. https://doi.org/10.1016/S1360-1385(03)00008-6

Peralta, G., Vázquez, D. P., Chacoff, N. P., Lomáscolo, S. B., Perry, G. L. W., and Tylianakis, J. M. 2020. Trait matching and phenological overlap increase the spatio-temporal stability and functionality of plant—pollinator interactions. *Ecology Letters* 23(7): 1107–1116. https://doi.org/10.1111/ele.13510

Perrings, C., Burgiel, S., Lonsdale, M., Mooney, H., and Williamson, M. 2010. International cooperation in the solution to trade-related invasive species risks. *Annals of the New York Academy of Sciences* 1195: 198–212. https://doi.org/10.1111/j.1749-6632.2010.05453.x

Pincebourde, S., and Casas, J. 2016. Hypoxia and hypercarbia in endophagous insects: Larval position in the plant gas exchange network is key. *Journal of Insect Physiology* 84: 137–153. https://doi.org/10.1016/j.jinsphys.2015.07.006

Post, E., Steinman, B. A., and Mann, M. E. 2018. Acceleration of phenological advance and warming with latitude over the past century. *Scientific Reports* 8(1): 3927. https://doi.org/10.1038/s41598-018-22258-0

Prentice, I. C., Farquhar, G. D., Fasham, M. J. R., Goulden, M. L., Heimann, M., Jaramillo, V. J., Kheshgi, H. S., Le Quéré, C., Scholes, R. J., Wallace, D. W., and Archer, D. 2001. The carbon cycle and atmospheric carbon dioxide. In *Climate change. Scientific basis* (pp. 185–237).

Pritchard, S. G. 2011. Soil organisms and global climate change. *Plant Pathology* 60(1): 82–99. https://doi.org/10.1111/j.1365-3059.2010.02405.x

Raven, P. H., and Wagner, D. L. 2021. Agricultural intensification and climate change are rapidly decreasing insect biodiversity. *Proceedings of the National Academy of Sciences of the United States of America* 118(2): e2002548117. https://doi.org/10.1073/pnas.2002548117

Renner, S. S., and Zohner, C. M. 2018. Climate change and phenological mismatch in trophic interactions among plants, insects, and vertebrates. *Annual Review of Ecology, Evolution, and Systematics* 49(1): 165–182. https://doi.org/10.1146/annurev-ecolsys-110617-062535

Renoz, F., Pons, I., and Hance, T. 2019. Evolutionary responses of mutualistic insect—bacterial symbioses in a world of fluctuating temperatures. *Current Opinion in Insect Science* 35: 20–26. https://doi.org/10.1016/j.cois.2019.06.006

Rivero, R. M., Sánchez, E., Ruiz, J. M., and Romero, L. 2003. Influence of temperature on biomass, iron metabolism and some related bioindicators in tomato and watermelon plants. *Journal of Plant Physiology* 160(9): 1065–1071. https://doi.org/10.1078/0176-1617-00907

Rogelj, J., Shindell, D., Jiang, K., Fifita, S., Forster, P., Ginzburg, V., and Zhou, W. 2018. *Mitigation pathways compatible with 1.5 C in the context of sustainable development* (pp. 93–174). Intergovernmental Panel on Climate Change.

Román-Palacios, C., and Wiens, J. J. 2020. Recent responses to climate change reveal the drivers of species extinction and survival. *Proceedings of the National Academy of Sciences of the United States of America* 117(8): 4211–4217. https://doi.org/10.1073/pnas.1913007117

Rosenzweig, C., Major, D. C., Demong, K., Stanton, C., Horton, R., and Stults, M. 2007. Managing climate change risks in New York City's water system: Assessment and adaptation planning. *Mitigation and Adaptation Strategies for Global Change* 12(8): 1391–1409. https://doi.org/10.1007/s11027-006-9070-5

Rubalcaba, J. G., Verberk, W. C. E. P., Hendriks, A. J., Saris, B., and Woods, H. A. 2020. Oxygen limitation may affect the temperature and size dependence of metabolism in aquatic ectotherms. *Proceedings of the National Academy of Sciences of the United States of America* 117(50): 31963–31968. https://doi.org/10.1073/pnas.2003292117

Samal, I., Bhoi, T. K., Majhi, P. K., Murmu, S., Pradhan, A. K., Kumar, D., Saini, V., Paschapur, A. U., Raj, M. N., Ankur, Manik, S., Behera, P. P., Mahanta, D. K., Komal, J., Alam, P., and Balawi, T. A. 2022. Combatting insects mediated biotic stress through plant associated endophytic entomopathogenic fungi in horticultural crops. *Frontiers in Plant Science* 13: 1098673. https://doi.org/10.3389/fpls.2022.1098673

Schleuning, M., Neuschulz, E. L., Albrecht, J., Bender, I. M. A., Bowler, D. E., Dehling, D. M., Fritz, S. A., Hof, C., Mueller, T., Nowak, L., Sorensen, M. C., Böhning-Gaese, K., and Kissling, W. D. 2020. Trait-based assessments of climate-change impacts on interacting species. *Trends in Ecology and Evolution* 35(4): 319–328. https://doi.org/10.1016/j.tree.2019.12.010

Schowalter, T. D. 2013. *Insects and sustainability of ecosystem services*. CRC Press.

Schowalter, T. D., Noriega, J. A., and Tscharntke, T. 2018. Insect effects on ecosystem services—Introduction. *Basic and Applied Ecology* 26: 1–7. https://doi.org/10.1016/j.baae.2017.09.011

Sentis, A., Hemptinne, J. L., and Brodeur, J. 2013. Effects of simulated heat waves on an experimental plant-herbivore-predator food chain. *Global Change Biology* 19(3): 833–842. https://doi.org/10.1111/gcb.12094

Shah, A. A., Dillon, M. E., Hotaling, S., and Woods, H. A. 2020. High elevation insect communities face shifting ecological and evolutionary landscapes. *Current Opinion in Insect Science* 41: 1–6. https://doi.org/10.1016/j.cois.2020.04.002

Shah, A. A., Funk, W. C., and Ghalambor, C. K. 2017. Thermal acclimation ability varies in temperate and tropical aquatic insects from different elevations. *Integrative and Comparative Biology* 57(5): 977–987. https://doi.org/10.1093/icb/icx101

Sharma, H. C., War, A. R., Pathania, M., Sharma, S. P., Akbar, S. M., and Munghate, R. S. 2016. Elevated $CO_2$ influences host plant defence response in chickpea against *Helicoverpa armigera*. *Arthropod-Plant Interactions* 10(2): 171–181. https://doi.org/10.1007/s11829-016-9422-3

Shrestha, S. 2019. Effects of climate change in agricultural insect pest. *Acta Scientific Agriculture* 3(12): 74–80. https://doi.org/10.31080/ASAG.2019.03.0727

Singh, S., Bhoi, T. K., Khan, I., Vyas, V., Athulya, R., Rathi, A., and Samal, I. 2023. Climate change drivers and soil microbe-plant interactions. In *Climate change and microbiome dynamics* (pp. 157–176). Springer. https://doi.org/10.1007/978-3-031-21079-2_11

Skendžić, S., Zovko, M., Živković, I. P., Lešić, V., and Lemić, D. 2021. The impact of climate change on agricultural insect pests. *Insects* 12(5): 440. https://doi.org/10.3390/insects12050440

Staley, J. T., Hodgson, C. J., Mortimer, S. R., Morecroft, M. D., Masters, G. J., Brown, V. K., and Taylor, M. E. 2007. Effects of summer rainfall manipulations on the abundance and vertical distribution of herbivorous soil macro-invertebrates. *European Journal of Soil Biology* 43(3): 189–198. https://doi.org/10.1016/j.ejsobi.2007.02.010

Steffen, W., Richardson, K., Rockström, J., Cornell, S. E., Fetzer, I., Bennett, E. M., Biggs, R., Carpenter, S. R., de Vries, W., de Wit, C. A., Folke, C., Gerten, D., Heinke, J., Mace, G. M., Persson, L. M., Ramanathan, V., Reyers, B., and Sörlin, S. 2015. Sustainability. Planetary boundaries: Guiding human development on a changing planet. *Science* 347(6223): 1259855. https://doi.org/10.1126/science.1259855

Stoeckli, S., Felber, R., and Haye, T. 2020. Current distribution and voltinism of the brown marmorated stink bug, *Halyomorpha halys*, in Switzerland and its response to climate change using a high-resolution CLIMEX model. *International Journal of Biometeorology* 64(12): 2019–2032. https://doi.org/10.1007/s00484-020-01992-z

Sunday, J. M., Bates, A. E., Kearney, M. R., Colwell, R. K., Dulvy, N. K., Longino, J. T., and Huey, R. B. 2014. Thermal-safety margins and the necessity of thermoregulatory behavior across latitude and elevation. *Proceedings of the National Academy of Sciences of the United States of America* 111(15): 5610–5615. https://doi.org/10.1073/pnas.1316145111

Sutherst, R. W., Baker, R. H. A., Coakley, S. M., Harrington, R., Kriticos, D. J., and Scherm, H. 2007. Pest under global change—Meeting your future landlords? In J. G. Canadell, D. E. Pataki, L. F. Pitelka, et al. (Eds.), *Terrestrial ecosystems in a changing world* (pp. 211–226). Springer. https://doi.org/10.1007/978-3-540-32730-1_17

Sutherst, R. W., Constable, F., Finlay, K. J., Harrington, R., Luck, J., and Zalucki, M. P. 2011. Adapting to crop pest and pathogen risks under a changing climate. *WIREs Climate Change* 2(2): 220–237. https://doi.org/10.1002/wcc.102

Sweeney, B. W., Funk, D. H., Camp, A. A., Buchwalter, D. B., and Jackson, J. K. 2018. Why adult mayflies of *Cloeon dipterum* (Ephemeroptera:Baetidae) become smaller as temperature warms. *Freshwater Science* 37(1): 64–81. https://doi.org/10.1086/696611

Tenllado, F., and Canto, T. 2020. Effects of a changing environment on the defenses of plants to viruses. *Current Opinion in Virology* 42: 40–46. https://doi.org/10.1016/j.coviro.2020.04.007

Tewksbury, J. J., Huey, R. B., and Deutsch, C. A. 2008. Ecology. Putting the heat on tropical animals. *Science* 320(5881): 1296–1297. https://doi.org/10.1126/science.1159328

Thakur, M. P. 2020. Climate warming and trophic mismatches in terrestrial ecosystems: The green–brown imbalance hypothesis. *Biology Letters* 16(2): 20190770. https://doi.org/10.1098/rsbl.2019.0770

Thierry, M., Pardikes, N. A., Lue, C. H., Lewis, O. T., and Hrček, J. 2021. Experimental warming influences species abundances in a drosophila host community through direct effects on species performance rather than altered competition and parasitism. *PLOS ONE* 16(2): e0245029. https://doi.org/10.1371/journal.pone.0245029

Thomas, M. B., and Blanford, S. 2003. Thermal biology in insect-parasite interactions. *Trends in Ecology and Evolution* 18(7): 344–350. https://doi.org/10.1016/S0169-5347(03)00069-7

Thomson, L. J., MacFadyen, S., and Hoffmann, A. A. 2010. Predicting the effects of climate change on natural enemies of agricultural pests. *Biological Control* 52(3): 296–306. https://doi.org/10.1016/j.biocontrol.2009.01.022

Tougeron, K., Brodeur, J., Le Lann, C., and Van Baaren, J. 2019. How climate change affects the seasonal ecology of insect parasitoids. *Ecological Entomology* 45(2): 167–181. https://doi.org/10.1111/een.12792

Tougeron, K., and Iltis, C. 2022. Impact of heat stress on the fitness outcomes of symbiotic infection in aphids: A meta-analysis. *Proceedings. Biological Sciences* 289(1971): 20212660. https://doi.org/10.1098/rspb.2021.2660

Tougeron, K., Le Lann, C., Brodeur, J., and van Baaren, J. 2017. Are aphid parasitoids from mild winter climates losing their winter diapause? *Oecologia* 183(3): 619–629. https://doi.org/10.1007/s00442-016-3770-7

Tylianakis, J. M., Didham, R. K., Bascompte, J., and Wardle, D. A. 2008. Global change and species interactions in terrestrial ecosystems. *Ecology Letters* 11(12): 1351–1363. https://doi.org/10.1111/j.1461-0248.2008.01250.x

Tylianakis, J. M., and Morris, R. J. 2017. Ecological networks across environmental gradients. *Annual Review of Ecology, Evolution, and Systematics* 48(1): 25–48. https://doi.org/10.1146/annurev-ecolsys-110316-022821

Van Dievel, M., Stoks, R., and Janssens, L. 2017. Beneficial effects of a heat wave: Higher growth and immune components driven by a higher food intake. *Journal of Experimental Biology* 220(21): 3908–3915. https://doi.org/10.1242/jeb.158899

van Dyck, H., Bonte, D., Puls, R., Gotthard, K., and Maes, D. 2015. The lost generation hypothesis: Could climate change drive ectotherms into a developmental trap? *Oikos* 124(1): 54–61. https://doi.org/10.1111/oik.02066

Venter, O., Sanderson, E. W., Magrach, A., Allan, J. R., Beher, J., Jones, K. R., Possingham, H. P., Laurance, W. F., Wood, P., Fekete, B. M., Levy, M. A., and Watson, J. E. 2016. Sixteen years of change in the global terrestrial human footprint and implications for biodiversity conservation. *Nature Communications* 7: 12558. https://doi.org/10.1038/ncomms12558

Verberk, W. C. E. P., Atkinson, D., Hoefnagel, K. N., Hirst, A. G., Horne, C. R., and Siepel, H. 2021. Shrinking body sizes in response to warming: Explanations for the temperature—size rule with special emphasis on the role of oxygen. *Biological Reviews of the Cambridge Philosophical Society* 96(1): 247–268. https://doi.org/10.1111/brv.12653

Vercelli, M., Novelli, S., Ferrazzi, P., Lentini, G., and Ferracini, C. A. 2021. A qualitative analysis of beekeepers' perceptions and farm management adaptations to the impact of climate change on honey bees. *Insects* 12(3): 228. https://doi.org/10.3390/insects12030228

Vilà, M., Beaury, E. M., Blumenthal, D. M., Bradley, B. A., Early, R., Laginhas, B. B., Trillo, A., Dukes, J. S., Sorte, C. J. B., and Ibáñez, I. 2021. Understanding the combined impacts of weeds and climate change on crops. *Environmental Research Letters* 16(3): 034043. https://doi.org/10.1088/1748-9326/abe14b

Wagner, D. L. 2020. Insect declines in the anthropocene. *Annual Review of Entomology* 65: 457–480. https://doi.org/10.1146/annurev-ento-011019-025151

Wagner, D. L., Grames, E. M., Forister, M. L., Berenbaum, M. R., and Stopak, D. 2021. Insect decline in the anthropocene: Death by a thousand cuts. *Proceedings of the National Academy of Sciences of the United States of America* 118(2): e2023989118. https://doi.org/10.1073/pnas.2023989118

Wang, Y., Yan, J., Sun, J., Shi, W., Harwood, J. D., Monticelli, L. S., Tan, X., and Chen, J. 2021. Effects of field simulated warming on feeding behavior of *Sitobion avenae* (Fabricius) and host defense systems. *Entomologia Generalis* 41(6): 567–578. https://doi.org/10.1127/entomologia/2021/1271

Warren, M. S., Maes, D., van Swaay, C. A. M., Goffart, P., Van Dyck, H., Bourn, N. A. D., Wynhoff, I., Hoare, D., and Ellis, S. 2021. The decline of butterflies in Europe: Problems, significance, and possible solutions. *Proceedings of the National Academy of Sciences of the United States of America* 118(2): e2002551117. https://doi.org/10.1073/pnas.2002551117

Warren, R., Price, J., Graham, E., Forstenhaeusler, N., and VanDerWal, J. 2018. The projected effect on insects, vertebrates, and plants of limiting global warming to 1.5°C Rather than 2°C. *Science* 360(6390): 791–795. https://doi.org/10.1126/science.aar3646

Weldegergis, B. T., Zhu, F., Poelman, E. H., and Dicke, M. 2015. Drought stress affects plant metabolites and herbivore preference but not host location by its parasitoids. *Oecologia* 177(3): 701–713. https://doi.org/10.1007/s00442-014-3129-x

Wellenreuther, M., Dudaniec, R. Y., Neu, A., Lessard, J. P., Bridle, J., Carbonell, J. A., Diamond, S. E., Marshall, K. E., Parmesan, C., Singer, M. C., Swaegers, J., Thomas, C. D., and Lancaster, L. T. 2022. The importance of eco-evolutionary dynamics for predicting and managing insect range shifts. *Current Opinion in Insect Science* 52: 100939. https://doi.org/10.1016/j.cois.2022.100939

Williams, C. M., Henry, H. A., and Sinclair, B. J. 2015. Cold truths: How winter drives responses of terrestrial organisms to climate change. *Biological Reviews of the Cambridge Philosophical Society* 90(1): 214–235. https://doi.org/10.1111/brv.12105

Williams, C. M., Marshall, K. E., MacMillan, H. A., Dzurisin, J. D. K., Hellmann, J. J., and Sinclair, B. J. 2012. Thermal variability increases the impact of autumnal warming and drives metabolic depression in an overwintering butterfly. *PLOS ONE* 7(3): e34470. https://doi.org/10.1371/journal.pone.0034470

Woodward, A., Smith, K. R., Campbell-Lendrum, D., Chadee, D. D., Honda, Y., Liu, Q., Olwoch, J., Revich, B., Sauerborn, R., Chafe, Z., Confalonieri, U., and Haines, A. 2014. Climate change and health: On the latest IPCC report. *Lancet* 383(9924): 1185–1189. https://doi.org/10.1016/S0140-6736(14)60576-6

Yadav, S., Stow, A., and Dudaniec, R. Y. 2021. Microgeographical adaptation corresponds to elevational distributions of congeneric montane grasshoppers. *Molecular Ecology* 30(2): 481–498. https://doi.org/10.1111/mec.15739

Yamamura, K., Yokozawa, M., Nishimori, M., Ueda, Y., and Yokosuka, T. 2006. How to analyze long-term insect population dynamics under climate change: 50-year data of three insect pests in paddy fields. *Population Ecology* 48(1): 31–48. https://doi.org/10.1007/s10144-005-0239-7

Yonow, T., Kriticos, D. J., Kirichenko, N., and Ota, N. 2018. Considering biology when inferring range-limiting stress mechanisms for agricultural pests: A case study of the beet armyworm. *Journal of Pest Science* 91(2): 523–538. https://doi.org/10.1007/s10340-017-0938-9

Yoro, K. O., and Daramola, M. O. 2020. $CO_2$ emission sources, greenhouse gases, and the global warming effect. In M. R. Rahimpour, M. Farsi, and M. A. Makarem (Eds.), *Advances in carbon capture* (1st ed., pp. 3–28). Woodhead Publishing.

Zhu, L., Ives, A. R., Zhang, C., Guo, Y., and Radeloff, V. C. 2019. Climate change causes functionally colder winters for snow cover-dependent organisms. *Nature Climate Change* 9(11): 886–893. https://doi.org/10.1038/s41558-019-0588-4

Ziska, L. H., Blumenthal, D. M., Runion, G. B., Hunt, E. R., Jr., and Diaz-Soltero, H. 2011. Invasive species and climate change: An agronomic perspective. *Climatic Change* 105(1–2): 13–42. https://doi.org/10.1007/s10584-010-9879-5

# 5 Climate Change and Insect Pathogens

*Raman Jasrotia, Neha Jamwal, and Seema Langer*

## 5.1 INTRODUCTION

The climate of the earth is never static. It does not imply change in a day, rather it is a long-term average of weather events of a region and has changed countless times since the earth's existence, due to natural events like sun spots, ice age glaciations and so on. Climate change is a measurable effect of continual warming trends and is usually measured in major shifts in rainfall, snow, temperature, wind patterns, etc. continuing for eras or more. But whenever a debate is heeded on 'climate change', all that is emphasised is the change in climate over the past few years credited to human beings directly or indirectly. Much of the share of climate change by humans is attributed to the burning of fossil fuels, deforestation, urbanisation, industrialisation, overpopulation, irrational use of natural resources, etc. A varied array of events like the plunging of the Arctic Sea ice, rise in ocean heat, and shifting of the temperature-sensitive organisms towards poles provide irrefutable sign of planetary-scale warming.

The earth has warmed at an unprecedented rate, and the average air temperature of the earth's surface has risen by about 1°C (1.8°F) within a century where at least half of the rise has been witnessed since the mid-1970s. Since 1992, each year has been one of the warmest years on record. An upsurge in extreme weather events like wildfires, strong tropical storms, heat waves, etc. led the year 2016 to be the hottest year so far. An increase in global temperature of about 1°C–4.5°C has been estimated in the coming century (IPCC, 2014) owing to variable precipitation, rise in carbon dioxide levels, increased temperature of the earth's surface and their overall interactions (Nayak et al., 2020; Jasrotia et al., 2022). Global warming and continental climate shift have created worldwide chaos as the habitats of various organisms are on a shift, invasiveness of many organisms (plants, animals and pathogens) has accelerated, indigenous organisms are getting vulnerable, diseases have become more powerful, and so on. Global warming impacts the planet in numerous ways (Figure 5.1) which endanger the existence of many organisms including man and occur due to numerous reasons both natural and human induced, but much of it is the cause of human meddling with the earth's plan (Figure 5.2). Much of the effects of climate change and global warming are taken out upon the agriculture-driven economies due to their sole dependence on rain water and temperature for high yields. Loss of efforts and economy are what leave a great void between developing and developed countries.

Arthropods have inhabited the earth since 480 million years ago with an average number of species varying between 1,170,000 and 10,000,000, making up around 80% population among the animal species (Lovett and St. Leger, 2017). Of all the arthropods, class Insecta is the chief subgroup which comprises the maximum species-rich members of any biological guild in land-living and freshwater situations. Insects have developed extensively based on the type and structure of environment they live in, thereby leading to their massive population across the globe and adaptations which favour their presence even in the coldest regions. Just because insects can survive in any given condition does not mean that there is no check on their population. In the past 400–450 million years, various organisms (often termed pathogens) have coevolved along with insects (Odegaard, 2000; Vega and Blackwel, 2005) in order to keep their population in check, along with other natural enemies. Similar to human beings and other vertebrates, insects are also vulnerable to


DOI: 10.1201/9781003382089-5

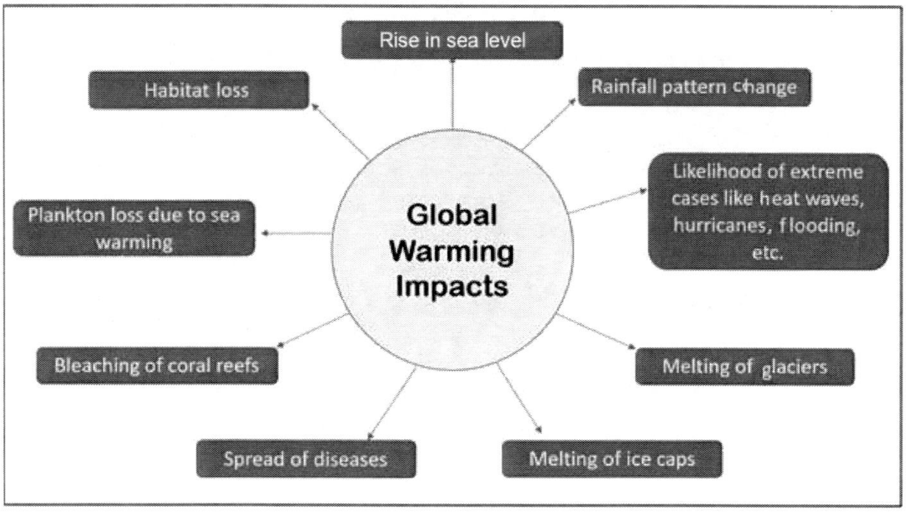


**FIGURE 5.1** Impacts of global warming.

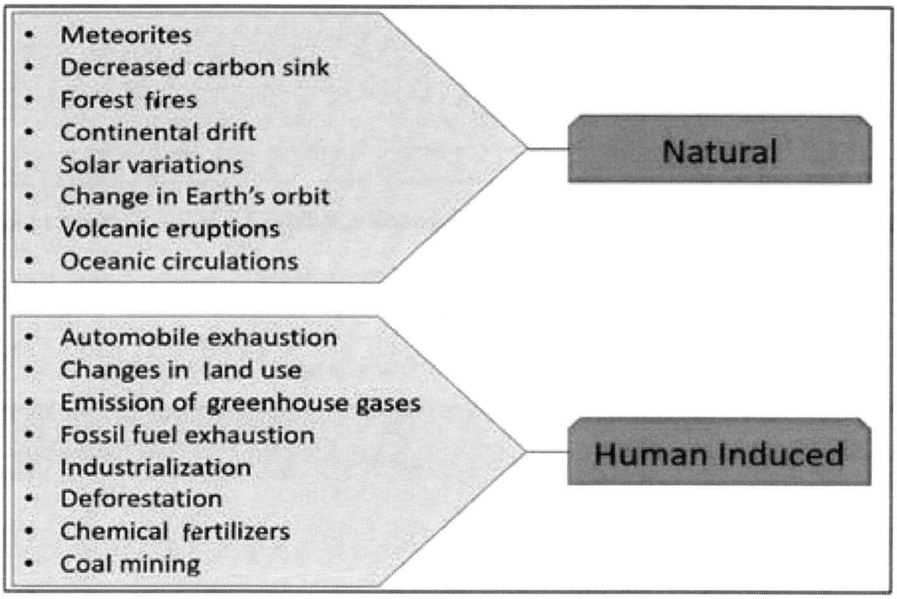


**FIGURE 5.2** Natural and Human induced causes of global warming.

these bountiful disease-causing entities which mostly include fungi, bacteria, protozoa, nematodes and viruses. Insects are extremely responsive to even slight differences in the temperature as various life processes are governed by prevailing climatic conditions. Climate change alters the insect population directly (Parmesan, 2007; Jasrotia et al., 2022) or indirectly (Harrington et al., 2001; Jasrotia et al., 2022) and therefore incessant monitoring of these sensitive species gives an account of the varying biodiversity patterns around the globe (Gregory et al., 2009). Climate change has a very negative effect on insects (Figure 5.3).

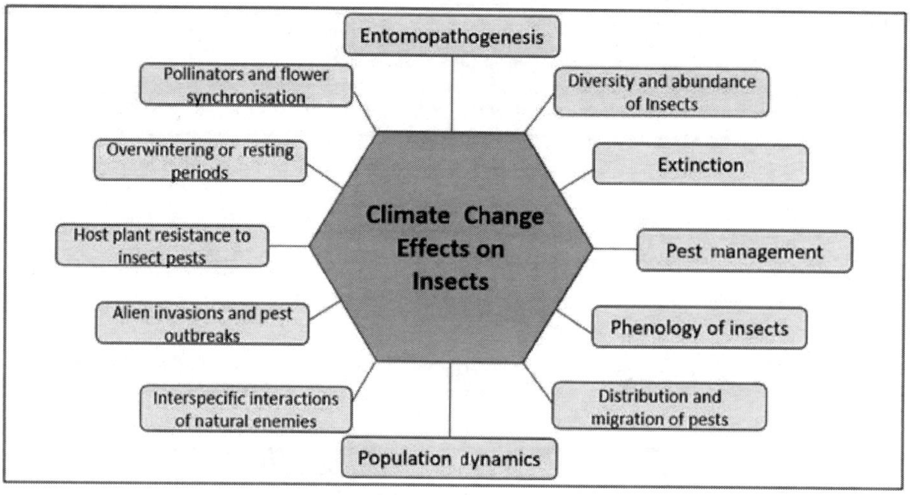


**FIGURE 5.3** Effect of climate change on insects.

### 5.1.1 Entomopathogenesis

A recently organized discipline as vocation leads to a new field of invertebrate pathology, the roots of which date back to ancient history where solutions for preventing honeybee and silkworm diseases were sought (Lacey et al., 2001). The early pioneers in this field, such as Louis Pasteur, Agostinio Bassi and Elie Matchnikoff, had suggested the application of microbes to control insect pest populations, and since then the events of microbial trails began. Many species have, thus far, been employed as potentials for biocontrol in forestry, stockpiled production, orchards, lawns and turfs, horticulture and for decline in pest and vector insects of veterinary and medical importance (Burges, 1981; Tanada and Kaya, 1993; Lacey and Kaya, 2000; Lacey et al., 2001). With numerous advantages over chemical pesticides, entomopathogenecity includes safe consumption of food and other resources by humans and nontarget organisms, increased biodiversity in managed ecosystems, reduction of pesticide residues in food and preservation of other natural enemies (Lacey et al., 2001). Fungi are the most common disease-causing agents in insects which have been used as a brilliant chemical alternative (Lovett and St. Leger, 2017), but it was not until the development of bacterium *Bacillus thuringiensis* Berliner, that this field started gaining momentum (Lacey et al., 2001). Discovery of novel varieties of *B. thuringiensis* Berliner against a wide array of insect orders like Diptera, Coleoptera and Lepidoptera has heightened its utility.

The use of these pathogens for effective insect control leads to the same strategies as followed by other biological control agents (Harper, 1987). They may be used for augmentation, conservation, inoculative release or inundative release. In this chapter we discuss some achievements in the field of entomopathogenecity and future speculations.

### 5.1.2 Naturally Occurring Pathogens

There are ample natural insect enemies that employ an extensive control of target insect populace. Several naturally occurring fungi and viruses exhibit epizootics, which is often responsible for a significant crash in an insect population (McCoy et al., 1988). Epizootics produced by nucleopolyhedroviruses (NPVs) of gypsy moth (*Lymantria dispar*) and sawflies (*Neodiprion* spp. and *Gilpinia hercyniae*) are highly credited (Kaya, 1976; Woods and Elkinton, 1987). Fungi mostly attack the phytophagous sucking insects because they conquer the host via the exoskeleton making them remarkable entomopathogens (Latge and Papierok, 1988). Very little is known about epizootics by protozoans, but its significance in the regulation of pest populations cannot be overlooked (Maddox, 1987; Brooks, 1988).

#### 5.1.2.1 Fungi

About 700 species falling under 90 genera are known to cause entomopathogenecity (St. Leger et al., 2011), but for insect control management strategies, only about ten species are currently being developed (Hajek and St. Leger, 1994). Except for basidiomycetes, these are virtually spread in every chief fungal taxonomic group (Roberts and Humber, 1981). The species of group Entomophthorales are among the most promising entomopathogenic fungi, responsible for epizootics often successfully regulating the insect pest population (Lacey et al., 2001). The biggest achievement was the introduction of Entomophthoralean fungi *Entomophaga maimaiga* introduced in the northwest United States in as early as 1910 and 1911 for control of gypsy moth, *L. dispar*, but unfortunately they could not be recovered. Later, a virulent strain of it was isolated and described from the moth in Japan (Soper et al., 1988). A variety of fungi-like species of *Metarhizium, Isaria, Lecanicillium* and *Beauveria* are produced commercially as they can be mass produced easily and can be used as inundative insecticides relative to classical biocontrol. Various species of Hyphomycetes, despite having a broad host range (Goettel et al., 1995), have been validated for activities that are highly specific against many insect pests which are the main candidates for commercial production and can prove beneficial against homopterous pests. These groups of fungi have been examined for their role as broad-spectrum insect pest management tools against grasshoppers, locusts, termites, whiteflies, beetles, thrips, aphids, etc. (McCoy et al., 1988; Fargues and Maniania, 1992; Zimmermann, 1993; Milner and Prior, 1994; Goettel et al., 1995; Keller et al., 1997). The season-long control of *Ostrinia nubilalis* in corn, is a result of conjugation with inundative application of the endophytic nature of *Beauveria bassiana* and incur a winning effect on the larvae that are overwintering (Anderson and Lewis, 1991). A multifaceted set of interacting processes is essential for or inhibitory to development of epizootics instigated by the entomopathogenic fungi. These processes can be both environmental and biotic in nature, including solar radiation sensitivity, behaviour of the host, vigor and age of pathogen, microbial antagonists, pesticide presence, physiological conditions and age, humidity, temperature and thresholds of the inoculum (Ferron et al., 1991; Lacey and Goettel, 1995). Highly effective methods of genetic engineering, environmental manipulation, culture formulation and methods, accurate dosage and timing (owing to availability of host stages) are critical for these entomopathogens to exhibit full epizootic nature. An amalgam of several techniques can be made in order to expect and improve the pathogenecity of fungi such as use of semiochemicals to trap the adult insects and infect them with the fungal spores which will aid in insect biocontrol as well as autodispersion of the fungi into difficult larval habitats such as soil (Klein and Lacey, 1999), cabbage heads (Furlong et al., 1995; Quintela and McCoy, 1997), corn ears (Vega et al., 1995) and other habitats. Similarly, control of the citrus root weevil, *Diaprepes abbreviatus*, larvae has been done using an amalgam of *M. anisopliae* and *B. bassiana* together with imidacloprid (a chloronictinyl insecticide) (Lacey et al., 2001). Although scientists have improved the effect of these fungi steadily in past decades via genetic engineering, when compared to the use of bacteria, they still need a lot of manoeuvring to escape the lag to produce new variants of fungi with enhanced virulence (Charley et al., 1997). They have achieved moderate success as biocontrol agents in several developing countries, and a significant improvement was achieved as the production of a strain of *Metarhizium* that produces a single-chain antibody fragment to block the transmission of malaria (Fang et al., 2011).

#### 5.1.2.2 Bacteria

Spore-forming bacilli are among the major bacteria species that account for the infection and killing of uncompromised, healthy insects. Of these spore-forming bacilli, the major groups include *B. thuringiensis* and *B. sphaericus* from which protoxins are released at the time of sporulation (Aronson et al., 1986). There are about 410 registered formulations of *B. thuringiensis* and 6 registered formulations of *B. lentimorbus* and *B. popillae* meant to be used against insect pests in the

United States (National Pesticide Information Retrieval System, Purdue University, West Lafayette, IN). The isolates of *B. thuringiensis* have been produced commercially against numerous insect pests belonging to Coleoptera, Diptera and Lepidoptera (Shah and Goettel, 1999). As of 1998, only 200 isolates of *B. thuringiensis* were registered for use in the United States (Schnepf et al., 1998). The use of these bacilli has both pros and cons. The advantage lies in the production of stable spores that can be formulated readily for effective pest management and are also curiously harmless to humans and other mammals (nontarget fauna) (Burges, 1982). The disadvantage of using this bacillus, on the other hand, is its non-inducting epizootics, except for cases where insects are restricted or live in jam-packed parts like insect-rearing facilities, beehives, or stored grain bins (Dulmage et al., 1981). Also, they provide short-term pest protection from a single application in cases where these are used and applied to the vegetation and get washed away by rain showers or get disabled by sunlight (Aronson et al., 1986). It has been discussed that many insect species developed resistance to *B. thuringiensis* but *Plutella xylostella* (Diamondback moth) is the only known species to develop so (Ferre et al., 1991). This can potentially be overcome by high doses, non-treated refugia, seed mixtures (non-transformed cultivars and transgenics) and rotation of toxins of *B. thuringiensis* (Gelernter, 1997).

There are numerous ways to deliver the *B. thuringiensis* insecticidal proteins which include the use of formulated products that are produced from naturally or conjugated strains and transgenic plants, which are developed as a result of incorporation of toxic genes that produce toxin in the plant genome. Sometimes, some alternative methods can also be implemented to deliver the *B. thuringiensis* genes, like using endophytic bacteria (*Clavibacter xyli or B. cereus*) to correctly deliver the toxins to the target insects. Also, certain other plant-colonising bacteria like *Pseudomonas cepacia, P. fluorescens, Rhizobium leguminosarum* and *Azosporillium* spp. are also used to administer the *B. thuringiensis* toxin in specific target insects (Obukowicz et al., 1986a, 1986b; Udayasuriyan et al., 1995). There has also been an attempt to control the mosquito larvae using this technique by building certain specific delivery systems onto hosts residing in aquatic systems which include cyanobacterium *Agmellenum quadruplicatum*, *Synechococcus* spp. and *Caulobacter crescentus* (Lacey et al., 2001).

#### 5.1.2.3 Virus

In order to fight the detrimental effects of Palm rhinoceros beetle (*Oryctes rhinoceros*) in the coconut and oil palms, a non-occluded virus was used for long-term beetle control (Bedford, 1980, 1981). Here the reservoirs turned out to be the adult beetles that were chronically infected and dissipated the virus and contaminated the whole colony of beetles via oral contact with the substrates infested with the virus. The highest potential as microbial control agents of insects are the Baculoviruses (Baculoviridae: Nucleopolyhedroviruses [NPVs] and granuloviruses [GVs]) (Hunter-Fujita et al., 1998). With limited knowledge about the host insects, baculoviruses are known to infect Hymenoptera and Lepidoptera mostly. Some of the registered baculoviruses for use in insect control are *Spodoptera* spp., *Plodia interpunctella* and *Helicoverpa/Heliothis* spp. (Lacey et al., 2001).

#### 5.1.2.4 Nematodes

A surfeit of about 30 families of nematodes is linked with the insects and other invertebrates (Kaya and Gaugler, 1993), but for research and development, seven families of nematodes have been a major focus due to their probability as biocontrol agents of insects. These include Mermithidae, Tetradonematidae, Allantonematidae, Phaenopsitylenchidae, Sphaerulariidae, Steinernematidae and Heterorhabditidae. Of all these families, Tetradonematidae, Allantonematidae and Sphaerulariidae are hard to mass produce owing to their complex life cycle, and they seem to be density independent, while Steinernematidae and Heterorhabditidae are receiving the most concern in this field against soil-borne insects. The reason for their growing concern is their ability to act analogous to parasitoids as they bear chemoreceptors and actively engage in searching their hosts (Gaugler et al., 1997);

**TABLE 5.1**
**Microbes Pathogenic to Insects**

| Group | Representative Species | Susceptible Insects | References |
|---|---|---|---|
| Viruses | *Autographa californica* Nuclear polyhedrosis virus (NPV) | Hemiptera, Lepidoptera, Hymenoptera, Coleoptera, Orthoptera, Trichoptera, Neuroptera, etc. | Faulkner, 1981; Granados, 1981; Kelly, 1981; Payne, 1981 |
| Rickettsiae | *Rickettsiella melolonthae* | Diptera, Coleoptera, Orthoptera | Krieg, 1971 |
| Fungi | *Beauvaria bassiana* *Metarhizium anisopliae* | Hymenoptera, Lepidoptera, Coleoptera, Diptera, Homoptera | Roberts and Yendol, 1971; Sweeney, 1981; Zacharuk, 1981 |
| Bacteria | *Bacillus popilliae* | Scarabaeidae beetles | Bulla et al., 1978; Burges and Hussey, 1971; Burges, 1982 |
| | *Bacillus thuringiensis* | Diptera, Lepidoptera | Burges and Hussey, 1971; Burges, 1982 |
| | *Bacillus sphaericus* | Mosquitoes | Burges and Hussey, 1971; Burges, 1982 |
| | *Clostridium malacosoma* | Tent caterpillars (*Malacoma* spp.) | Bucher, 1961 |
| | *Pseudomonas aeruginosa* | Opportunistic pathogen with susceptible species in major insect taxa | Bucher, 1967 |
| | *Xenorhabdus nematophilus* | Insects susceptible to the nematode *Steinernema feltiae* | Poinar, 1971, 1979 |

also, they have an association with the mutualistic bacteria *Xenorhabdus* (Steinernematidae) and *Photorhabdus* (Heterorhabditidae). This mutualistic cooperation is extremely virulent as it causes mortality of its host within 48 hours with no danger to plants, many invertebrates and vertebrates (Kaya and Gaugler, 1993; Lacey et al., 2001).

#### 5.1.2.5 Protozoa

The infections produced by protozoans in insects are commonly host specific and act slowly, ultimately leading to chronic infections. These pathogens in insects are pervasive and comprise an important regulatory check on insect populations (Maddox, 1987; Brooks, 1988). They exhibit a very complex nature as they tend to develop in live hosts with a few requiring intermediate hosts as well. Microsporidia are the most commonly noted protozoans, and with very few successful agents the only species to have been registered and commercially developed is the grasshopper pathogen called *Nosema locustae* (Henry and Oma, 1981; Lacey et al., 2001). Their main drawback is the requirement of *in vivo* production and low level of immediate mortality, in the inundative microbial control.

Some of the microorganisms that are pathogenic to insects have been summarised in Table. 5.1.

## 5.2 CONCLUSION

Naturally there has been a mechanism to control booming insect pest populations across the globe, but due to climate fluctuations, it needs certain manoeuvring. As discussed earlier, a vast variety of insect pathogens are sensitive to temperature fluctuations and cannot tolerate even a slight variability in temperature. With the earth's rising temperature, it is not only the insect pests that are vulnerable but also the pathogens that assist in natural biological control become equally endangered. The right time and adequate dosage with the appropriate virulent are necessary for biological control to happen, but due to global climatic shift, insect invasions to non-native places, extension of their geographical reach, host shifting, acquisition of multiple hosts, etc. has led to ambiguity in pathogen selection and implementation. This may result in challenges, greater than expected, in combatting the control of insect pests which will ultimately lead to challenges in food security and safety and will compromise the health of organisms inhabiting the earth. So, global warming and climatic shift are hazardous, creating a physical disruption of the planet and changing the ecology of the planet to a greater extent, which may be irreversible.

## REFERENCES

Anderson, L., and Lewis, L.C. 1991. Suppression of *Ostrinia nubilalis* (Hubner) (Lepidoptera: Pyralidae) by endophytic *Beauveria bassiana* (Balsamo) Vuillemin. *Environmental Entomology* 20: 1207–1211.

Aronson, A.I., Beckman, W., and Dunn, P. 1986. *Bacillus thringiensis* and related insect pathogens. *Microbiological Reviews* 50(1): 1–24.

Bedford, G.O. 1980. Biology, ecology and control of palm rhinoceros beetles. *Annual Review of Entomology* 25: 309–339.

Bedford, G.O. 1981. Control of rhinoceros beetle by baculovirus. In: Burges, H.D. (ed.), *Microbial Control of Pests and Plant Diseases 1970–1980.* Academic Press, London. pp. 409–426.

Brooks, W.M. 1988. Entomophagous protozoa. In: Ignoffo, C.M., and Mandava, N.B. (eds.), *Handbook of Natural Pesticides, Vol. V: Microbial Insecticides, Part A: Entomogenous Protozoa and Fungi.* CRC Press, Boca Raton, FL. pp. 1–149.

Bucher, G.E. 1961. Artificial culture of *Clostridium brevifaciens* n. sp. and *C. malacosoma* n. sp.; the causes of brachytosis of tent caterpillers. *Journal of Insect Pathology* 3: 439–445.

Bucher, G.E. 1967. Pathogens of tobacco and tomato hornworms. *Journal of Invertebrate Pathology* 9: 82–89.

Bulla, L.A., Jr., Costilow, R.N., and Sharpe, E.S. 1978. Biology of *Bacillus popilliae. Advances in Applied Microbiology* 22: 1–18.

Burges, H.D. (ed.). 1981. *Microbial Control of Pests and Plant Diseases 1970–1980.* Academic Press, London.

Burges, H.D. 1982. Control of insects by bacteria. *Parasitology* 84: 79–117.

Burges, H.D., and Hussey, N.W. 1971. *Microbial control of insects and mites.* Academic Press, Inc., New York.

Charley, A.K., Cobb, B., and Clarkson, J.M. 1997. Towards the improvement of fungal insecticides. In: Evans, H.F. (Chair.), *Microbial Insecticides: Novelty or Necessity? BCPC Symposium Proceedings No. 68.* British Crop Protection Enterprises, pp. 115–126.

Dulmage, H.T., Soper, R.S., and Smith D.B. 1981. The use of bacteria and fungi in insect control. In: *Applications of biorational substances and natural enemies. 1st Japan/USA Symposium on IPM,* Tsukuba, Japan. Secretariat of Agriculture, Forestry and Fisheries Research Council and National Institute of Agricultural Science, Tokyo. pp. 112–127.

Fang, W., Vega-Rodriguez, J., Ghosh, A.K., et al., 2011. Development of transgenic fungi that kill human malaria parasites in mosquitoes. *Science* 331: 1074–1077.

Fargues, J., and Maniania, N.K. 1992. Variabilite de la sensibilite de *Spodopteralittoralis* (Lepidoptera: Noctuidae) a l'hyphomyceteentomopathogene*Nomuraearileyi. Entomophaga* 37: 545–554.

Faulkner, P. 1981. Baculovirus. In: Davidson, E.W. (ed.), *Pathogenesis of Invertebrate Microbial Diseases.* Allenheld, Osmun Publishers, Totowa, NJ. pp. 3–37.

Ferre, J., Real, M.D., Van Rie, J., et al., 1991. Resistance to the *Bacillus thuringiensis* bioinsecticide in a field population of *Plutella xylostellan* is due to change in midgut membrane receptor. *Proceedings of the National Academy of Sciences of the United States of America* 88: 5119–5123.

Ferron, P., Fargues, J., and Riba, G. 1991. Fungi as microbial insecticides against pests. In: Arora, D.K., Ajelio, L., and Mukerji, K.G. (eds.), *Handbook of Applied Mycology.* Dekker, New York, Vol 2. pp. 665–706.

Furlong, M.J., Pell, J.K., Choo, O.P., and Rahman, S.A. 1995. Field and laboratory evaluation of a sex pheromone trap for the auto-dissemination of the fungal entomopathogen *Zoophthora radicans* (Entomophthorales) by the diamond-back moth, *Plutella xylostella* (Lepidoptera: Yponomeutidae). *Bulletin of Entomological Research* 85: 331–337.

Gaugler, R., Lewis, E., and Stuart, R.J. 1997. Ecology in the service of biological control: The case of entomopathogenic nematodes. *Oecologia* 109: 483–489.

Gelernter, W.D. 1997. Resistance to microbial insecticides: The scale of the problem and how to manage it. In: Evans, H.F. (Chair.), *Microbial Insecticides: Novelty or Necessity? BCPC Symposium Proceedings No. 68.* British Crop Protection Enterprises, pp. 201–212.

Goettel, M.S., Johnson, D.L., and Inglis, G.D. 1995. The role of fungi in the biological control of grasshoppers. *Canadian Journal of Botany* 73: S1–S75.

Granados, R.R. 1981. Entomopoxvirus infections in insects. In: Davidson, E.W. (ed.), *Pathogenesis of Invertebrate Microbial Disease.* Allenheld, Osmun Publishers, Totowa, NJ. pp. 101–126.

Gregory, P.J., Johnson, S.N., Newton, A.C., and Ingram, J.S. 2009. Integrating pests and pathogens into the climate change/food security debate. *Journal of Experimental Botany* 60(10): 2827–2838. https://doi.org/10.1093/jxb/erp080

Hajek, A.E., and St. Leger, R.J. 1994. Interactions between fungal pathogens and insect hosts. *Annual Review of Entomology* 39: 293–322.

Harper, J.D. 1987. Applied epizootiology: Microbial control of insects. In: Fuxa, J.R., and Tanada, Y. (eds.), *Epizootiology of Insect Diseases*. Wiley, New York. pp. 473–496.

Harrington, R., Fleming, R.A., and Woiwod, I.P. 2001. Climate change impacts on insect management and conservation in temperate regions: Can they be predicted? *Agricultural and Forest Entomology* 3(4): 233–240. https://doi.org/10.1046/j.1461-9555.2001.00120.x

Henry, J.E., and Oma, E.A. 1981. Pest control of *Nosema locustae*, a pathogen of grasshoppers and crickets. In: Burges, H.D. (ed.), *Microbial Control of Pests and Plant Diseases 1970–1980*. Academic Press, London. pp. 573–586.

Hunter-Fujita, F.R., Entwistle, P.F., Evans, H.F., and Crook, N.E. 1998. *Insect Viruses and Pest Management*. Wiley, Chichester.

IPCC. 2014. *Climate Change 2014: Synthesis Report. Contribution of Working Groups I, II and III to the Fifth Assessment Report of the Intergovernmental Panel on Climate Change*. [Core Writing Team, R. K. Pachauri and L. A. Meyer (eds.)]. IPCC, 151 pp.

Jasrotia, R., Koul, M., Jamwal, N., and Langer, S. 2022. Climate change and global insect dynamics. In: Ahmed, M. (ed.), *Global Agricultural Production: Resilience to Climate Change*. Springer Nature. pp. 335–352.

Kaya, H.K. 1976. Insect pathogens in natural and microbial control of forest defoliators. In: Anderson, J.F., and Kaya, H.K. (eds.), *Perspectives in Forest Entomology*. Academic Press, New York. pp. 251–263.

Kaya, H.K., and Gaugler, R. 1993. Entomopathogenic nematodes. *Annual Review of Entomology* 38: 181–206.

Keller, S., Schweizer, C., Keller, E., and Brenner, H. 1997. Control of white grubs (*Melolontha melolontha* L.) by treating adults with the fungus *Beauveria brongniartii*. *Biocontrol Science and Technology* 7: 105–116.

Kelly, D.C. 1981. Non-occluded viruses. In: Davidson, E.W. (ed.), *Pathogenesis of invertebrate microbial disease*. Allenheld, Osmun Publishers, Totowa, NJ. pp. 39–60.

Klein, M.G., and Lacey, L.A. 1999. An attractant trap for autodissemination of entomopathogenic fungi into populations of Japanese beetle, *Popillia japonica* (Coleoptera: Scarabidae). *Biocontrol Science and Technology* 9: 151–158.

Krieg, A. 1971. Possible use of rickettsiae for microbial control of insects. In: Burges, H.D., and Hussey, N.W. (eds.), *Microbial Control of Insects and Mites*. Academic Press, Inc., New York. pp. 459–468.

Lacey, L.A., Frutos, R., Kaya, H.K., and Vail, P. 2001. Insect pathogens as biological control agents: Do they have a future? *Biological Control* 21: 230–248.

Lacey, L.A., and Goettel, M. 1995. Current developments in microbial control of insect pests and prospects for the early 21st century. *Entomophaga* 40: 3–27.

Lacey, L.A., and Kaya, H.K. (eds.). 2000. *Field Manual of Techniques in Invertebrate Pathology: Application and Evaluation of Pathogens for Control of Insects and Other Invertebrate Pests*. Kluwer Academic, Dordrecht.

Latge, J.P., and Papierok, B. 1988. Aphid pathogens. In: Minks, A.K., and Harrewijn, P. (eds.), *Aphids: Their Biology, Natural Enemies and Control*. Elsevier, Amsterdam. Vol. B. pp. 323–335.

Lovett, B., and St. Leger, R.J. 2017. The insect pathogens. *Microbiology Spectrum* 5(2): FUNK-0001-2016.

Maddox, J.V. 1987. Protozoan diseases. In: Fuxa, J.R., and Tanada, Y. (eds.), *Epizootiology of Insect Diseases*. Wiley, New York. pp. 417–452.

McCoy, C.W., Samsin, R.A., and Boucias, D.G. 1988. Entomogenous fungi. In: Ignoffo, C.M., and Mandava, N.B. (eds.), *Handbook of Natural Pesticides. Vol. V: Microbial Insecticides, Part A: Entomogenous Protozoa and Fungi*. CRC Press, Boca Raton, FL. pp. 151–236.

Milner, R.J., and Prior, C. 1994. Susceptibility of Australian plague locust, *Chortoicetes terminifera*, and wingless grasshopper, *Phaulacridium vittatum*, to the fungi *Metarhizium* spp. *Biological Control* 4: 132–137.

Nayak, S.B., Rao, K.S., and Ramalakshmi, V. 2020. Impact of climate change on insect pests and their natural enemies. *International Journal of Ecology and Environmental Sciences* 2: 579–584.

Obukowicz, M.G., Perlak, F.J., Kusano-Kretzmer, K., et al., 1986a. Tn5-mediated integration of the delta-endotoxin gene from *Bacillus thringiensis* into the chromosome of root-colonizing pseudomonads. *Journal of Bacteriology* 168: 982–989.

Obukowicz, M.G., Perlak, F.J., Kusano-Kretzmer, K., et al., 1986b. Integration of the delta-endotoxin gene of *Bacillus thuringiensis* into the chromosome of root-colonizing pseudomonads using Tn5. *Gene* 45: 327–331.

Odegaard, F. 2000. How many species of arthropods? Erwin's estimate revised. *Biological Journal of Linnean Society* 71: 583–597.

Parmesan, C. 2007. Influences of species, latitudes and methodologies on estimates of phenological response to global warming. *Global Change Biology* 13(9): 1860–1872.

Payne, C.C. 1981. Cytoplasmic polyhedrosis viruses. In: Davidson, E.W. (ed.), *Pathogenesis of Invertebrate Microbial Disease*. Allenheld, Osmun Publishers, Totowa, NJ. pp. 61–100.

Poinar, G.O., Jr. 1971. Use of nematodes for biological control of insects. In: Burges, H.D., and Hussey, N.W. (eds.), *Microbial Control of Insects and Mites*. Academic Press, Inc., New York. pp. 181–203.

Poinar, G.O., Jr. 1979. *Nematodes for Biological Control of Insects*. CRC Press, Boca Raton, FL.

Quintela, E.D., and McCoy, C.W. 1997. Pathogenicity enhancement of *Metarhizium anisopliae* and *Beauveria basiana* to first instars of *Diapreps abbreviatus* (Coleoptera: Curculionidae) with sublethal doses of imidacloprid. *Environmental Entomology* 26: 1173–1182.

Roberts, D.W., and Humber, R.A. 1981. Entomogenous fungi. In: Cole, G.T., and Kendrick, W.B. (eds.), Biology of Conidial Fungi: Academic Press, Inc., New York. vol. 2. pp. 201–236.

Roberts, D.W., and Yendol, W.G. 1971. Use of fungi for microbial control of insects. In: Burges, H.D., and Hussey, N.W. (eds.), *Microbial Control of Insects and Mites*. Academic Press, Inc., New York. pp. 125–149.

Schnepf, E., Crickmore, N., Van Rie, J., et al., 1998. *Bacillus thuringiensis* and its pesticidal proteins. *Microbiology and Molecular Biology Reviews* 62: 775–806.

Shah, P.A., and Goettel, M.S. 1999. *Directory of Microbial Control Products*. Society of Invertebrate Pathology, Division of Microbial Control. www.sipweb.org/directory.htm

Soper, R.S., Shimazu, M., Humber, R.A., et al., 1988. Isolation and characterization of *Entomophaga maimaiga* sp. Nov., a fungal pathogen of gypsy moth, *Lymantria dispar*, from Japan. *Journal of Invertebrate Pathology* 51: 229–241.

St. Leger, R.J., Wang, C., and Fang, W. 2011. New perspectives on insect pathogens. *Fungal Biology Reviews* 25: 84–88.

Sweeney, A.W. 1981. Fungal pathogens of mosquito larvae. In: Davidson, E.W. (ed.), *Pathogenesis of Invertebrate Microbial Disease*. Allenheld, Osmun Publishers, Totowa, NJ. pp. 403–424.

Tanada, Y., and Kaya, H.K. 1993. *Insect Pathology*. Academic Press, New York.

Udayasuriyan, V., Nakamura, A., Masaki, H., and Uozomi, T. 1995. Transfer of an insecticidal protein gene of *Bacillus thuringiensis* into plant-colonizing *Azospirillum*. *World Journal of Microbiology and Biotechnology* 11: 163–167.

Vega, F.E., and Blackwel, M. 2005. *Insect-Fungal Associations: Ecology and Evolution*. Oxford University Press, Oxford.

Vega, F.E., Dowd, P.F., and Bartelt, R.J. 1995. Dissemination of microbial agents using an autoinoculating device and several insect species as vectors. *Biological Control* 5: 545–552.

Woods, S.A., and Elkinton, J.S. 1987. Bimodal patterns of mortality from nuclear polyhedrosis virus in gypsy moth (*Lymantria dispar*) populations. *Journal of Invertebrate Pathology* 50: 151–157.

Zacharuk, R.Y. 1981. Fungal diseases on terrestrial insects. In: Davidson, E.W. (ed.), *Pathogenesis of Invertebrate Microbial Disease*. Allenheld, Osmun Publishers, Totowa, NJ. pp. 367–402.

Zimmermann, G. 1993. The entomopathogenic fungus *Metarhizium anisopliae* and its potential as a biocontrol agent. *Journal of Pesticide Science* 37: 375–379.

# 6 Climate Change and Agricultural Intensification Influence on Insect Biodiversity

*Asif Sajjad, Roughaina Tahir, Muhammad Yasin, and Shoaib Fareed*

## 6.1 INTRODUCTION

Insects hold the distinction of being the most extensive and diverse group of organisms in the animal kingdom. They serve numerous crucial roles in ecosystems and offer various benefits to humanity. Insects play a vital part in providing, regulating, and shaping a wide range of ecosystem services. In terms of provisioning services, insects provide alternative sources of nutrition, contribute to economic activities, enhance food chains, support industrial production, and offer medicinal services (Weisser and Siemann, 2013; Schowalter, 2013). Regulating services offered by insects include regulating and mitigating the effects of climate change, controlling and suppressing pathogens, preserving habitat genetic diversity, facilitating network services, providing pest control, regulating population dynamics, enhancing pollination, maintaining soil fertility, managing soil nutrients, regulating spatial variability of soil nutrients, and preventing soil erosion (Schowalter, 2013; Ameixa et al., 2018). Insects also provide various essential supporting ecosystem services such as protecting biodiversity, facilitating decomposition, removing dung, influencing hydrological soil properties, promoting mineralization, accumulating nutrients, facilitating nutrient flow, recycling matter, dispersing seeds, improving soil structure, and enhancing soil water infiltration (Weisser and Siemann, 2013; Schowalter, 2013).

Despite their beneficial contributions, insects can also negatively impact human well-being by acting as pests in agricultural crops or as carriers of diseases (Schaeckermann et al., 2015). However, it is important to recognize that insects deliver ecosystem services that have significant economic value. According to Losey and Vaughan (2006), the estimated annual worth of these beneficial services in the United States alone is at least $57 billion. Moreover, the economic value of insect pollination on a global scale is estimated to range from $235 billion to $577 billion per year (Potts et al., 2016).

Climate change is expected to impact agricultural systems globally in both positive and negative ways. However, the negative impacts are predicted to outweigh the positive ones (Mueller, 2013), leading to an overall negative impact on world agriculture. This is mainly due to factors such as changing weather patterns, rising temperatures, and changes in precipitation, which can negatively affect crop yields and livestock production, etc. The intensification of agriculture has led to simplified and specialized agroecosystems by declining landscape heterogeneity, increased use of agrochemicals, and abandonment of less fertile lands. This has resulted in decreased agrobiodiversity by eroding the quality and quantity of habitats for crops and associated species such as pollinators, pests, competitors, and predators (Tscharntke et al., 2012).

The loss of natural habitats like hedgerows, trees, and other landscape elements, coupled with the frequent application of chemical fertilizers and pesticides, has an adverse effect on insect

DOI: 10.1201/9781003382089-6

diversity (Tilman et al., 2001). Additionally, pesticides have a detrimental impact, either directly through insecticide drifting (Geiger et al., 2010) or indirectly by eliminating crucial nectar sources and larval food plants with the use of herbicides. This results in a trade-off between weed control and insect survival. The decline in biological diversity is leading to the loss of ecosystem services and functioning (Díaz et al., 2019), mainly by decreasing the species richness (number of species) and altering the balance in species evenness (relative abundance of species) (Chapin et al., 2000; Matson et al., 1997).

## 6.2 IMPACT ON POLLINATORS

About 73% of the world's cultivated crops depend on bees for pollination, followed by flies (19%), bats (6.5%), wasps (5%), beetles (5%), birds (4%), butterflies (4%), and moths (4%) (Abrol, 2009). However, human-induced global change such as land use and climate change disrupt the plant-pollinator interactions (Ricketts et al., 2004). These human-induced changes impact the distribution and phenology of species leading to mismatches between mutualistic partners. These changes may also have an impact on pollinator behaviors, floral resources, nesting sites, and species abundance, which further make plant-pollinator interactions more complex (Hegland et al., 2009). Consequently, mutualistic plant-pollinator interactions are more vulnerable to global change and may serve as an "early warning" signal before basic biodiversity indicators such as species richness and composition (Tylianakis et al., 2008). Due to change in temperature, plants and their pollinators may experience an alteration in the timing of critical development stages such as first flowering in plants and first emergence date in pollinators. The rate of spring advancement in phenology was observed to be three times faster in certain butterflies than their host plants, indicating a growing mismatch between the two groups. This phenological miss matching may lead to reduced pollinator visitation to the plants that rely on them.

Climate change poses a major threat to pollination services (Hegland et al., 2009; Schweiger et al., 2010). Increased temperatures are believed to be the most significant impact of climate change on plant-pollinator interactions (Kjøhl et al., 2011). In response to climate change, species may adapt to new environments, migrate to more suitable areas, or face extinction. Protecting pollinators and their habitats through conservation efforts and sustainable land-use practices, as well as reducing greenhouse gas emissions, are essential to maintain the delicate balance between plants and their pollinators.

Honeybees can be impacted by climate change in various ways. These impacts can be direct, affecting their behavior and physiology, or indirect, such as changes in the floral environment that affect their colony's development and harvesting capacity (Le Conte and Navajas, 2008). Extreme temperatures can cause behavioral responses in bees that negatively affect pollination services. Moreover, increasing temperatures may affect the pollen removal and deposition efficiency of pollinators, putting them at risk of overheating (Reddy et al., 2012).

Honeybees face numerous threats from pathogens, parasites, and pests. Climate change can exacerbate the spread and virulence of these diseases and parasites. Pathogens often have different haplotypes that vary in their virulence, and climate change can facilitate the transfer of these haplotypes between honeybee populations (Le Conte and Navajas, 2008). Additionally, climate change may expose different species and races of honeybees with the new pathogens that they have not coevolved with. For example, the interaction between *Varroa destructor* and *Apis mellifera* is one such case (Solignac et al., 2005).

Insect pollinators are experiencing a decline, primarily driven by changes in land cover, land-use intensity, and climate (Potts et al., 2010; Steffan-Dewenter and Westphal, 2008). The decline in pollinators extends to various groups, including wild bees, butterflies, honeybees, moths, hoverflies, and flower-visiting wasps (Ollerton, 2017). The decline in populations of wild bees and butterflies is associated with historic changes in the landscape (Burkle et al., 2013; Bommarco et al., 2014; Senapathi et al., 2015), resulting in a loss of important floral resources, nesting sites, and foraging areas (Goulson et al., 2005; Biesmeijer et al., 2006; Potts et al., 2010; Scheper et al., 2014; Baude et al., 2016). Habitat loss and their degradation has resulted in decreased fruit set in both the wild plants (Aguilar et al., 2006; Batáry et al., 2013; Clough et al., 2014) and the insect-pollinated crops (Klein et al., 2002, 2012) by reducing the abundance and diversity of pollinators.

Pesticides such as neonicotinoids have a close association with declining bee populations (Goulson, 2013; Woodcock et al., 2016), honeybee health (Sanchez-Bayo and Goka, 2014), and bumblebee behavior (Crall et al., 2018). Other chemical inputs such as fungicides and herbicides tend to have indirect effects on pollinator biodiversity by increasing pesticide toxicity and reducing floral diversity (Millard et al., 2021). A high reliance on nitrogen-based fertilizers contributes to a decrease in plant species diversity and eliminates the need for clover field crop rotation, thereby reducing floral resources available for pollinators (Millard et al., 2021).

## 6.3 IMPACT ON NATURAL ENEMIES

Climate change and global warming can impact higher trophic levels in different ways. They may directly alter the behavior of natural enemies or indirectly impact the physiological features of host plants and behavioral characteristics of herbivorous insects (Skendžić et al., 2021). Climate change may influence the interactions between predator and prey through several means. First, climate change can directly affect species by altering their behavior and distribution. Second, climate change may alter predator-prey interactions by influencing direct consumptive and indirect nonconsumptive predator effects (Laws, 2017).

Climate change can impact various characteristics of species, including their development time (Culler et al., 2015), oviposition rates (Puentes et al., 2015), and behavioral traits (Bodlah et al., 2016), which can ultimately impact the interactions between predator and prey. Changes in the traits of plants can potentially affect interactions between predators and prey. Climate change is expected to modify the quality of plants in various ways, such as altering nutrient content and the ratio of protein to carbohydrates (Zvereva and Kozlov, 2006) as well as changing the concentrations of secondary compounds in plants (Ode et al., 2014). These changes can influence the foraging behaviors of herbivores and their nutrient content, which can ultimately impact interactions between predators and prey (Ainsworth and Long, 2005).

Climate change can influence species interactions by changing the effectiveness of chemical signaling between and within species. Ode et al. (2014) conducted a study to evaluate the impact of greenhouse gases on signaling and chemical composition of plants. Their research indicated that $CO_2$ can overwhelm the jasmonic acid signaling pathway and increase the salicylic acid signaling pathway. This may increase the vulnerability of plants to leaf-chewing herbivores.

Recent experiments have revealed that increased levels of carbon dioxide ($CO_2$) can have direct effects on the consumptive predation effects. Gao et al. (2010) found that lacewings grown in high $CO_2$ conditions were less effective at killing prey than those grown in ambient conditions. Lacewings raised under high $CO_2$ conditions mature more quickly but have lower body mass because aphids grown under these conditions have less available soluble proteins. These physiological changes may lead to a reduction in the predation rates of lacewings. Bezemer et al. (1998) conducted a study to examine the efficacy of *Aphidius picipes* Nee, a parasitoid wasp dining on the English grain aphid (*Sitobion avenae* F.) and found that parasitism increased in association with increasing $CO_2$, but the same high $CO_2$ level led to reduced fecundity of the wasps.

Climate change may alter the consumptive effects of predators. Increased temperature may lead to a higher predation rate. Warming may enhance the predation rates of rotifers and dyctisid beetles feeding on mosquito larvae (Seifert et al., 2015; Culler et al., 2015). Hance et al. (2007) noticed that an early and warm spring may cause a natural enemy to develop quicker than its prey (such as aphids). This early emergence leads to the death of predators due to lack of food. If that phenomenon is repeated over several years, this may result in the extinction of natural enemies.

Various life-history features of parasitoid insects can be significantly influenced by exposure to extreme temperature which can be either lethal or sublethal. Exposure to cold temperatures can cause either metabolic ailments such as osmotic stress or anoxia or physical harm from ice crystals, both of which can be fatal to parasitoids (Turnock and Fields, 2005). When insects are exposed to temperatures lower than their base temperature, their development can be ceased. Low temperatures can also affect the number of larval instars, e.g., *Ephestia kuehniella* undergoes four instars at 18°C compared to five instars at 25°C. On the other hand, for *Galleria mellonella*, a cold shock during the final larval instar leads the larva to complete an additional instar (Denlinger and Lee, 2019).

Through a variety of mechanisms, low temperatures can alter the adult size of parasitoids. According to the temperature-size rule, ectotherms tend to become larger at lower temperatures. This is because colder temperatures lead to slower growth rate, longer developmental time, and eventually, larger adult size in insects (Sibly and Atkinson, 1994). On the other hand, cold temperatures can also lead certain species to stop their feeding earlier than expected, such as *Sarcophaga bullata*, a dipteran whose adult size is demonstrated during the last stage of larval development. *Trichogramma carverae* exposed to temperatures of 10°C or below in the early pupal stage likewise show a reduction in size following cold exposure.

Exposing immature parasitoids to low temperatures generally shortens the lifespan of surviving adults (Pandey and Johnson, 2005). Similarly, Foerster and Doetzer (2006) discovered that adult Scelionidae, when exposed to low temperature and later returned to normal temperature, experienced a decline in their lifespan. Likewise, the temperature can affect the sex allocation of female parasitoids, particularly in arrhenotokous hymenopteran parasitoids. Typically, exposure of either the immature or adult stages to low temperature can disrupt the normal reproductive processes, leading to the development of more males (Denlinger and Lee, 2019).

The intensification in agriculture significantly impacts the biological control by distorting the balance of natural enemy communities and favoring only a few dominant species (Chapin III et al., 2000; Matson et al., 1997). It can also destabilize the interspecific relationships that lead to biodiversity loss. Zhao et al. (2015) evaluated the response of natural enemies at two agriculture intensification scales, i.e., cropland expansion and increasing nitrogen input. The response of natural enemies was highly species-specific, i.e., predator-pest ratio decreased and percentage of parasitism increased. Crop diversification at landscape scale is a promising tool for ecological intensification, whereby biological control is improved and pesticide applications reduced (Redlich et al., 2018). In addition, as a pest management strategy, there is a need to balance the cost and benefit ratio of agriculture intensification and restore biological control service through strengthening the role of natural enemies (Zhao et al., 2015).

## 6.4 IMPACT ON INSECT PESTS

Climate change has substantial effects on agriculture, as well as on insect pests that damage crops. The changing global climate directly influences pests in terms of their reproduction, development, survival, and dispersal. Indirectly, it also affects the interactions between pests, their environment, and other insect species, including vectors, natural enemies, competitors, and mutualists (Prakash et al., 2014). As a result of climate change, new ecological niches emerge, creating favorable conditions for insect pests to establish themselves and expand into new geographic regions, often shifting from one area to another (FAO, 2020).

Changes in temperature have a significant impact on insect physiology, as their metabolic rate tends to increase by approximately twofold with every 10°C rise in temperature (Dukes et al., 2009). The effects of elevated temperatures are more pronounced for insects that primarily inhabit aboveground environments compared to those that spend a significant portion of their life cycle in the soil. This is because the soil acts as a thermal insulator, buffering temperature fluctuations and reducing their impact (Bale et al., 2002). The rise in temperature can impact the growth rates, fecundity (reproductive capacity), consumption rates, and population densities of insect pests (Awmack et al.,

1997). For example, whitefly populations are primarily regulated by temperature, precipitation, and humidity. High temperatures, coupled with high humidity, positively correlate with the buildup of whitefly populations (Pathania et al., 2020).

Ezcurra et al. (1978) established that the distribution of insect pests may be influenced by several factors, including natural biogeography, distribution of crops, agricultural practices (such as monocultures, irrigation, fertilizers, and pesticides), climate change, trade, and cultural patterns. As per FAO (2020), climate change can create novel ecological niches, which can offer opportunities for insect pests to establish themselves and spread to new regions, as well as shift from one location to another.

An increase in temperature has a direct impact on the reproduction, survival, dissemination, and population dynamics of pests, as well as the interactions that exist between them, their environment, and their natural enemies (Prakash et al., 2014). The increase in atmospheric $CO_2$ levels can impact the performance, abundance, and distribution of herbivorous insects. Increased concentration of atmospheric $CO_2$ affects the feeding behavior of certain pest groups (Bezemer et al., 1998), causing them to consume plants at a higher rate. This response is commonly observed in insects that feed on leaves, such as caterpillars, miners, and chewers (Hamilton et al., 2005).

Hamilton et al. (2005) grew soybeans under increased atmospheric $CO_2$ concentrations. During the early stage of growth, the soybeans showed 57% more damage by the Japanese beetle (*Popilia japonica* Newman), potato leafhopper (*Empoasca fabae* Harris), Mexican bean beetle (*Epilachna varivestis* Mulsant), and western corn rootworm (*Diabrotica virgifera virgifera* Le Conte) compared to soybeans grown under normal atmospheric conditions.

The survival and diapause of insect species that overwinter in soil can be negatively impacted by excessive rainfall, which may cause flooding and the prolonged stagnation of water (Skendžić et al., 2021). Heavy rainfall and flooding can also wash away insect eggs and larvae. Small pests such as aphids, mites, jassids, whiteflies, etc., are particularly vulnerable and can be washed away during heavy rain (Pathak et al., 2012). Therefore, variable rainfall patterns can significantly influence insect populations. Staley et al. (2007) investigated how increased summer rainfall and drought affect the population growth of the soil-dwelling wireworm (*Agriotes lineatus* L.) in grassland plots. He concluded that wireworm populations grew rapidly in the upper part of the soil when exposed to increased summer rainfall, in contrast to ambient and drought conditions.

The presence of herbivorous insects and microbial pathogens can significantly affect plant productivity in agricultural and natural ecosystems. The decrease in plant species richness that occurs due to intensified agricultural practices leads to alterations in the communities of herbivorous insects, their natural predators and parasites, and the microbial community that attacks crops. The reduced diversity in monocultural agricultural systems typically results in more significant crop losses from a less diverse but more abundant insect pest complex (Andow, 1991).

## 6.5 IMPACT ON INVASIVE SPECIES

The concept of invasive species may be exaggerated by climate change as some species that were once invasive may become less problematic, while others that were noninvasive before may become invasive, and many native species may move into new areas where they were not previously found (Hellmann et al., 2008). The globalization of trade, agriculture, and international travel has led to a rapid increase in the spread of species beyond their native ranges over the last millennium (Ricciardi et al., 2013). According to the Millennium Ecosystem Assessment, biological invasions are among the top five causes for the reduction in biodiversity and the subsequent decrease in ecosystem services worldwide. The occurrence of biological invasions has rapidly increased in the last 50 years, leading to a decline in global biodiversity (McNeely, 2001; Pyšek and Richardson, 2006; McGeoch et al., 2010).

Insect physiology exhibits a high degree of sensitivity to temperature and climate change. Consequently, climate change is anticipated to have predominantly positive effects on invasive insects, at least in a direct manner (Bale et al., 2002; Deutsch et al., 2008). Parmesan and Yohe (2003) reported that climate change is causing insects species to expand at the average rate of 6.1 km per decade. Increase in temperature has enabled insect species to survive in the areas where they previously could not thrive (Wolfe et al., 2008;). Rising temperatures generally result in increased insect activity, including movement, survival, growth rate, dispersal, and feeding rate (Bale et al., 2002). Insects exhibit swift responses to climate change in terms of their distribution and abundance. This rapid response is attributed to their high mobility, short generation time, physiological sensitivity to temperature, and high reproductive potential (Weed et al., 2013).

*Diabrotica virgifera virgifera* (western corn rootworm) has successfully migrated from North America to Europe. This invasive pest has caused considerable damage to corn production in both continents. Increasing air temperature and fluctuations in phenology have facilitated the upper limit of this invasive pest to expand toward northern regions (Aragón and Lobo, 2012). Likewise, distribution of *Adelges tsugae* (invasive hemlock woolly adelgid) is currently restricted within the Eastern United States due to cold winter temperatures. However, future projections suggest that the predicted warming could facilitate the hemlock woolly adelgid to proliferate freely across the entire range of hemlock (*Tsuga* spp.) in North America.

Extremely high temperatures, however, can significantly decrease insect fitness (Deutsch et al., 2008). Warming may have detrimental impacts on invasive insects by disturbing their developmental synchronization with their host plants (Bale et al., 2002). Additionally, decrease in protective snow cover, due to warming temperatures, can have detrimental effects by exposing overwintering life stages to freezing conditions (Bale and Hayward, 2010). Sobek-Swant et al. (2012) have presented evidence that supreme winter warm spells followed by supreme cold can lead to the mortality of emerald ash borer (*Agrilus planipennis*) larvae.

Elevated levels of carbon dioxide ($CO_2$) do not have a direct impact on insects (Guerenstein and Hildebrand, 2008), however, insects, particularly herbivores, may be indirectly influenced by the effects of $CO_2$ on plants. For example, plants raise under elevated $CO_2$ conditions often exhibit higher carbon-to-nitrogen ratios in their tissues. This leads to a decrease in nutrient content, particularly nitrogen, which is typically a growth-limiting factor for insects (Coviella and Trumble, 1999). Consequently, insects need to consume more plant tissue to acquire sufficient nutrients for their growth and development (Dermody et al., 2008; Johnson and McNicol, 2010).

Nevertheless, there are cases in certain forests where elevated levels of carbon dioxide ($CO_2$) can lead to a reduction in insect feeding (Knepp et al., 2005). However, many climate change factors, such as elevated $CO_2$, drought, ozone ($O_3$), ultraviolet (UV) light, and temperature can affect the plant secondary chemistry that plays a crucial role in regulating the growth and survival of herbivores (Burkle and Runyon, 2016; Jamieson et al., 2017; Yuan et al., 2009). Furthermore, climate change can also alter a plant's ability to produce chemical defenses in response to attacks by invasive herbivores (Zavala et al., 2008). During drought, dry plants may become more attractive to insects which could result in an increase in crop pests (Ziska et al., 2011).

Three main physical and biochemical processes, including thermal tolerance, insecticide resistance, and immune priming, help invasive insect species adapt to the changing climatic conditions. Thermal tolerance enables insects to adapt to both high and low temperatures, enabling them to expand their range and outcompete other species that occupy similar ecological niches. Many invasive insect species have shown many-fold resistance against several insecticides belonging to different classes (Wan and Yang (2016). The immune priming refers to the ability of invertebrates to develop immune memory and become more resistant to pathogen infections after being exposed to them previously (Rowley and Powell, 2007). It is characterized by an enhanced level of protection against pathogens or immune stimulants following prior exposure (Tidbury et al., 2012). Many studies have demonstrated the occurrence of immune priming in insect species like *Tribolium castaneum*, *Bombus terrestris*, and *Plodia interpunctella* (Rowley and Powell, 2007; Tidbury et al., 2011, 2012; Roth et al., 2009). The presence of immune priming in this species could potentially enhance its ability to adapt and disperse in new regions.

## 6.6 CONCLUSION

Climate change is impacting agricultural systems globally mainly due to changing weather patterns, rising temperatures, and changes in precipitation, which are negatively affecting crop yields and livestock production. In addition, the intensification of agriculture has increased the use of agrochemicals, expanded specialized agroecosystems, and decreased the variability of the landscape. In this chapter, we discussed how the biodiversity of insects is being, directly and indirectly, affected by the physical and biological factors of climate change and agriculture intensification. How the important ecosystem services, delivered by different functional groups of insects, are being affected due to loss of their biodiversity is also discussed. These include pollinators, natural enemies (predators and parasitoids), insect pests, and invasive species.

## REFERENCES


Abrol, D. P. 2009. Plant-pollinator interactions in the context of climate change—an endangered mutualism. *Journal of Palynology* 45: 1–25.

Aguilar, R., Ashworth, L., Galetto, L., and Aizen, M. A. 2006. Plant reproductive susceptibility to habitat fragmentation: Review and synthesis through a meta-analysis. *Ecology Letters* 9(8): 968–980.

Ainsworth, E. A., and Long, S. P. 2005. What have we learned from 15 years of free-air $CO_2$ enrichment (FACE)? A meta-analytic review of the responses of photosynthesis, canopy properties and plant production to rising $CO_2$. *New Phytologist* 165(2): 351–372.

Ameixa, O. M. C. C., Soares, A. O., Soares, A. M. V. M., and Lillebø, A. I. 2018. Ecosystem services provided by the little things that run the world. In *Selected Studies in Biodiversity*. Intechopen, 267–302.

Andow, D. A. 1991. Vegetational diversity and arthropod population response. *Annual Review of Entomology* 36(1): 561–586.

Aragón, P., and Lobo, J. M. 2012. Predicted effect of climate change on the invasibility and distribution of the Western corn root-worm. *Agricultural and Forest Entomology* 14(1): 13–18.

Awmack, C., Woodcock, C., and Harrington, R. 1997. Climate change may increase vulnerability of aphids to natural enemies. *Ecological Entomology* 22(3): 366–368.

Bale, J. S., and Hayward, S. A. L. 2010. Insect overwintering in a changing climate. *Journal of Experimental Biology* 213(6): 980–994.

Bale, J. S., Masters, G. J., Hodkinson, I. D., Awmack, C., Bezemer, T. M., Brown, V. K., . . . Whittaker, J. B. 2002. Herbivory in global climate change research: Direct effects of rising temperature on insect herbivores. *Global Change Biology* 8(1): 1–16.

Batáry, P., Sutcliffe, L., Dormann, C. F., and Tscharntke, T. 2013. Organic farming favours insect-pollinated over non-insect pollinated forbs in meadows and wheat fields. *PLOS ONE* 8(1): e54818.

Baude, M., Kunin, W. E., Boatman, N. D., Conyers, S., Davies, N., Gillespie, M. A., . . . Memmott, J. 2016. Historical nectar assessment reveals the fall and rise of floral resources in Britain. *Nature* 530(7588): 85–88.

Bezemer, T. M., Jones, T. H., and Knight, K. J. 1998. Long-term effects of elevated $CO_2$ and temperature on populations of the peach potato aphid *Myzus persicae* and its parasitoid *Aphidius matricariae*. *Oecologia* 116: 128–135.

Biesmeijer, J. C., Roberts, S. P., Reemer, M., Ohlemuller, R., Edwards, M., Peeters, T., . . . Kunin, W. E. 2006. Parallel declines in pollinators and insect-pollinated plants in Britain and the Netherlands. *Science* 313(5785): 351–354.

Bodlah, M. A., Zhu, A. X., and Liu, X. D. 2016. Host choice, settling and folding leaf behaviors of the larval rice leaf folder under heat stress. *Bulletin of Entomological Research* 106(6): 809–817.

Bommarco, R., Lindborg, R., Marini, L., and Öckinger, E. 2014. Extinction debt for plants and flower-visiting insects in landscapes with contrasting land use history. *Diversity and Distributions* 20(5): 591–599.

Burkle, L. A., Marlin, J. C., and Knight, T. M. 2013. Plant-pollinator interactions over 120 years: Loss of species, co-occurrence, and function. *Science* 339(6127): 1611–1615.

Burkle, L. A., and Runyon, J. B. 2016. Drought and leaf herbivory influence floral volatiles and pollinator attraction. *Global Change Biology* 22(4): 1644–1654.

Chapin III, F. S., Zavaleta, E. S., Eviner, V. T., Naylor, R. L., Vitousek, P. M., Reynolds, H. L., . . . Díaz, S. (2000). Consequences of changing biodiversity. *Nature* 405(6783): 234–242.

Clough, Y., Ekroos, J., Báldi, A., Batáry, P., Bommarco, R., Gross, N., . . . Smith, H. G. 2014. Density of insect-pollinated grassland plants decreases with increasing surrounding land-use intensity. *Ecology Letters* 17(9): 1168–1177.

Coviella, C. E., and Trumble, J. T. 1999. Effects of elevated atmospheric carbon dioxide on insect-plant interactions. *Conservation Biology* 13(4): 700–712.

Crall, J. D., Switzer, C. M., Oppenheimer, R. L., Ford Versypt, A. N., Dey, B., Brown, A., . . . de Bivort, B. L. 2018. Neonicotinoid exposure disrupts bumblebee nest behavior, social networks, and thermoregulation. *Science* 362(6415): 683–686.

Culler, L. E., Ayres, M. P., and Virginia, R. A. 2015. In a warmer Arctic, mosquitoes avoid increased mortality from predators by growing faster. *Proceedings of the Royal Society B: Biological Sciences* 282(1815): 20151549.

Denlinger, D. L., and Lee, R. E. 2019. Physiology of cold sensitivity. In *Temperature Sensitivity in Insects and Application in Integrated Pest Management*. CRC Press, 55–95.

Dermody, O., O'Neill, B. F., Zangerl, A. R., Berenbaum, M. R., and DeLucia, E. H. 2008. Effects of elevated $CO_2$ and $O_3$ on leaf damage and insect abundance in a soybean agroecosystem. *Arthropod-Plant Interactions* 2: 125–135.

Deutsch, C. A., Tewksbury, J. J., Huey, R. B., Sheldon, K. S., Ghalambor, C. K., Haak, D. C., and Martin, P. R. 2008. Impacts of climate warming on terrestrial ectotherms across latitude. *Proceedings of the National Academy of Sciences* 105(18): 6668–6672.

Díaz, S. M., Settele, J., Brondízio, E., Ngo, H., Guèze, M., Agard, J., and Zayas, C. 2019. The global assessment report on biodiversity and ecosystem services: Summary for policy makers. https://scholar.google.com/scholar?hl=en&as_sdt=0%2C5&q=The+global+assessment+report+on+biodiversity+and+ecosystem+services%3A+Summary+for+policy+makers&btnG=

Dukes, J. S., Pontius, J., Orwig, D., Garnas, J. R., Rodgers, V. L., Brazee, N., . . . Ayres, M. 2009. Responses of insect pests, pathogens, and invasive plant species to climate change in the forests of northeastern North America: What can we predict? *Canadian Journal of Forest Research* 39(2): 231–248.

Ezcurra, E., Rapoport, E. H., and Marino, C. R. 1978. The geographical distribution of insect pests. *Journal of Biogeography* 149–157.

FAO. 2020. Climate Related Transboundary Pests and Diseases. Available online: www.fao.org/3/a-ai785e.pdf (accessed on 19 December 2020).

Foerster, L. A., and Doetzer, A. K. 2006. Cold storage of the egg parasitoids *Trissolcus basalis* (Wollaston) and *Telenomus podisi* Ashmead (Hymenoptera: Scelionidae). *Biological Control* 36(2): 232–237.

Gao, F., Chen, F., and Ge, F. 2010. Elevated $CO_2$ lessens predation of *Chrysopa sinica* on *Aphis gossypii*. *Entomologia experimentalis et applicata* 135(2): 135–140.

Geiger, F., Bengtsson, J., Berendse, F., Weisser, W. W., Emmerson, M., Morales, M. B., . . . Inchausti, P. (2010). Persistent negative effects of pesticides on biodiversity and biological control potential on European farmland. *Basic and Applied Ecology* 11(2): 97–105.

Goulson, D. 2013. An overview of the environmental risks posed by neonicotinoid insecticides. *Journal of Applied Ecology* 50(4): 977–987.

Goulson, D., Hanley, M. E., Darvill, B., Ellis, J. S., and Knight, M. E. 2005. Causes of rarity in bumblebees. *Biological Conservation* 122(1): 1–8.

Guerenstein, P. G., and Hildebrand, J. G. 2008. Roles and effects of environmental carbon dioxide in insect life. *Annual Review of Entomology* 53: 161–178.

Hamilton, J. M., Maddison, D. J., and Tol, R. S. 2005. Effects of climate change on international tourism. *Climate Research* 29(3): 245–254.

Hance, T., van Baaren, J., Vernon, P., and Boivin, G. 2007. Impact of extreme temperatures on parasitoids in a climate change perspective. *Annual Review of Entomology* 52: 107–126.

Hegland, S. J., Nielsen, A., Lázaro, A., Bjerknes, A. L., and Totland, Ø. 2009. How does climate warming affect plant-pollinator interactions? *Ecology Letters* 12(2): 184–195.

Hellmann, J. J., Byers, J. E., Bierwagen, B. G., and Dukes, J. S. 2008. Five potential consequences of climate change for invasive species. *Conservation Biology* 22(3): 534–543.

Jamieson, M. A., Burkle, L. A., Manson, J. S., Runyon, J. B., Trowbridge, A. M., and Zientek, J. 2017. Global change effects on plant-insect interactions: The role of phytochemistry. *Current Opinion in Insect Science* 23: 70–80.

Johnson, S. N., and McNicol, J. W. 2010. Elevated $CO_2$ and aboveground-belowground herbivory by the clover root weevil. *Oecologia* 162: 209–216.

Kjøhl, M., Nielsen, A., and Stenseth, N. C. 2011. *Potential Effects of Climate Change on Crop Pollination*. Food and Agriculture Organization of the United Nations (FAO).

Klein, A. M., Brittain, C., Hendrix, S. D., Thorp, R., Williams, N., and Kremen, C. 2012. Wild pollination services to California almond rely on semi-natural habitat. *Journal of Applied Ecology* 49(3): 723–732.

Klein, A. M., Steffan-Dewenter, I., Buchori, D., and Tscharntke, T. 2002. Effects of land-use intensity in tropical agroforestry systems on coffee flower-visiting and trap-nesting bees and wasps. *Conservation Biology* 16(4): 1003–1014.

Knepp, R. G., Hamilton, J. G., Mohan, J. E., Zangerl, A. R., Berenbaum, M. R., and DeLucia, E. H. 2005. Elevated $CO_2$ reduces leaf damage by insect herbivores in a forest community. *New Phytologist* 167(1): 207–218.

Laws, A. N. 2017. Climate change effects on predator—prey interactions. *Current Opinion in Insect Science* 23: 28–34.

Le Conte, Y., and Navajas, M. 2008. Climate change: Impact on honey bee populations and diseases. *Revue Scientifique et Technique-Office International des Epizooties* 27(2): 499–510.

Losey, J. E., and Vaughan, M. 2006. The economic value of ecological services provided by insects. *Bioscience* 56(4): 311–323.

Matson, P. A., Parton, W. J., Power, A. G., and Swift, M. J. 1997. Agricultural intensification and ecosystem properties. *Science* 277(5325): 504–509.

McGeoch, M. A., Butchart, S. H., Spear, D., Marais, E., Kleynhans, E. J., Symes, A., . . . Hoffmann, M. 2010. Global indicators of biological invasion: Species numbers, biodiversity impact and policy responses. *Diversity and Distributions* 16(1): 95–108.

McNeely, J. 2001. Invasive species: A costly catastrophe for native biodiversity. *Land Use and Water Resources Research* 1 (1732-2016-140260).

Millard, J., Outhwaite, C. L., Kinnersley, R., Freeman, R., Gregory, R. D., Adedoja, O., . . . Newbold, T. 2021. Global effects of land-use intensity on local pollinator biodiversity. *Nature Communications* 12(1): 2902.

Mueller, C. 2013. African lessons on climate change risks for agriculture. *Annual Review of Nutrition* 33: 395–411.

Ode, P. J., Johnson, S. N., and Moore, B. D. 2014. Atmospheric change and induced plant secondary metabolites—are we reshaping the building blocks of multi-trophic interactions? *Current Opinion in Insect Science* 5: 57–65.

Ollerton, J. 2017. Pollinator diversity: Distribution, ecological function, and conservation. *Annual Review of Ecology, Evolution, and Systematics* 48: 353–376.

Pandey, R. R., and Johnson, M. W. 2005. Effects of cool storage on *Anagyrus ananatis* Gahan (Hymenoptera: Encyrtidae). *Biological Control* 35(1): 9–16.

Parmesan, C., and Yohe, G. 2003. A globally coherent fingerprint of climate change impacts across natural systems. *Nature* 421(6918): 37–42.

Pathak, H., Aggarwal, P. K., and Singh, S. D. 2012. *Climate Change Impact, Adaptation and Mitigation in Agriculture: Methodology for Assessment and Applications.* Indian Agricultural Research Institute, 302.

Pathania, M., Verma, A., Singh, M., Arora, P. K., and Kaur, N. 2020. Influence of abiotic factors on the infestation dynamics of whitefly, *Bemisia tabaci* (Gennadius 1889) in cotton and its management strategies in North-Western India. *International Journal of Tropical Insect Science* 40: 969–981.

Potts, S. G., Biesmeijer, J. C., Kremen, C., Neumann, P., Schweiger, O., and Kunin, W. E. 2010. Global pollinator declines: Trends, impacts and drivers. *Trends in Ecology & Evolution* 25(6): 345–353.

Potts, S. G., Ngo, H. T., Biesmeijer, J. C., Breeze, T. D., Dicks, L. V., Garibaldi, L. A., . . . Vanbergen, A. 2016. The assessment report of the Intergovernmental Science-Policy Platform on Biodiversity and Ecosystem Services on pollinators, pollination and food production. https://scholar.google.com/scholar?hl=en&as_sdt=0%2C5&q=The+assessment+report+of+the+Intergovernmental+Science-Policy+Platform+on+Biodiversity+and+Ecosystem+Services+on+pollinators%2C+pollination+and+food+production&btnG=

Prakash, A., Rao, J., Mukherjee, A. K., Berliner, J., Pokhare, S. S., Adak, T., . . . Shashank, P. R. 2014. *Climate Change: Impact on Crop Pests.* Applied Zoologists Research Association (AZRA), Central Rice Research Institute.

Puentes, A., Torp, M., Weih, M., and Björkman, C. 2015. Direct effects of elevated temperature on a tri-trophic system: Salix, leaf beetles and predatory bugs. *Arthropod-Plant Interactions* 9: 567–575.

Pyšek, P., and Richardson, D. M. 2006. The biogeography of naturalization in alien plants. *Journal of Biogeography* 33(12): 2040–2050.

Raza, M. M., Khan, M. A., Arshad, M., Sagheer, M., Sattar, Z., Shafi, J., . . . Sattar, A. 2015. Impact of global warming on insects. *Archives of Phytopathology and Plant Protection* 48(1): 84–94.

Reddy, P. V., Verghese, A., and Rajan, V. V. 2012. Potential impact of climate change on honeybees (*Apis* spp.) and their pollination services. *Pest Management in Horticultural Ecosystems* 18(2): 121–127.

Redlich, S., Martin, E. A., and Steffan-Dewenter, I. 2018. Landscape-level crop diversity benefits biological pest control. *Journal of Applied Ecology* 55(5): 2419–2428.

Ricciardi, A., Hoopes, M. F., Marchetti, M. P., and Lockwood, J. L. 2013. Progress toward understanding the ecological impacts of nonnative species. *Ecological Monographs* 83(3): 263–282.

Ricketts, T. H., Daily, G. C., Ehrlich, P. R., and Michener, C. D. 2004. Economic value of tropical forest to coffee production. *Proceedings of the National Academy of Sciences* 101(34): 12579–12582.

Roth, O., Sadd, B. M., Schmid-Hempel, P., and Kurtz, J. 2009. Strain-specific priming of resistance in the red flour beetle, *Tribolium castaneum*. *Proceedings of the Royal Society B: Biological Sciences* 276(1654): 145–151.

Rowley, A. F., and Powell, A. 2007. Invertebrate immune systems—specific, quasi-specific, or nonspecific? *Journal of Immunology* 179(11): 7209–7214.

Sanchez-Bayo, F., and Goka, K. 2014. Pesticide residues and bees—a risk assessment. *PLOS ONE* 9(4): e94482.

Schaeckermann, J., Pufal, G., Mandelik, Y., and Klein, A. M. 2015. Agro-ecosystem services and dis-services in almond orchards are differentially influenced by the surrounding landscape. *Ecological Entomology* 40: 12–21.

Scheper, J., Reemer, M., van Kats, R., Ozinga, W. A., van der Linden, G. T., Schaminée, J. H., . . . Kleijn, D. 2014. Museum specimens reveal loss of pollen host plants as key factor driving wild bee decline in The Netherlands. *Proceedings of the National Academy of Sciences* 111(49): 17552–17557.

Schowalter, T. D. 2013. *Insects and Sustainability of Ecosystem Services*. CRC Press.

Schweiger, O., Biesmeijer, J. C., Bommarco, R., Hickler, T., Hulme, P. E., Klotz, S., . . . and Settele, J. 2010. Multiple stressors on biotic interactions: How climate change and alien species interact to affect pollination. *Biological Reviews* 85(4): 777–795.

Seifert, L. I., Weithoff, G., Gaedke, U., and Vos, M. 2015. Warming-induced changes in predation, extinction and invasion in an ectotherm food web. *Oecologia* 178: 485–496.

Senapathi, D., Carvalheiro, L. G., Biesmeijer, J. C., Dodson, C. A., Evans, R. L., McKerchar, M., . . . Potts, S. G. 2015. The impact of over 80 years of land cover changes on bee and wasp pollinator communities in England. *Proceedings of the Royal Society B: Biological Sciences* 282(1806): 20150294.

Sibly, R. M., and Atkinson, D. 1994. How rearing temperature affects optimal adult size in ectotherms. *Functional Ecology* 486–493.

Skendžić, S., Zovko, M., Živković, I. P., Lešić, V., and Lemić, D. 2021. The impact of climate change on agricultural insect pests. *Insects* 12(5): 440.

Sobek-Swant, S., Crosthwaite, J. C., Lyons, D. B., and Sinclair, B. J. 2012. Could phenotypic plasticity limit an invasive species? Incomplete reversibility of mid-winter deacclimation in emerald ash borer. *Biological Invasions* 14: 115–125.

Solignac, M., Cornuet, J. M., Vautrin, D., Le Conte, Y., Anderson, D., Evans, J., . . . Navajas, M. 2005. The invasive Korea and Japan types of *Varroa destructor*, ectoparasitic mites of the Western honeybee (*Apis mellifera*), are two partly isolated clones. *Proceedings of the Royal Society B: Biological Sciences* 272(1561): 411–419.

Staley, J. T., Hodgson, C. J., Mortimer, S. R., Morecroft, M. D., Masters, G. J., Brown, V. K., and Taylor, M. E. 2007. Effects of summer rainfall manipulations on the abundance and vertical distribution of herbivorous soil macro-invertebrates. *European Journal of Soil Biology* 43(3): 189–198.

Steffan-Dewenter, I., and Westphal, C. 2008. The interplay of pollinator diversity, pollination services and landscape change. *Journal of Applied Ecology* 737–741.

Tidbury, H. J., Best, A., and Boots, M. 2012. The epidemiological consequences of immune priming. *Proceedings of the Royal Society B: Biological Sciences* 279(1746): 4505–4512.

Tidbury, H. J., Pedersen, A. B., and Boots, M. 2011. Within and transgenerational immune priming in an insect to a DNA virus. *Proceedings of the Royal Society B: Biological Sciences* 278(1707): 871–876.

Tilman, D., Fargione, J., Wolff, B., D'antonio, C., Dobson, A., Howarth, R., . . . Swackhamer, D. 2001. Forecasting agriculturally driven global environmental change. *Science* 292(5515): 281–284.

Tscharntke, T., Clough, Y., Wanger, T. C., Jackson, L., Motzke, I., Perfecto, I., . . . Whitbread, A. 2012. Global food security, biodiversity conservation and the future of agricultural intensification. *Biological Conservation* 151(1): 53–59.

Turnock, W. J., and Fields, P. G. 2005. Winter climates and coldhardiness in terrestrial insects. *European Journal of Entomology* 102(4): 561.

Tylianakis, J. M., Didham, R. K., Bascompte, J., and Wardle, D. A. 2008. Global change and species interactions in terrestrial ecosystems. *Ecology Letters* 11(12): 1351–1363.

Wan, F. H., and Yang, N. W. 2016. Invasion and management of agricultural alien insects in China. *Annual Review of Entomology* 61: 77–98.

Weed, A. S., Ayres, M. P., and Hicke, J. A. 2013. Consequences of climate change for biotic disturbances in North American forests. *Ecological Monographs* 83(4): 441–470.
Weisser, W. W., and Siemann, E. (eds.). 2013. *Insects and Ecosystem Function* (Vol. 173). Springer Science & Business Media.
Wolfe, D. W., Ziska, L., Petzoldt, C., Seaman, A., Chase, L., and Hayhoe, K. 2008. Projected change in climate thresholds in the Northeastern US: Implications for crops, pests, livestock, and farmers. *Mitigation and Adaptation Strategies for Global Change* 13: 555–575.
Woodcock, B. A., Isaac, N. J., Bullock, J. M., Roy, D. B., Garthwaite, D. G., Crowe, A., and Pywell, R. F. 2016. Impacts of neonicotinoid use on long-term population changes in wild bees in England. *Nature Communications* 7(1): 12459.
Yuan, J. S., Himanen, S. J., Holopainen, J. K., Chen, F., and Stewart, C. N. 2009. Smelling global climate change: Mitigation of function for plant volatile organic compounds. *Trends in Ecology & Evolution* 24(6): 323–331.
Zavala, J. A., Casteel, C. L., DeLucia, E. H., and Berenbaum, M. R. 2008. Anthropogenic increase in carbon dioxide compromises plant defense against invasive insects. *Proceedings of the National Academy of Sciences* 105(13): 5129–5133.
Zhao, Z. H., Hui, C., He, D. H., and Li, B. L. 2015. Effects of agricultural intensification on ability of natural enemies to control aphids. *Scientific Reports* 5(1): 8024.
Ziska, L. H., Blumenthal, D. M., Runion, G. B., Hunt Jr, E. R., and Diaz-Soltero, H. 2011. Invasive species and climate change: An agronomic perspective. *Climatic Change* 105(1–2): 13–42.
Zvereva, E. L., and Kozlov, M. V. 2006. Consequences of simultaneous elevation of carbon dioxide and temperature for plant-herbivore interactions: A meta-analysis. *Global Change Biology* 12(1): 27–41.

# 7 Potential Impacts of Climate Change on Insect Behaviour

*Muhammad Yasin, Hafiz Muhammad Bilal Yousuf, Asif Sajjad, Mirza Abdul Qayyum, Habib Ali, Asim Abbasi, Muhammad Anjum Aqueel, Abu Bakar Muhammad Raza, and Sajjad Ali*

## 7.1 INTRODUCTION

Climate change (CC) is a significant and well-documented global phenomenon with far-reaching impacts on ecosystems and human and animal societies. Anthropogenic CC is caused by the release of greenhouse gases (GHGs) into the atmosphere, primarily from human activities such as burning fossil fuels and deforestation. Rising concentrations of atmospheric carbon dioxide ($CO_2$) and increasing temperatures are two of the most prominent effects of CC. Other notable impacts include ozone ($O_3$) depletion and altered precipitation patterns. Like other organisms, CC has significant impacts on insect behaviour, including changes in migration patterns, reproductive behaviour, population dynamics, and their inter- and intra-species interactions. The concentration of atmospheric $CO_2$ has increased from approximately 280 parts per million (ppm) before the industrial era to over 400 ppm today, primarily due to human activities such as burning fossil fuels and deforestation (IPCC, 2021). This increase in $CO_2$ has led to changes in temperature, precipitation, and other environmental factors, which have had significant impacts on insect behaviour. One notable effect of rising $CO_2$ levels is the alteration of plant chemistry. Plants grown under elevated $CO_2$ levels tend to have lower nitrogen content and higher carbon content, which can affect the nutritional quality of the plants for herbivorous insects. This, in turn, can lead to changes in the behaviour of insect herbivores, including altered feeding behaviour, growth rates, and reproduction (Zavala et al., 2013).

Increasing temperatures associated with CC have also been shown to have significant effects on insect behaviour. Many insects are sensitive to temperature changes and exhibit altered activity patterns, changes in feeding behaviour, and altered developmental rates under elevated temperatures (Kingsolver et al., 2013). Warmer temperatures can accelerate the development of insect larvae, leading to earlier emergence as adults (Hoffmann et al., 2010). This can affect the timing of plant-insect interactions, potentially disrupting pollination and herbivory processes (Memmott et al., 2007). Likewise, increasing temperatures can lead to earlier egg hatching and larval development in butterflies, which can have implications for population dynamics and interactions with other species (Deutsch et al., 2008). Male crickets use chirping to attract females, but increasing temperatures can alter the frequency and timing of their chirps, potentially making it harder for males to find mates (Bailey et al., 2018). Similarly, warmer temperatures can cause female mosquitoes to lay more eggs and increase the frequency of blood meals, which can lead to higher rates of disease transmission (Paaijmans et al., 2010). Some insect species may shift their geographic range or alter their migration patterns. The painted lady butterfly has expanded its range northward by approximately 300 km over the past two decades, likely due to warming temperatures (Roy et al., 2019). Also, the range of the southern pine beetle has expanded northward in the eastern United States due to warming temperatures (Lesk et al., 2017).

$O_3$ depletion is another consequence of CC change that can also have significant impacts on insect behaviour. Ozone depletion can increase ultraviolet (UV) radiation exposure, which can affect the physiology and behaviour of insects (Ziska et al., 2016). The UV radiation exposure can


DOI: 10.1201/9781003382089-7

lead to altered feeding behaviour in grasshoppers, potentially impacting population dynamics and ecosystem processes (Nevo et al., 2016). Changes in precipitation patterns associated with CC can also have significant impacts on insect behaviour. Alterations in the timing, frequency, and intensity of precipitation can affect insect activity patterns, feeding behaviour, and reproductive success (Parmesan, 2007). Increased precipitation can lead to increased activity and reproduction in mosquitoes, potentially increasing the risk of vector-borne disease transmission (Zhan et al., 2018). Overall, these examples illustrate the complex and varied ways in which CC can impact insect behaviour. As global temperatures continue to rise, it is important to understand and anticipate these changes in order to mitigate their effects on both insects and the ecosystems they inhabit.

## 7.2 PROJECTED CLIMATIC CHANGES

Long-term changes in the earth's temperature, precipitation patterns, sea levels, and weather events are caused by an increase in GHGs such as $CO_2$, methane ($CH_4$), and nitrous oxide ($N_2O$) (Ciais et al., 2013; EPA, 2021). These gases trap heat in the earth's atmosphere, leading to a rise in temperature and other climate-related changes (Willetts et al., 2010). Over the past few decades, there has been a significant increase in the concentration of GHGs, primarily $CO_2$, in the earth's atmosphere (Hegerl et al., 2018). This increase in GHGs is causing the earth's climate to change in ways that will have far-reaching impacts on the environment and human societies (Christensen et al., 2007). The scientific consensus is that human activities, such as burning fossil fuels, deforestation, and industrial processes, are the primary drivers of CC. Factors involved in CC are discussed the following sections.

### 7.2.1 Increase in Temperature

One of the most significant impacts of CC is the increase in temperature, which has been linked to a range of negative consequences, including extreme weather events, rising sea levels, and food insecurity (IPCC, 2018). Global temperatures are projected to rise by 1.5°C–4.5°C by the end of the century, depending on future emissions scenarios (Otto et al., 2015). The increase in temperature has resulted in several effects on the environment, including the melting of glaciers and ice caps, rising sea levels, and extreme weather events. This will have far-reaching effects on the planet and its inhabitants (Solomon et al., 2007). Apart from the GHGs and burning fossil fuels, other contributing factors that lead to an increase in global temperature are deforestation and land-use changes, which are responsible for absorbing $CO_2$ through photosynthesis, so when forests are cut down or burned, $CO_2$ is released into the atmosphere, which resultantly causes a rise in the earth's temperature (Pan et al., 2011). In addition, agricultural practices such as the use of fertilizers and irrigation also contribute to the emission of GHGs (Lal, 2004). This could lead to more frequent and severe heat waves, droughts, floods, and other extreme weather events, which would have devastating impacts on human health, the environment, and the global economy (IPCC, 2014).

### 7.2.2 Increase in $CO_2$ Level

$CO_2$ is playing a critical role in regulating the planet's temperature and is essential for the growth and survival of plants and animals (Ainsworth and Long, 2005). The earth's atmosphere has been experiencing significant changes due to the increase in the concentration of GHGs. Among them, $CO_2$ is the most critical factor for global warming, and its concentration in the atmosphere has been rising rapidly (Solomon et al., 2007). This increase in $CO_2$ levels due to CC is one of the most severe environmental problems of our time, with a variety of potential consequences for ecosystems, weather patterns, and human health (EPA, 2020). The primary source of $CO_2$ in the atmosphere is the burning of fossil fuels, including coal, oil, and gas, which releases carbon into the atmosphere in the form of $CO_2$ (Friedlingstein et al., 2021). However, since the Industrial Revolution, human activities have significantly increased the concentration of $CO_2$ in the atmosphere. Global energy-related

$CO_2$ emissions reached a record high of 33.1 gigatons in 2019 (IEA, 2020). Other human activities such as deforestation, industrial processes, and agricultural practices such as livestock rearing and crop burning also contribute to the increase in $CO_2$ concentrations (FAO, 2016; IPCC, 2018). Rice paddies and cows produce $CH_4$ during digestion, and their manure produces $N_2O$ (EPA, 2021). Moreover, volcanic eruptions release large amounts of $CO_2$ and other gases into the atmosphere.

### 7.2.3 $O_3$ Depletion

The $O_3$ layer plays a vital role in protecting the earth from harmful UV radiation from the sun. It absorbs UV radiation and converts it into heat, protecting living organisms from its damaging effects (WMO, 2021). However, since the 1970s, the depletion of the $O_3$ layer has become a major environmental concern (Solomon et al., 2016). $O_3$ depletion refers to the thinning of the $O_3$ layer in the stratosphere, which leads to an increase in the amount of UV radiation that reaches the earth's surface (Newman et al., 2006; Liang et al., 2018). Human activities have led to the depletion of the $O_3$ layer, primarily through the release of ozone-depleting substances such as chlorofluorocarbons (CFCs), halons, and hydrochlorofluorocarbons (HCFCs) into the atmosphere, which have been found to destroy $O_3$ molecules. CFCs are stable compounds that can persist in the atmosphere for decades (Solomon et al., 2016). Once released into the atmosphere, CFCs can rise to the stratosphere, where they are broken down by UV radiation (Montzka et al., 2018). This process releases chlorine and bromine atoms, which can then react with $O_3$ molecules, breaking them down into oxygen molecules. This destroys $O_3$ molecules and leads to a depletion of the $O_3$ layer (Ravishankara et al., 2009). The depletion of the $O_3$ layer has significant environmental and health consequences, including increased risks of skin cancer, cataracts, and other health problems.

In addition to human-made chemicals, there are also natural sources of chlorine and bromine that can contribute to $O_3$ depletion, such as volcanic eruptions that can release chlorine and bromine gases into the atmosphere, which can then react with $O_3$ molecules (Oppenheimer, 2003). Similarly, solar activity can have a significant impact on the $O_3$ layer. During periods of high solar activity, the amount of $O_3$ in the stratosphere can decrease by up to 1.7%, and these fluctuations in $O_3$ levels could have implications for CC and human health (Hood and Soukharev, 2012). Another natural factor that can affect the $O_3$ layer is the movement of air masses. The stratosphere is a dynamic system, with air masses constantly moving around the planet. These air masses can carry pollutants and other substances that can affect the $O_3$ layer, such as $N_2O$ and $CH_4$ (Polvani and Waugh, 2004).

### 7.2.4 Change in Precipitation Patterns and Floods

A change in precipitation patterns is one of the most severe threats to human societies and ecosystems. Floods are a critical manifestation of such changes, and they represent the most widespread and costliest natural disasters worldwide (Kundzewicz et al., 2019). They affect millions of people, cause severe economic damage, and have long-term impacts on ecosystems, agriculture, and public health (Arnell, 2016). Floods can lead to significant loss of life and property damage. For example, floods in Pakistan between 2010 and 2020 led to the loss of over 10,000 lives, affected over 52 million people, and caused over $19 billion in property damage (NDMA, 2021). Changes in precipitation patterns can alter the timing and quantity of water availability, affecting plant growth and ecosystem productivity. For instance, in the Amazon rainforest, changes in precipitation patterns have led to an increase in the frequency and intensity of droughts, which have caused tree mortality and reduced carbon uptake (Brienen et al., 2015). Precipitation patterns have changed significantly in recent decades due to human activities, natural variability, and CC. These changes are due to a range of factors, including changes in atmospheric temperature, atmospheric moisture content, and atmospheric circulation patterns (Zhang et al., 2011). Due to CC, some regions are experiencing more precipitation, while others are experiencing less precipitation. These changes in precipitation patterns are causing an increase in the frequency and severity of flooding events (Hirabayashi et al., 2013).

### 7.2.5 Wildfires

Wildfires, also known as forest fires, are a type of fire that occurs in natural areas such as forests or grasslands. In recent years, the frequency and intensity of wildfires have increased due to a combination of factors, and CC is believed to be one of the primary causes. Wildfires have significant impacts on both natural and human systems, including loss of life and property, environmental damage, economic impacts, and health impacts (Giglio et al., 2013; Bowater et al., 2020). These impacts can be long-lasting, even for decades. Wildfires could be due to lightning strikes, spontaneous combustion, volcanic eruptions, campfires, cigarette smoking, sparking of power lines, fireworks, etc. (Abatzoglou and Kolden, 2013; Abatzoglou and Williams, 2016; Millington et al., 2018; Marzano and Chiesa, 2019).

## 7.3 EFFECTS OF CLIMATE CHANGE ON AGRICULTURE AND FOREST ECOSYSTEMS

CC is having significant impacts on agriculture and forest ecosystems worldwide. The consequences of CC on these sectors are diverse and complex, and the potential risks and impacts of these changes are still not fully understood. These changes in temperature, precipitation, and other climate variables can manifest in a variety of ways, from reduced crop yields and forest productivity to changes in species distribution and biodiversity loss. CC is already having significant impacts on global agriculture, and these impacts are expected to become more severe in the coming decades. Rising temperatures, changes in precipitation patterns, and extreme weather events are all contributing to reduced crop yields and food security in many parts of the world. One of the most significant impacts of CC on agriculture is the reduction in crop yields. As temperatures rise, crops like wheat, maize, and rice are expected to experience yield declines, with some projections suggesting reductions of up to 25% by 2050 (Lobell et al., 2011).

Droughts, floods, and other extreme weather events are also expected to become more common, further impacting crop yields and food security. CC also affects the distribution and spread of pests and diseases, which can have significant impacts on crop productivity. Warmer temperatures can lead to increased pest prevalence and range, suggesting that global warming may increase the range of the coffee berry borer by up to 50% by 2050. Diseases, such as wheat rust, are also expected to become more widespread, further reducing crop yields and food security.

Similarly, CC has significant impacts on forest ecosystems worldwide. It is affecting the health, distribution, and productivity of forests in many regions (IPCC, 2019). One of the most significant impacts of CC on forests is the increased risk of wildfires. Rising temperatures and changes in precipitation patterns can create dry conditions that increase the likelihood of wildfires (Flannigan et al., 2016). CC change could increase the area burned by wildfires by up to 125% by the end of the century (Wang et al., 2017). It is also affecting the distribution and composition of forest ecosystems. As temperatures rise, some tree species may not be able to adapt, leading to changes in forest composition (IPCC, 2019). For example, some tree species are likely to shift by up to 500 km by 2100 (Iverson et al., 2014). These changes in forest composition can have significant impacts on biodiversity, ecosystem services, and forest productivity. Forests are important carbon sinks, absorbing and storing $CO_2$ from the atmosphere. However, as forests die back and decline, their capacity to sequester carbon is reduced, leading to a positive feedback loop that exacerbates CC (Pan et al., 2011).

## 7.4 EFFECTS OF LONG-TERM CLIMATE CHANGE ON INSECT BEHAVIOUR

CC is having a significant impact on the earth's biodiversity, with many species being forced to adapt or face extinction (Bellard et al., 2012). One group of animals that is particularly sensitive to CC is insects which play a crucial role in ecosystem functioning, from pollinating plants to providing food for other animals. Changes in the climate have altered their behaviour in various ways, with some effects being more severe than others (Potts et al., 2016a). For instance, the Asian tiger mosquito, which can transmit dengue fever and chikungunya, has expanded its range into Europe and the Americas as temperatures have warmed (Rafferty and Ives, 2011). Insect behaviours that are affected by CC are discussed in the following sections.

### 7.4.1 Effect of Climate Change on Insect Orientation

Insects are an essential part of the ecosystem and play an important role in pollination, decomposition, and nutrient cycling. CC has a significant impact on their orientation or navigation, which is crucial for their survival and reproduction (Basset et al., 2013). Changes in temperature, precipitation, and other environmental factors can alter insect orientation and affect their ability to find food, mates, and suitable habitats (Table 7.1). This disruption can have serious consequences for ecosystems and human societies, including impacts on pollination, pest control, and food security (Wiley and

**TABLE 7.1**
**Effect of Long-Term Changes in the Climate on Insect Orientation**

| Insect Species | Effect of Climate Change | References |
|---|---|---|
| | **Wildfires** | |
| *Drosophila melanogaster* | Reduced olfactory-guided behaviour and reduced mating success | Gadenne et al., 2016 |
| *Trichoptera species* | Changes in larval habitat preference and distribution | Jacobsen et al., 2020 |
| *Neodiprion sertifer* | Reduced flight activity and altered flight patterns | Cote et al., 2021 |
| *Polyergus breviceps* | Reduced nest relocation success and increased mortality | Helms and Herbers, 2003 |
| *Cicindela hirticollis* | Reduced habitat availability and abundance, leading to population decline | Knisley and Hill, 1992 |
| | **Increase in temperature** | |
| *Apis mellifera* | Reduced ability to navigate | Lindauer, 1952 |
| *Danaus plexippus* | Altered migratory behaviour | Saunders et al., 1987 |
| *Drosophila melanogaster* | Increased sensitivity to light | Kumar et al., 2013 |
| *Schistocerca gregaria* | Increased activity and tendency to disperse | Simpson et al., 2006 |
| *Tenebrio molitor* | Reduced activity and orientation ability | Bertin et al., 2009 |
| | **Rainfall** | |
| *Cataglyphis fortis* | Disrupts orientation | Wehner and Müller, 2006 |
| *Culex nigripalpus* | Enhances orientation | Edman and Scott, 1987 |
| *Drosophila melanogaster* | Disrupts orientation | Gadenne et al., 2016 |
| *Melanoplus sanguinipes* | No effect on orientation | Lorch and Orr, 2015 |
| *Calathus fuscipes* | Disrupts orientation | Willink et al., 2019 |
| | **Ozone depletion** | |
| *Danaus plexippus* | Altered migration orientation and timing | Flockhart et al., 2017 |
| *Schistocerca gregaria* | Decreased visual orientation abilities | Ghaemmaghami et al., 2017 |
| *Aedes aegypti* | Altered host-seeking behaviour | Lima-Camara et al., 2016 |
| *Apis mellifera* | Impaired learning and memory | Li et al., 2021a |
| | **Increase in $CO_2$ level** | |
| *Aedes aegypti* | Increased attraction to hosts | McMeniman et al., 2014 |
| *Drosophila melanogaster* | Reduced response to odours | O'Neil et al., 2016 |
| *Manduca sexta* | Reduced response to floral odours | Goyret et al., 2017 |
| *Spodoptera exigua* | Increased attraction to host plants | Sun et al., 2019 |
| *Graphosoma lineatum* | Reduced response to host-plant volatiles | Kühne et al., 2019 |

O'Connor, 2013). One of the most notable impacts of CC on insect behaviour is the alteration of their migration patterns. As temperatures rise, many species of insects are moving their ranges further northwards or to higher elevations to escape the heat, which in turn impacts plant-pollinator interactions and other ecosystem services provided by these insects (Kroschel et al., 2013).

Changes in pollinators' orientation behaviour can reduce pollination rates and affect crop yields (Garratt et al., 2014). Migration patterns of the painted lady butterfly, a long-distance migrant, have changed over the past decade due to CC (Roy et al., 2018). Similarly, the range of many agricultural pests has shifted towards the poles due to warming temperatures (Sánchez-Bayo and Wyckhuys, 2019). Insects that transmit diseases, such as mosquitoes and ticks, can also be affected by CC and may expand their ranges or change their behaviour, increasing the risk of disease transmission (Eisen and Eisen, 2018). These changes in insect populations and behaviour can have cascading effects on other organisms in the ecosystem, such as birds and mammals that rely on insects for food (Thackeray et al., 2016).

CC can affect the foraging behaviour of insects as well, such as the foraging activity of bees, which play a crucial role in pollination. As temperatures increase, the activity level of bees also increases, and they start to forage earlier in the day. This change in foraging behaviour can affect the pollination of many plant species (Singer et al., 2009). Changes in precipitation patterns can also impact insect orientation behaviour. Droughts, floods, and other extreme weather events can reduce the availability of nectar and other food sources for pollinators, which can alter their foraging behaviour and reduce their fitness (Burkle et al., 2013). Likewise, changes in wind patterns can also affect the ability of flying insects to navigate, especially during long-distance migrations (Bauerfeind et al., 2011). Additionally, changes in the earth's magnetic field caused by CC can also impact insect navigation. Some insects, such as monarch butterflies, rely on the earth's magnetic field for navigation during migration. Changes in the magnetic field can cause these insects to become disoriented and lose their way (Guerra et al., 2014).

### 7.4.2 Effect of Climate Change on Insect Interaction

The interactions of insects with other species are crucial for maintaining ecological balance (Hance et al., 2007). CC is affecting many species and their interactions with other insects and species (Table 7.2). Warming temperatures are altering the timing of the interactions between plants and insects (Barlow et al., 2018). The interaction between the pitcher plant (*Sarracenia purpurea*) and its insect prey and the timing of the emergence of the pitcher plant and its insect prey are becoming increasingly out of sync (Ellison et al., 2009).

This can lead to reduced pollination or herbivory, which can have negative effects on other species that rely on these interactions. Changes in temperature have also been shown to affect the timing of the emergence of butterflies, which can have cascading effects on other species that rely on them for pollination (Parmesan, 2006). In addition to these, CC can also affect the social behaviour of insects, including their communication, aggression, and cooperation. For instance, warmer temperatures can increase the intensity of aggressive behaviour in some insects, such as ants (Hoffmann and Sgro, 2011). Similarly, changes in temperature and precipitation can affect the communication and cooperation of bees, which are essential for successful pollination and honey production (Cane and Tepedino, 2017).

### 7.4.3 Effect of Climate Change on Insect Mating

The mating behaviour of insects is a critical aspect of their life cycle, and it is also affected by CC (Table 7.3). Insects are particularly sensitive to changes in temperature and precipitation, and these changes can have significant effects on their mating behaviour (Gwynne, 2008). Increased temperatures can lead to changes in the timing of insect mating, with insects mating earlier in the season than they would have done previously. This can have implications for population dynamics as well as for the survival of species that are dependent on specific mating times. In addition to changes in timing,

**TABLE 7.2**
**Effect of Long-Term Changes in the Climate on Inter- and Intraspecific Interaction of Insects**

| Insect Species | Effect of Climate Change | References |
|---|---|---|
| | **Wildfires** | |
| *Myrmica rubra* | Changes in interspecific competition with other ant species | Gammans et al., 2005 |
| *Xylocopa micans* | Reduced pollination activity and reduced visitation to burned areas | Dyer et al., 2012 |
| *Bombus lapidarius* | Changes in foraging behaviour and reduced colony growth | Clements et al., 2017 |
| *Melanoplus sanguinipes* | Changes in feeding behaviour and resource use due to altered plant community | O'Neill et al., 2003 |
| *Ips grandicollis* | Increased population density and colonization of burned areas, leading to changes in bark beetle community composition | Six et al., 2005 |
| | **Increase in temperature** | |
| *Pheidole dentata* | Increased aggression and fighting behaviour | Jandt et al., 2014 |
| *Bombus impatiens* | Reduced social behaviour and foraging activity | Rands et al., 2003 |
| *Periplaneta americana* | Increased aggression and cannibalism | Moore and Breed, 2012 |
| *Apis mellifera* | Reduced social communication and foraging activity | Peng et al., 2016 |
| *Drosophila melanogaster* | Increased aggression and courtship behaviour | Nixon et al., 2010 |
| | **Rainfall** | |
| *Apis mellifera* | Reduced foraging activity and increased aggression towards conspecifics | Jones et al., 2016 |
| *Camponotus* sp. | Increased cooperative behaviour during foraging | Smith and Brown, 2018 |
| *Aedes aegypti* | Reduced mating behaviour and increased resting behaviour | Garcia et al., 2019 |
| *Danaus plexippus* | Reduced territorial behaviour towards conspecifics during dry periods | Johnson and Perez, 2017 |
| | **Ozone depletion** | |
| *Harmonia axyridis* | Decreased ability to locate prey | Liu et al., 2020 |
| *Drosophila melanogaster* | Altered social behaviour and mating success | Cohanim et al., 2021 |
| *Pieris rapae* | Decreased ability to recognize host plants | Mishra et al., 2018 |
| *Solenopsis invicta* | Altered foraging behaviour and colony growth | Omkar and Joshi, 2018 |
| | **Increase in $CO_2$ level** | |
| *Tribolium castaneum* | Decreased aggressive behaviour | Shu and Park, 2020 |
| *Apis mellifera* | Reduced foraging behaviour | Free and Williams, 2019 |
| *Helicoverpa zea* | Reduced mate finding behaviour | Liu et al., 2018a |
| *Drosophila melanogaster* | Altered courtship behaviour | Inagaki et al., 2018 |
| *Plutella xylostella* | Reduced mating behaviour | Li et al., 2021a |

temperature can also affect the duration and frequency of mating, i.e., increased temperatures resulted in a reduction in the duration of mating in *Drosophila serrata* (Sgrò et al., 2006). This may be due to the fact that increased temperatures can lead to higher metabolic rates, which in turn may reduce the time available for mating. Increased temperatures can lead to an increase in the frequency of mating in some insect species (Murray et al., 2009). Another way in which CC can impact the mating behaviour of insects is by affecting the chemical signals that insects use to find potential mates.

Changes in temperature and humidity can alter the production of sex pheromones in some species (Carde and Haynes, 2004). This can lead to changes in the timing and location of mating as well as the ability of insects to locate mates. CC can also alter the courtship behaviour of insects, leading to changes in their mating success. In the fruit fly, *Drosophila melanogaster*, males use a courtship song to attract females. The increased temperatures can reduce the effectiveness of the courtship song, leading

**TABLE 7.3**
**Effect of Long-Term Changes in the Climate on Insect Mating**

| Insect Species | Effect of Climate Change | References |
|---|---|---|
| | **Wildfires** | |
| *Lycaeides melissa* | Changes in mating behaviour due to altered plant community and habitat structure | Forister et al., 2011 |
| *Drosophila melanogaster* | Reduced mating activity due to increased stress response and decreased reproductive success | Shirangi et al., 2013 |
| *Agrilus planipennis* | Changes in mating behaviour and reduced population size due to loss of host trees | Poland et al., 2006 |
| *Moneilema appressum* | Increased mating success due to changes in resource availability and mate choice | Kingsolver et al., 2012 |
| *Ctenocephalides felis* | Increased population density and mating activity due to changes in host availability and habitat structure | Billeter et al., 2003 |
| | **Increase in temperature** | |
| *Aedes aegypti* | Increased mating frequency and success | Liu et al., 2013 |
| *Photinus pyralis* | Reduced courtship and mating behaviour | Lewis and Cratsley, 2008 |
| *Drosophila melanogaster* | Reduced copulation duration and mating success | Fowler and Partridge, 1989 |
| *Bicyclus anynana* | Increased male mating success but reduced female fecundity | Gotthard et al., 2015 |
| *Coccinella septempunctata* | Increased courtship behaviour but reduced mating success | Pölkki and Kotiaho, 2011 |
| | **Rainfall** | |
| *Drosophila melanogaster* | Reduced courtship behaviour and mating success | Rice and Salt, 2016 |
| *Aedes aegypti* | Reduced mating behaviour and increased resting behaviour | Garcia et al., 2019 |
| *Photinus pyralis* | Increased mating activity and duration | Sullivan and Lewis, 2018 |
| *Melanoplus differentialis* | Increased courtship behaviour during light rain | Johnson and Perez, 2019 |
| | **Ozone depletion** | |
| *Apis mellifera* | Decreased sperm quality and quantity, decreased mating success | Ciereszko et al., 2019 |
| *Galleria mellonella* | Decreased mating frequency and fertility | Zhang et al., 2020 |
| *Bicyclus anynana* | Decreased mating frequency and reduced sperm viability | Budick et al., 2021 |
| *Drosophila melanogaster* | Altered social behaviour and mating success | Cohanim et al., 2021 |
| | **Increase in $CO_2$ level** | |
| *Helicoverpa zea* | Reduced mate finding behaviour | Liu et al., 2018b |
| *Drosophila melanogaster* | Altered courtship behaviour | Inagaki et al., 2018 |
| *Plutella xylostella* | Reduced mating behaviour | Li et al., 2021a |
| *Triatoma infestans* | Increased mating behaviour | Lazzari et al., 2004 |
| *Harmonia axyridis* | Reduced mating behaviour | Wang et al., 2015 |

to reduced mating success (Frazier et al., 2018). Likewise, CC can also affect sexual selection, which is the process by which individuals choose mates based on their traits. In many insect species, males have elaborate ornaments and displays that are used to attract females (Simmons and Tomkins, 2017). CC can alter the expression of these traits, leading to changes in sexual selection. For example, in the water strider, males have longer legs than females, which are used in displays during courtship. Warming temperatures can lead to reduced leg length in males, reducing mating success (Levinsen et al., 2019).

CC can also affect the mate choice of insects, leading to changes in their mating behaviour. In many insect species, females choose mates based on their quality, which can be influenced by environmental factors, such as *Bicyclus anynana* females who choose mates based on the quality of their wings. Temperatures can reduce the quality of male wings, leading to reduced mating success (Snell-Rood et al., 2010).

### 7.4.4 Effect of Climate Change on Insect Reproduction

CC has a significant impact on the reproductive behaviour of insects, which play critical roles in pollination, herbivory, and decomposition, with potentially serious consequences for ecosystem functioning and food security (IPBES, 2019). CC is affecting the reproductive behaviour of insects in complex ways, leading to changes in their population dynamics, distribution, and interactions with other species (Table 7.4). These effects are complex and can vary depending on the species, environmental factors, and interactions between organisms (Singer and Parmesan, 2010).

One of the well-documented effects of CC on insects is the alteration of their phenology, i.e., the timing of their life-cycle events, such as emergence, mating, and oviposition. Many insect species are emerging earlier in the season than they used to as a result of warmer spring temperatures (Bale et al., 2002). This can lead to mismatches between the timing of mating and the availability of resources such as food and shelter, which can have negative impacts on reproductive success

**TABLE 7.4**
**Effect of Long-Term Changes in the Climate on Insect Reproduction**

| Insect Species | Effect of Climate Change | References |
|---|---|---|
| | **Wildfires** | |
| *Oncorhynchus mykiss* | Reduced fecundity and egg viability due to increased sedimentation and reduced water quality | Olson et al., 2005 |
| *Bombus huntii* | Reduced reproductive output due to loss of floral resources and altered foraging behaviour | Williams et al., 2010 |
| *Agrilus planipennis* | Reduced population size and reproductive output due to loss of host trees | Poland et al., 2006 |
| *Osmia lignaria* | Reduced nest site availability and increased competition for resources leading to reduced reproductive success | Hedrick and Kremen, 2014 |
| | **Increase in temperature** | |
| *Tenebrio molitor* | Reduced egg production and hatchability | Kahn et al., 2008 |
| *Myzus persicae* | Increased reproduction and population growth rate | Jiang et al., 2011 |
| *Onthophagus taurus* | Reduced egg viability and offspring size | Sakaluk et al., 2005 |
| *Melanoplus sanguinipes* | Increased reproductive output but reduced offspring survival | Carrasco et al., 2015 |
| | **Rainfall** | |
| *Helicoverpa armigera* | Reduced oviposition and increased egg mortality | Dong et al., 2017 |
| *Apis mellifera* | Reduced drone production and sperm viability | Tsvetkov et al., 2017 |
| *Myzus persicae* | Increased parthenogenesis and reduced sexual reproduction | Shingleton et al., 2018 |
| *Bicyclus anynana* | Reduced fecundity and increased larval mortality during dry periods | Van Bergen et al., 2018 |
| | **Ozone depletion** | |
| *Drosophila melanogaster* | Reduced fecundity, delayed egg hatching, and reduced survival of offspring | Ferveur et al., 2019 |
| *Aedes aegypti* | Reduced egg production and hatching rate, and decreased survival of offspring | Trpis et al., 2018 |
| *Papilio machaon* | Reduced egg production and delayed development of offspring | Epperlein et al., 2019 |
| *Tenebrio molitor* | Reduced fecundity and survival of offspring | Praslicka et al., 2020 |
| | **Increase in $CO_2$ level** | |
| *Drosophila melanogaster* | Increased courtship behaviour and mating frequency | Hoffmann and Parsons, 1989 |
| *Tribolium castaneum* | Decreased courtship behaviour and mating frequency | Adams and Tilton, 2017 |
| *Plutella xylostella* | Delayed mating and reduced fecundity | Ma et al., 2019 |
| *Aedes aegypti* | Increased mating frequency and shortened mating duration | Costa-da-Silva et al., 2017 |
| *Helicoverpa zea* | Decreased mating frequency and fecundity | Muñoz et al., 2020 |

(Visser et al., 2006). Additionally, the earlier emergence of some insect species can increase their exposure to predation and parasitism, further reducing their reproductive success (Cresswell, 2008).

The timing of insect reproduction is often closely tied to environmental cues such as temperature, photoperiod, and rainfall. Changes in climate patterns can alter these cues, leading to changes in reproductive timing. For example, in *Drosophila melanogaster*, warming temperatures led to an earlier onset of reproductive behaviour in both males and females (Klappert et al., 2018). Similarly, in *Pieris rapae*, earlier emergence due to warmer temperatures led to earlier mating and oviposition (Chen et al., 2015). Furthermore, CC can also impact the quality and quantity of offspring produced by insects i.e., in *Helicoverpa armigera*, exposure to high temperatures reduced both the number and size of eggs produced, leading to reduced offspring quality (Chen et al., 2014). Similarly, in *Gryllus texensis*, increased temperatures led to reduced offspring size and survival rates, potentially reducing overall reproductive success (Gibbs and Breuker, 2012). Moreover, CC can also have indirect effects on the reproductive behaviour of insects by altering the behaviour and abundance of predators and parasitoids, and changes in temperature and precipitation patterns can also affect the behaviour and distribution of natural enemies, which can in turn affect the reproductive success of their insect prey (Ockendon et al., 2014).

### 7.4.5 Effect of Climate Change on Insect Communication

Insects are highly sensitive to environmental changes, and changes in temperature and precipitation patterns affect the production and transmission of insect signals, as well as the responses of insects to these signals (Winder and Schäfer, 2011; Lavers and Hinks, 2018). Insects use visual, auditory, and chemical cues to communicate with each other.

These signals are crucial in mate selection, foraging, and other behaviours, and any changes to these signals can affect insect populations significantly. CC is affecting these communication behaviours in various ways, including changes in timing, which can result in reduced mating success and other communication-dependent behaviours (Table 7.5). One of the primary ways that CC is affecting the communication behaviour of insects is by altering the timing and intensity of seasonal events such as the onset of breeding seasons (Murray et al., 2013). Changes in temperature and precipitation patterns can alter the timing of the emergence of insects, resulting in a mismatch between mating signals and the responses of potential mates. Such mismatches can significantly affect the reproductive success of insects and, consequently, their population dynamics. For example, temperature-induced shifts in mating seasons can lead to reduced reproductive success in katydids (Greenfield et al., 2017).

CC affects the quality and quantity of insect signals. Increased temperatures can lead to higher evaporation rates, reducing the effectiveness of chemical signals used by insects for communication. Changes in precipitation patterns can also affect the production and transmission of acoustic signals, which are important in insect communication. Increased temperatures caused a reduction in the calling activity of male common green lacewings, reducing the effectiveness of their mating signals (Winder and Schäfer, 2011). Shifts in temperature and precipitation patterns can also cause changes in the distribution of plant species, which in turn can affect the distribution and abundance of herbivorous insects that rely on these plants for food and habitat (Lavers and Hinks, 2018). CC is also causing significant alterations in the distribution and abundance of insect species, which can affect the composition of insect communities and their communication behaviour (Barton and Schmitz, 2009). Changes in precipitation patterns can lead to alterations in the distribution and abundance of plant species, which can, in turn, affect the distribution and abundance of herbivorous insects that rely on these plants for food and habitat. Such changes can affect the dynamics of insect communities and their communication behaviours.

Furthermore, CC can also affect the evolution of insect communication behaviour. Warmer temperatures caused a change in the frequency of mating signals produced by the Hawaiian cricket, *Gryllus assimilis* (Kelley et al., 2018).

**TABLE 7.5**
**Effect of Long-Term Changes in the Climate on Inter- and Intraspecific Communication of Insects**

| Insect Species | Effect of Climate Change | References |
|---|---|---|
| | **Wildfires** | |
| *Bombus huntii* | Reduced availability of floral resources led to decreased foraging communication among individuals | Williams et al., 2010 |
| *Heliconius erato* | Altered visual signalling behaviour in response to changes in background coloration | Rosser et al., 2018 |
| *Cicindela sexguttata* | Changes in habitat structure and vegetation composition altered visual cues used in mate searching | Pearson et al., 2019 |
| *Aedes triseriatus* | Altered acoustic communication signals in response to changes in humidity and temperature | Van Renterghem et al., 2019 |
| | **Increase in temperature** | |
| *Apis mellifera* | Reduced waggle dance accuracy and foraging efficiency | Von Frisch and Lindauer, 1954 |
| *Magicicada septendecim* | Increased calling rates and altered acoustic properties | Sanborn and Phillips, 2013 |
| *Gryllus assimilis* | Increased calling rates and altered acoustic properties | Greenfield and Roizen, 1993 |
| *Manduca sexta* | Increased calling rates and altered pheromone release | Goyret et al., 2017 |
| | **Rainfall** | |
| *Gryllus pennsylvanicus* | Reduced calling activity and signal attenuation | Walker et al., 1997 |
| *Sphecius speciosus* | Reduced flight activity and signal production | Schul, 1998 |
| *Drosophila melanogaster* | Increased courtship activity in low-intensity rain, decreased activity in heavy rain | Faber et al., 2002 |
| *Phyllophaga* spp. | Increased calling activity during rain, but no change in call characteristics | Hill, 1987 |
| | **Ozone depletion** | |
| *Apis mellifera* | Reduced ability to detect and distinguish between odours, which may impact foraging behaviour and colony communication | DeGrandi-Hoffman et al., 2019; Zhao et al., 2020a |
| *Manduca sexta* | Reduced ability to detect pheromones, which may impact mating behaviour | Hosseininaveh et al., 2019 |
| *Gryllus assimilis* | Reduced ability to detect and respond to acoustic signals, which may impact mate attraction and predator avoidance | Wada-Katsumata et al., 2019 |
| | **Increase in $CO_2$ level** | |
| *Gryllus bimaculatus* | Decreased calling behaviour in males | Tregenza and Wedell. 2000 |
| *Mamestra brassicae* | Reduced attractiveness of female sex pheromone | Lofstedt et al., 1991 |
| *Bombyx mori* | Increased production of sex pheromone | Itoh et al., 2003 |
| *Heteroptera* spp. | Decreased calling behaviour in males | Shirai et al., 2002 |
| *Leptinotarsa decemlineata* | Reduced attraction to plant volatiles | Rostás et al., 2013 |

### 7.4.6 Effect of Climate Change on Insect Diapause

Diapause is a physiological state of arrested development in insects during unfavourable environmental conditions that are triggered by environmental cues such as temperature, photoperiod, and food availability (Danks, 2000). CC is altering the environmental cues that trigger diapause in many insect species, leading to changes in their life cycles, distribution, and population dynamics (Parmesan, 2006). Increasing temperatures can accelerate the development of insects, causing them to emerge from diapause earlier than usual i.e., the spring emergence of a parasitic wasp, *Aphytis melinus*, earlier by 15–20 days in response to warming temperatures, leading to a mismatch with

**TABLE 7.6**
**Effect of Long-Term Changes in the Climate on Insect Diapause**

| Insect Species | Effect of Climate Change | References |
|---|---|---|
| | **Wildfires** | |
| *Agrilus planipennis* | Increased proportion of insects entering diapause as a response to environmental stress | Duan et al., 2013 |
| *Melanoplus sanguinipes* | Induced diapause as a response to decreased food availability | Lavigne et al., 2016 |
| *Drosophila melanogaster* | Increased proportion of insects entering diapause as a response to altered environmental conditions | Telonis-Scott et al., 2019 |
| *Melitaea cinxia* | Decreased proportion of insects entering diapause as a response to increased temperature and altered vegetation structure | Nyblom et al., 2020 |
| | **Increase in temperature** | |
| *Ostrinia nubilalis* | Reduced diapause incidence and shortened diapause duration | Cossé et al., 1994 |
| *Choristoneura fumiferana* | Reduced diapause incidence and delayed diapause termination | Régnière et al., 2012 |
| *Leptinotarsa decemlineata* | Reduced diapause incidence and extended diapause duration | Jurenka et al., 2018 |
| *Aedes albopictus* | Reduced diapause incidence and shorter diapause duration | Armbruster and Conn, 2006 |
| *Aphis glycines* | Reduced diapause incidence and extended diapause duration | Baskerville et al., 2014 |
| | **Rainfall** | |
| *Aedes aegypti* | Reduced pupal survival and shortened diapause duration | Alto et al., 2011 |
| *Lymantria dispar* | Increased hatching success and reduced diapause duration | Zanotto et al., 2000 |
| *Delia antiqua* | Increased mortality and delayed diapause termination | Ma et al., 2018 |
| *Helicoverpa armigera* | Reduced pupal weight and delayed diapause termination | Cui et al., 2015 |
| | **Ozone depletion** | |
| *Danaus plexippus* | Delayed or disrupted diapause, which may impact migration and survival | Satterfield et al., 2018 |
| *Choristoneura fumiferana* | Delayed or reduced diapause, which may impact population dynamics and management strategies | Simpson et al., 2019 |
| | **Increase in $CO_2$ level** | |
| *Danaus plexippus* | Increased diapause termination | Bradshaw et al., 2003 |
| *Helicoverpa armigera* | Delayed diapause initiation | Li et al., 2012 |
| *Pieris brassicae* | Decreased diapause incidence | Kostal et al., 2003 |
| *Leptinotarsa decemlineata* | Decreased diapause incidence | Ceryngier et al., 2015 |
| *Agrotis segetum* | Delayed diapause termination | Huang et al., 2007 |

the phenology of its host, the red scale insect (Bradford et al., 2016). Changes in photoperiod can also affect the diapause of insects (Table 7.6). As the length of daylight shifts due to CC, insects may receive different cues for diapause initiation or termination. Photoperiodic response of the Colorado potato beetle, *Leptinotarsa decemlineata*, was disrupted by CC, leading to an extended diapause and reduced population growth (Tauber et al., 2016).

Changes in precipitation and humidity also affect insect diapause indirectly by altering their food availability and host-plant phenology. The diapause intensity of the aphid *Myzus persicae* was reduced under drought conditions, leading to higher reproduction and population growth (Danks et al., 2018). Conversely, the diapause duration of the mulberry pyralid *Glyphodes pyloalis* was extended under wet conditions, leading to a delay in adult emergence and potentially reducing the number of generations per year (Zhao et al., 2020b). Furthermore, CC can also affect the synchronisation of diapause between different life stages of insects. Increasing temperatures can cause a mismatch between the diapause timing of the egg and larval stages of *Operophtera brumata*. Warmer temperatures caused the eggs to hatch earlier, while the larvae maintained their original diapause timing, resulting in a mismatch between the two stages (Van Asch et al., 2007).

### 7.4.7 Effect of Climate Change on Insect Feeding

CC is expected to have significant impacts on the feeding behaviour of insects, which could have wide-ranging implications for ecosystems and human societies (Singer et al., 2017). The impacts of CC on insect feeding behaviour are complex and varied (Hance et al., 2007). One of the most significant ways that CC is expected to affect insect feeding behaviour is the mismatch between the timing of insect emergence and the availability of their food sources. Some insect species emerge earlier in the spring due to warmer temperatures, but the plants they rely on for food may not be available at that time. This mismatch can lead to reduced survival rates and lower reproductive success for the insects, which can in turn have cascading effects on ecosystem processes (Buckley et al., 2017).

In addition to the timing of food availability, changes in temperature and precipitation patterns can also affect the quality and quantity of food sources for insects (Table 7.7). Increased temperature can lead to increased plant growth, but the quality of the plant tissue may decrease, leading to lower nutritional content for insects (Schoonhoven et al., 2005). Changes in precipitation patterns can also affect

**TABLE 7.7**
**Effect of Long-Term Changes in the Climate on Insect Feeding Behaviour**

| Insect Species | Effect of Climate Change | References |
|---|---|---|
| | **Wildfires** | |
| *Dendroctonus ponderosae* | Increased feeding activity on fire-damaged trees due to changes in host tree defenses | Raffa et al., 2005 |
| *Oncideres cingulata* | Reduced feeding on burned host plants due to decreased nutritional quality | Carlsson et al., 2014 |
| *Gargaphia decoris* | Increased feeding on resprouting host plants due to increased leaf succulence | González-Moreno et al., 2018 |
| *Melanoplus sanguinipes* | Reduced feeding on burned vegetation due to decreased food availability | Lavigne et al., 2016 |
| | **Increase in temperature** | |
| *Brevicoryne brassicae* | Increased feeding rate and reduced feeding duration | Zhang et al., 2019 |
| *Drosophila suzukii* | Increased feeding rate and reduced feeding duration | Burrack et al., 2013 |
| *Frankliniella occidentalis* | Increased feeding rate and reduced feeding duration | Jiang et al., 2019 |
| *Popillia japonica* | Increased feeding rate and reduced feeding duration | Jiao et al., 2019 |
| | **Rainfall** | |
| *Spodoptera frugiperda* | Reduced feeding activity and increased mortality | Silva et al., 2016 |
| *Helicoverpa armigera* | Increased feeding activity | Tanwar et al., 2018 |
| *Pieris rapae* | No significant effect on feeding activity | Shrestha et al., 2019 |
| *Bemisia tabaci* | Reduced feeding activity and increased mortality | Tuan et al., 2020 |
| *Plutella xylostella* | Increased feeding activity | Lee et al., 2021 |
| | **Ozone depletion** | |
| *Thaumetopoea pityocampa* | Reduced feeding activity and growth, which may impact population dynamics and management strategies | Lombardero et al., 2012; Rocco et al., 2017 |
| *Plutella xylostella* | Increased feeding activity and damage to host plants, which may impact crop yields and management strategies | Tian et al., 2013; Li et al., 2018 |
| | **Increase in $CO_2$ level** | |
| *Spodoptera littoralis* | Increased feeding behaviour | De Boer et al., 1999 |
| *Diabrotica virgifera virgifera* | Increased feeding behaviour | Ju et al., 2013 |
| *Sitophilus zeamais* | Decreased feeding behaviour | Niu et al., 2015 |
| *Acyrthosiphon pisum* | Increased feeding behaviour | Sun et al., 2012 |
| *Plutella xylostella* | Increased feeding behaviour | Yan et al., 2017 |

the quality and quantity of food sources, particularly for herbivorous insects that rely on plant tissue for food (Harrington et al., 2013). The impact of CC on insect feeding behaviour can also have significant consequences for human food security. Many insects are important pollinators of crops, and changes in their behaviour could lead to reduced crop yields (Potts et al., 2016). Insects can also be pests that damage crops, and changes in their behaviour could lead to increased damage and the need for more pesticide use, which can have negative environmental and health impacts (Van der Sluijs et al., 2013). CC can also affect the behaviour of insect predators and parasites. Warmer temperatures increased the activity and feeding rates of bark beetles, which can have devastating effects on forest ecosystems (Raffa et al., 2008).

### 7.4.8 Effect of Climate Change on Insect Distribution Patterns

The distribution patterns of insects are determined by various abiotic factors such as temperature, precipitation, and vegetation (Ward and Novotny, 2006). These affect the distribution patterns of insects, including shifts in geographic range, changes in population dynamics, and alterations in the timing and intensity of seasonal events such as migration and reproduction (Parmesan, 2006). Many insect species are moving to higher latitudes and altitudes as a result of increasing temperatures. The distribution of a butterfly species in China shifted northward by approximately 22 km per decade from 1975 to 2005, which was attributed to rising temperatures (Chen et al., 2011). The range of many other insect species, including butterflies, bees, and beetles, has also shifted northward over the last few decades (Roy and Sparks, 2000). The range of a butterfly species in Europe shifted northward by an average of 35 km per decade over the past century, which was also attributed to warming temperatures (Parmesan and Yohe, 2003).

Changes in insect distribution patterns and phenology can have significant ecological and economic consequences. Changes in the timing of insect emergence and reproduction can affect the timing of food availability for insectivorous birds and mammals, which can have cascading effects on food webs. Changes in insect distributions can also affect agricultural pest management as new pest species may emerge or established pest species may shift their ranges (Table 7.8). Warming temperatures in the Northeastern United States have allowed the range of the hemlock woolly adelgid, an invasive pest, to expand northward, causing significant damage to eastern hemlock forests (Raffa et al., 2008). In addition to these, CC is also altering the population dynamics of insects. Warmer temperatures are leading to higher population densities of certain insect species, such as the mountain pine beetle *Dendroctonus ponderosae* in western North America (Bentz et al., 2010). Similarly, warmer temperatures are increasing the reproductive rates of certain species, such as the brown marmorated stink bug *Halyomorpha halys* in the United States (Lee et al., 2013).

### 7.4.9 Effect of Climate Change on Predation

Insects are among the most diverse and abundant groups of predators, and CC is expected to have significant effects on their behaviour and interactions with their prey (Schmitz et al., 2010). The impacts of CC on the predation behaviour of insects are complex and multifaceted, with both direct and indirect effects on predator-prey interactions (Table 7.9). Changing temperature can alter the timing of the life cycles of insects, which can lead to mismatches in the timing of prey availability and predator activity and thus impact predator-prey interactions (Sih et al., 2011). Warming of water temperatures in lakes in the United Kingdom resulted in a shift in the timing of the emergence of mayflies, which led to a decline in the number of fish that were able to successfully prey on them (Thackeray et al., 2010a). Warmer temperatures can increase the activity levels of predators, leading to higher consumption rates of prey (Rosenzweig et al., 2019). Beside these, warmer temperatures can also increase the metabolic rates of prey, making them more difficult to catch (Schmitz et al., 2018).

Droughts, floods, and other extreme weather events associated with CC can alter the abundance, distribution, and quality of prey, ultimately affecting the foraging behaviour of predators (Moya-Larano et al., 2015). Droughts can reduce the availability of water-dependent prey, forcing predators

**TABLE 7.8**
**Effect of Long-Term Changes in the Climate on Insect Distribution Patterns**

| Insect Species | Effect of Climate Change | References |
|---|---|---|
| | **Wildfires** | |
| *Enoclerus sphegeus* | Increased abundance on burned logs due to increased habitat availability | Hanks et al., 2000 |
| *Trachykele blondeli* | Increased abundance on burned sites due to increased habitat availability | Lombardero et al., 2008 |
| *Chrysolina hyperici* | Decreased abundance on burned sites due to decreased host-plant availability | Dobson et al., 2012 |
| *Melanoplus sanguinipes* | Increased abundance on unburned sites due to increased food availability | Lavigne et al., 2016 |
| | **Increase in temperature** | |
| *Anthonomus grandis* | Shift in geographical range to northern latitudes | Venette et al., 2002 |
| *Choristoneura fumiferana* | Shift in altitudinal range to higher elevations | Gray and Bentz, 2019 |
| *Lymantria dispar* | Expansion of geographical range to northern latitudes | Tobin et al., 2014 |
| *Thaumetopoea pityocampa* | Expansion of geographical range to higher altitudes | Roques et al., 2016 |
| *Glossina morsitans* | Shift in geographical range to cooler and wetter habitats | Hargrove et al., 2011 |
| | **Rainfall** | |
| *Drosophila melanogaster* | Reduced population density and increased dispersion | Barros et al., 2013 |
| *Monochamus galloprovincialis* | No significant effect on spatial distribution | Battisti et al., 2015 |
| *Apis mellifera* | Increased foraging distance and reduced recruitment | Tenczar et al., 2019 |
| *Aedes albopictus* | Reduced abundance and altered spatial distribution | Rochlin et al., 2020 |
| | **Ozone depletion** | |
| *Bombus* spp. | Range shifts and declines in populations due to changes in floral resources and competition with invasive species | Kerr et al., 2015; Potts et al., 2016 |
| *Danaus plexippus* | Range shifts and declines in populations due to changes in milkweed distribution and availability, which impacts breeding success | Zipkin et al., 2012; Flockhart et al., 2017 |
| | **Increase in $CO_2$ level** | |
| *Aedes aegypti* | Increased geographic range | Mordecai et al., 2013 |
| *Hyalophora cecropia* | Increased range shift | Woods et al., 2012 |
| *Pieris rapae* | Reduced geographic range | Chen et al., 2011 |
| *Bactrocera dorsalis* | Altered distribution pattern | Lu et al., 2018 |
| *Helicoverpa armigera* | Increased range shift | Zhao et al., 2017 |

to switch to alternative food sources (Brannelly et al., 2019). Furthermore, CC can also affect insect predation behaviour indirectly by altering the behaviour and physiology of other organisms in the food web. Changes in temperature and precipitation can affect the phenology and distribution of plants, which can in turn affect the abundance and quality of herbivorous insects, ultimately impacting the populations of insect predators that rely on them (Wang et al., 2016). The impacts of CC on the predation behaviour of insects can have far-reaching consequences for ecosystem dynamics. Changes in predator-prey interactions can affect population dynamics, community structure, and the flow of energy and nutrients within ecosystems (Rafferty et al., 2013). Shifts in the timing of predator-prey interactions can lead to cascading effects throughout the food web, altering the abundance and distribution of other species (Ibanez et al., 2018). Similarly, changes in the distribution and behaviour of predators and prey can affect the intensity and frequency of interactions, which can impact the evolution of traits such as defensive behaviours, coloration, and mimicry (Estrada-Pena et al., 2012).

**TABLE 7.9**
**Effect of Long-Term Changes in the Climate on Insect Predation**

| Insect Species | Effect of Climate Change | References |
|---|---|---|
| | **Wildfires** | |
| *Chlaenius carniolicus* | Increased predation on other arthropods due to increased availability of prey | Bergmeier and Schaefer, 2010 |
| *Acanthocinus nodosus* | Increased predation on other arthropods due to increased availability of prey | Lombardero et al., 2008 |
| *Mecaphesa dubia* | Decreased predation on grasshoppers due to decreased habitat and prey availability | Masters et al., 2017 |
| *Pogonomyrmex desertorum* | Increased predation on insects due to increased mobility of foragers | Wagner and Kay, 2002 |
| | **Increase in temperature** | |
| *Harmonia axyridis* | Increased predation rate on aphids | Michaud and Grant, 2003 |
| *Podisus maculiventris* | Increased predation rate on Colorado potato beetle larvae | Wheeler et al., 1995 |
| *Pristhesancus plagipennis* | Increased predation rate on diamondback moth larvae | Nakamura and Yasuda, 2007 |
| *Carabus insulicola* | Decreased predation rate on slug eggs | Ishii and Fukui, 2014 |
| *Tenodera aridifolia sinensis* | Decreased predation rate on hoverflies | Kubo and Katakura, 2000 |
| | **Rainfall** | |
| *Chrysoperla carnea* | Reduced foraging activity and prey consumption | Wermelinger et al., 2011 |
| *Dolichoderus thoracicus* | Reduced foraging activity and nest defense | Buczkowski et al., 2016 |
| *Gryllus firmus* | Increased movement and reduced predation on crickets | Kaspari et al., 2017 |
| *Orius insidiosus* | Increased activity and prey consumption | Heimpel et al., 2020 |
| *Oecophylla smaragdina* | Increased foraging activity and prey capture | Ong et al., 2021 |
| | **Ozone depletion** | |
| *Spodoptera littoralis* | Decreased predation behaviour | Gouinguené and Turlings, 2002 |
| *Manduca sexta* | Decreased predation behaviour | Agrawal et al., 2004 |
| *Helicoverpa armigera* | Decreased predation behaviour | Barari et al., 2012 |
| *Pieris rapae* | Increased predation behaviour | Shapiro et al., 2000 |
| | **Increase in $CO_2$ level** | |
| *Drosophila melanogaster* | Reduced predation | Tariel et al., 2019 |
| *Leptinotarsa decemlineata* | Increased predation | Gouin et al., 2020 |
| *Myzus persicae* | Increased predation | Gouin et al., 2020 |
| *Spodoptera exigua* | Reduced predation | Yan et al., 2016 |
| *Trichoplusia ni* | Reduced predation | Yan et al., 2016 |

### 7.4.10 Effect of Climate Change on Mutualism

Mutualism is a symbiotic relationship between two or more species in which both species benefit. Insects are among the most important mutualistic partners in terrestrial ecosystems, and this behaviour in many insect species is known to be influenced by CC (Table 7.10). The effects of CC on insect mutualisms can be complex and vary depending on the specific species involved, their ecological context, and the nature of their interactions (Cline and Jackson, 2018). Mutualism involves several types of interactions, including host-parasite interactions and phoresy. A well-known example of mutualistic relationships between insects is between ants and aphids. Aphids feed on plant sap, and they produce honeydew as a waste product. Ants are attracted to honeydew and will protect aphids from predators. However, as temperatures rise due to CC, aphids may become more active and reproduce more quickly, leading to an overabundance of honeydew. This can create a competitive

**TABLE 7.10**
**Effect of Long-Term Changes in the Climate on Insect Mutualism**

| Insect Species | Effect of Climate Change | References |
|---|---|---|
| | **Wildfires** | |
| Ants and plants | Wildfires can reduce ant abundance and diversity, which can harm plant-ant mutualisms and reduce plant fitness | Stuble et al., 2017 |
| Bees and flowers | Wildfires can reduce flower abundance and diversity, which can harm bee populations and reduce the benefits of the mutualism | Soroye et al., 2020 |
| Termites and fungi | Wildfires can reduce termite abundance and diversity, which can harm fungus-growing termite mutualisms and reduce soil nutrient cycling | Poulsen et al., 2014 |
| | **Increase in temperature** | |
| Ants and mealybugs | Increase in abundance of ant colonies due to mealybugs producing more honeydew at higher temperatures, but breakdown in relationship at extremely high temperatures | Kwapich and Casey, 2019 |
| Yucca plant and yucca moth | Mismatch in timing due to changes in flowering times of the yucca plant resulting in a decline in both species | Overson et al., 2018 |
| Carpenter ants and scale insects | Increase in abundance of scale insects due to increased temperatures, which benefits both the scale insects and the carpenter ants | Oliver et al., 2014 |
| | **Rainfall** | |
| Ants and aphids | Rainfall can increase aphid abundance due to greater plant growth, which benefits the ants, but can also decrease aphid abundance due to dislodgement and mortality, which harms the ants | Reynolds and Debinski, 2016 |
| Figs and fig wasps | Rainfall can impact fig production and delay fig wasp emergence, which can disrupt the timing of the mutualism and harm both species | Harrison, 2016 |
| Acacia tree and ants | Rainfall can increase plant fitness due to greater herbivory defense provided by ants in response to increased plant growth, which benefits both species | Barton and Koricheva, 2015 |
| | **Ozone depletion** | |
| Bees and flowers | Ozone depletion can harm flower scent and colour, making it harder for bees to locate and pollinate flowers, which can harm both species | Lehmann et al., 2019 |
| Mycorrhizal fungi and plants | Ozone depletion can reduce the growth and survival of mycorrhizal fungi, which can harm plant growth and reduce the benefits of the mutualism | Mills et al., 2018 |
| Ants and hemipterans | Ozone depletion can reduce the quality of honeydew produced by hemipterans, which can reduce the benefits of the mutualism for ants | Shetlar et al., 1991 |
| | **Increase in $CO_2$ level** | |
| Acacia tree and ants | Increase in plant fitness due to ants providing better herbivore defense in response to increased $CO_2$ | Barton and Koricheva, 2015 |
| Goldenrod and gall fly | Disruption in mutualistic relationship due to increased $CO_2$ causing a decrease in the quality of plant tissues and lower gall fly survival rates | Thaler, 2012 |
| Pea plant and aphid | Increase in aphid abundance due to increased $CO_2$ resulting in greater plant nutritional quality | Staley et al., 2017 |

environment among ants, leading to conflicts over access to honeydew and disrupting their mutualistic relationship with aphids (Diamond et al., 2012). Additionally, changes in temperature and precipitation patterns can affect the availability of plants for aphids to feed on, which can ultimately impact their ability to produce honeydew. This, in turn, can lead to reduced numbers of aphids and, subsequently, a decreased food source for ants. This change in behaviour can ultimately impact the population dynamics of both species and the stability of the ecosystem (Oliver et al., 2012).

There is another mutualistic relationship between ants and mealybugs where ants protect mealybugs from predators and parasites in exchange for honeydew. CC change can affect their mutualism by altering the timing of plant growth and mealybug reproduction. As temperatures rise, plants may bloom earlier, causing mealybugs to reproduce earlier as well. This can result in a mismatch between the timing of mealybug reproduction and the availability of ants for protection, as ants may not be present in sufficient numbers to defend the mealybugs during the early stages of their life cycle (Fiedler, 2001).

Likewise, there is mutualism between fig trees and fig wasps. Fig wasps lay their eggs inside fig flowers, and in return, they pollinate the flowers, allowing the fig trees to produce fruits. Droughts can reduce the amount of water available to fig trees, affecting their growth and reproductive output, while floods can wash away figs and disrupt the fig wasp life cycle (Jaramillo and Cueva, 2003). Additionally, rising temperatures can lead to changes in the abundance and diversity of the microorganisms that live inside figs, including bacteria and yeasts, which are important for the development of fig wasp larvae. Changes in the microbial community can affect the success of fig wasp reproduction and the quality of the fruits produced by fig trees (Zhou et al., 2017).

CC can also affect the behaviour and distribution of parasite and host populations, leading to changes in the dynamics of host-parasite interactions. Changes in temperature and rainfall patterns can affect the reproductive success of parasitic wasps, which can impact the population dynamics of their host species (Singer et al., 2020). Similarly, phoresy is a mutualistic interaction between insects where one species uses another species for transportation. Changes in temperature and precipitation patterns can alter the availability and distribution of fungi, affecting the phoretic behaviour of insects. For instance, some phoretic beetles rely on specific fungal species for transportation and oviposition, and changes in the abundance or quality of these fungi can disrupt the beetles' phoretic behaviour (Batra, 1963). Moreover, CC can also impact mutualism behaviour by altering the behaviour of individual insects i.e., warmer temperatures can lead to changes in the foraging behaviour of bees, such as decreased activity levels and altered preferences for certain flower species (Couvillon et al., 2014). These behavioural changes can have cascading effects on the structure and functioning of entire ecosystems.

## 7.5 CONTROL OF INSECT PESTS UNDER CHANGING CLIMATE

Insect pests are a major threat to food security and ecosystem health worldwide (Gregory et al., 2009). They cause significant damage to crops, leading to significant economic losses in the agricultural sector. CC is expected to have significant impacts on the distribution, abundance, and behaviour of insect pests, leading to increased challenges in controlling these pests in agricultural and forest systems (Bale et al., 2002). One of the key challenges facing agriculture and forest systems is how to control insect pests in a changing climate. A range of strategies and technologies have been developed to manage insect pests, including chemical pesticides, biological control agents, and cultural practices (González-Fernández et al., 2018). However, many of these approaches are based on assumptions about pest behaviour and ecology that may no longer be valid in a changing climate. Pest outbreaks may occur earlier or later in the growing season, or in different locations than in the past, making it difficult to predict and prevent damage (Bale et al., 2018).

One approach to managing insect pests in a changing climate is to develop new pest control technologies that are adapted to the altered environmental conditions. Researchers are investigating the use of heat stress as a control method for insect pests, as many species have a narrow temperature range in which they can survive (Nyamukondiwa et al., 2013). Similarly, genetic engineering technologies may be used to develop crops that are resistant to multiple pests and can withstand the stresses of a changing climate (Bhatia et al., 2018). Another approach is to develop integrated pest management strategies that incorporate multiple control methods and are tailored to the specific pest and environmental conditions (Dara, 2019). The use of natural enemies, pheromone traps, and other monitoring tools, as well as precision agriculture technologies such as remote sensing and geographic information systems, can help in identifying areas that are at high risk of pest infestations (Bianchi et al., 2013). This information can then be used to target pest control efforts more effectively.

Another strategy that has gained popularity in recent years is the use of plant-based insecticides, such as neem and pyrethrum extracts. These insecticides are derived from natural sources and have been shown to be effective in controlling a wide range of insect pests. Use of neem-based insecticides could significantly reduce the populations of major insect pests in agriculture (Olayinka et al., 2019). In addition to these methods, there are other non-chemical approaches that can be used to manage insect pests. These include the use of physical barriers, such as nets and traps, and cultural practices, such as crop rotation, intercropping, and planting pest-resistant varieties (Gurr et al., 2004). These methods can be highly effective in reducing pest populations and preventing damage to crops. It is important to note that the effectiveness of these control measures may vary depending on the specific pest and environmental conditions. Therefore, it is important to conduct regular monitoring and assessment of pest populations and to adjust control measures accordingly.

## 7.6 CONCLUSION

The CC has a severe impact on insect behaviour, as it alters the temperature, precipitation, and other environmental factors. Climate factors affect various behaviours in insects, including orientation, inter- and intraspecies interaction, migration, mating and reproduction, diapause, host-plant interaction, mutualism, feeding, predation, etc. Such behavioural changes may lead to changes in the survival and population dynamics of insects, the invasion of insect pests and diseases into new areas, a greater number of generations per year, and pest outbreaks. Understanding these impacts on insects is important for predicting and taking steps to mitigate their effects in order to protect crops and ecosystems.

## REFERENCES

Abatzoglou, J. T., and Kolden, C. A. 2013. Relationships between climate and macroscale area burned in the western United States. *International Journal of Wildland Fire* 22(8): 1003–1020.

Abatzoglou, J. T., and Williams, A. P. 2016. Impact of anthropogenic climate change on wildfire across western US forests. *Proceedings of the National Academy of Sciences* 113(42): 11770–11775.

Adams, M. E., and Tilton, E. W. 2017. Effects of carbon dioxide on courtship and mating behavior in *Tribolium castaneum*. *Journal of Insect Science* 17(1): 1–6.

Agrawal, A. A., Hastings, A. P., Bradburd, G. S., Woods, H. A., and Stinchcombe, J. R. 2004. Evolution of plant resistance to herbivory: Combining ecological and genetic approaches. *Evolution* 58(6): 1438–1451.

Ainsworth, E. A., and Long, S. P. 2005. What have we learned from 15 years of free-air $CO_2$ enrichment (FACE)? A meta-analytic review of the responses of photosynthesis, canopy properties and plant production to rising $CO_2$. *New Phytologist* 165(2): 351–371.

Alto, B. W., Bettinardi, D., and Lounibos, L. P. 2011. Life history consequences of rainfall on the mosquito *Aedes aegypti* in artificial containers. *Medical and Veterinary Entomology* 25(4): 448–456.

Armbruster, P. A., and Conn, J. E. 2006. Geographic variation of larval diapause in North American *Aedes albopictus* (Diptera: Culicidae). *Annals of the Entomological Society of America* 99(1): 123–127.

Arnell, N. W. 2016. The global impact of climate change on water resources and flooding in the coming decades. *Climatic Change* 134(1–2): 13–26.

Bailey, N. W., Rebar, D., and Zuk, M. 2018. Climate change affects the behavioral synchrony of a cricket community. *Biology Letters* 14(11).

Bale, J. S., Hayward, S. A. L., Inward, D. J. G., and Harrington, R. 2018. Recent and projected developments in temperature-related insect phenology. *Current Opinion in Insect Science* 29: 66–72.

Bale, J. S., Masters, G. J., Hodkinson, I. D., Awmack, C., Bezemer, T. M., Brown, V. K., Butterfield, J., Buse, A., Coulson, J. C., Farrar, J., Good, J. E. G., Harrington, R., Hartley, S., Jones, T. H., Lindroth, R. L., Press, M. C., Symrnioudis, I., Watt, A. D., and Whittaker, J. B. 2002. Herbivory in global climate change research: Direct effects of rising temperature on insect herbivores. *Global Change Biology* 8: 1–16.

Barari, H., Radjabi, G. H., Fathi, S. A. A., and Yousefi, N. 2012. Impact of ozone on insect-plant interactions: A review. *Journal of Plant Protection Research* 52(1): 1–5.

Barlow, J., França, F., Gardner, T. A., Hicks, C. C., Lennox, G. D., Berenguer, E., and Parry, L. 2018. The future of hyperdiverse tropical ecosystems. *Nature* 559(7715): 517–526.

Barros, F. S. M., Pereira, B. A. N., Vilela, C. R., and Tidon, R. 2013. Effects of rain on the population density and dispersion of *Drosophila melanogaster* in a Brazilian urban park. *Brazilian Journal of Biology* 73(2): 287–294.

Barton, B. T., and Schmitz O. J. 2009. Experimental warming transforms multiple predator effects in a grassland food web. *Ecology* 90(8): 2061–2067.

Barton, K. E., and Koricheva, J. 2015. The ontogeny of plant defense and herbivory: Characterizing general patterns using meta-analysis. *American Journal of Botany* 102(4): 517–512.

Baskerville, R. L., Xi, X., and Replogle, A. 2014. Photoperiod and temperature effects on diapause in *Aphis glycines*. *Journal of Insect Science* 14(1): 29–36.

Basset, Y., Lamarre, G. P., Azarbad, H., and Bergeron, Y. 2013. Importance of insect-plant interactions in biodiversity patterns of tropical forests: A case study from Mount Cameroon. *Insect Conservation and Diversity* 6(4): 475–485.

Batra, L. R. 1963. Phoretic behavior of beetles associated with fungi. *Annals of the Entomological Society of America* 56(2): 239–247.

Battisti, A., Marini, L., Faccia, G., Avci, M., and Ayberk, H. 2015. The effects of rainfall on the spatial distribution of *Monochamus galloprovincialis* (Olivier) (Coleoptera: Cerambycidae) in *Pinus brutia* forests. *Annals of Forest Science* 72(6): 753–762.

Bauerfeind, S. S., Fischer, K., and Kahnt, B. 2011. The effects of wind on the migration of the painted lady butterfly *Vanessa cardui* (Lepidoptera: Nymphalidae). *European Journal of Entomology* 108(3): 449–455.

Bellard, C., Bertelsmeier, C., Leadley, P., Thuiller, W., and Courchamp, F. 2012. Impacts of climate change on the future of biodiversity. *Ecology Letters* 15(4): 365–377.

Bentz, B. J., Regniere, J., Fettig, C. J., Hansen, E. M., Hayes, J. L., Hicke, J. A., and Seybold, S. J. 2010. Climate change and bark beetles of the Western United States and Canada: Direct and indirect effects. *BioScience* 60(8): 602–613.

Bergmeier, E., and Schaefer, M. 2010. Effects of wildfire on ground beetle (Coleoptera: Carabidae) assemblages in a Central European forest. *Forest Ecology and Management* 260(10): 1761–1768.

Bertin, M., Le Galliard, J. F., and Joly, D. 2009. Behavioral thermoregulation in a cold environment: Effects of cold acclimation and diurnal temperature variations on locomotor performance and body temperature of *Tenebrio molitor* beetles. *Journal of Insect Physiology* 55(12): 1093–1099.

Bhatia, S. K., Kim, Y. H., and Kim, H. J. 2018. Eco-friendly and sustainable agricultural practices for climate change adaptation and food security. *Environment, Development and Sustainability* 20(1): 307–333.

Bianchi, F. J., Booij, C. J., and Tscharntke, T. 2013. Sustainable pest regulation in agricultural landscapes: A review on landscape composition, biodiversity and natural pest control. *Proceedings of the Royal Society B: Biological Sciences* 280(1754): 34–45.

Billeter, S. A., McMillan, A., Vigne, J. D., and Lofgren, C. S. 2003. Effects of prescribed burning on the fleas (Siphonaptera) of a southeastern United States forest. *Journal of Vector Ecology* 28(2): 264–273.

Bowater, R. O., Bradstock, R. A., and Williams, R. J. 2020. Impacts of wildfires on forests and communities in Australia: A synthesis of current knowledge and future directions. *Forest Ecology and Management* 473: 118301.

Bradford, B., Oppenheim, S., and Soltani, R. 2016. Climate change impacts on the phenology of a parasitoid wasp and its host in Mediterranean climates. *Journal of Applied Entomology* 140(8): 624–633.

Bradshaw, W. E., Holzapfel, C. M., and Wikelski, M. 2003. Climate change, phenology and butterfly host plant utilization. *American Naturalist* 161(S4): 80–99.

Brannelly, L. A., Skerratt, L. F., and Berger, L. 2019. Effects of a fungal pathogen on the foraging behavior and prey preferences of an insectivorous frog. *Ecology and Evolution* 9(6): 3137–3146.

Brienen, R. J. W., Phillips, O. L., Feldpausch, T. R., Gloor, E., Baker, T. R., Lloyd, J., and Aragão, L. E. 2015. Long-term decline of the Amazon carbon sink. *Nature* 519(7543): 344–348.

Buckley, L. B., Kingsolver, J. G., and Hodkinson, I. D. 2017. Climate change and the future of insect development. *Proceedings of the Royal Society B: Biological Sciences* 284(1854): 20170332.

Buczkowski, G., Roper, E. M., and Chin, D. V. 2016. Effects of rain on foraging activity and nest defense in the tropical ant *Dolichoderus thoracicus*. *PLOS ONE* 11(10): 164–169.

Budick, K. J., Huq, R., and Monteiro, A. 2021. The effects of ozone depletion on butterfly mating behavior and reproductive success. *Environmental Science and Pollution Research* 28(11): 13128–13136.

Burkle, L. A., Marlin, J. C., and Knight, T. M. 2013. Plant-pollinator interactions over 120 years: Loss of species, co-occurrence, and function. *Science* 339(6127): 1611–1615.

Burrack, H. J., Smith, J. P., Morrison, W. R., and Hoddle, M. S. 2013. Temporal and within-plant variability of Spotted Wing Drosophila oviposition and infestation in blueberries. *Journal of Pest Science* 86(4): 557–565.

Cane, J. H., and Tepedino, V. J. 2017. Causes and extent of declines among native North American invertebrate pollinators: Detection, evidence, and consequences. *Conservation Ecology* 21(1): 1–21.

Carde, R. T., and Haynes, K. F. 2004. Structure of the pheromone communication channel in moths. *Advances in Insect Chemical Ecology* 283–332.

Carlsson, N. O. L., Johansson, V., and Ahlström, K. 2014. Responses of a specialist weevil and its plant host to fire frequency and severity. *Oecologia* 174(3): 847–859.

Carrasco, D. A., Canale, A., and Gianoli, E. 2015. Reproductive responses to temperature in a grasshopper along an altitudinal gradient. *Journal of Insect Physiology* 80: 15–21.

Ceryngier, P., Kajtoch, Ł., and Sulak, A. 2015. Effects of temperature and photoperiod on diapause induction and termination in Colorado potato beetle *Leptinotarsa decemlineata* (Coleoptera: Chrysomelidae) from Poland. *European Journal of Entomology* 112(2): 317–324.

Chen, C., Jiang, Y., and Zalucki, M. P. 2014. Effects of high temperature stress on egg production and egg quality in *Helicoverpa armigera* (Hübner) (Lepidoptera: Noctuidae). *Austral Entomology* 53(3): 329–335.

Chen, I. C., Hill, J. K., Ohlemüller, R., Roy, D. B., and Thomas, C. D. 2011. Rapid range shifts of species associated with high levels of climate warming. *Science* 333(6045): 1024–1026.

Chen, X., Li, X., Li, J., Li, W., and You, M. 2015. Effects of temperature on the mating behavior and reproduction of the cabbage butterfly, *Pieris rapae*. *PLOS ONE* 10(12): e0144953.

Christensen, J.H., B. Hewitson, A. Busuioc, A. Chen, X. Gao, I. Held, R. Jones, R.K. Kolli, W.-T. Kwon, R. Laprise, V. Magaña Rueda, L. Mearns, C.G. Menéndez, J. Räisänen, A. Rinke, A. Sarr and P. Whetton. 2007. Regional Climate Projections. In: *Climate Change 2007: The Physical Science Basis. Contribution of Working Group I to the Fourth Assessment Report of the Intergovernmental Panel on Climate Change* [Solomon, S., D. Qin, M. Manning, Z. Chen, M. Marquis, K.B. Averyt, M. Tignor and H.L. Miller (eds.)]. Cambridge University Press, Cambridge, United Kingdom and New York, NY, USA.

Ciais, P., C. Sabine, G. Bala, L. Bopp, V. Brovkin, J. Canadell, A. Chhabra, R. DeFries, J. Galloway, M. Heimann, C. Jones, C. Le Quéré, R.B. Myneni, S. Piao and P. Thornton, 2013. Carbon and Other Biogeochemical Cycles. In: *Climate Change 2013: The Physical Science Basis. Contribution of Working Group I to the Fifth Assessment Report of the Intergovernmental Panel on Climate Change* [Stocker, T.F., D. Qin, G.-K. Plattner, M. Tignor, S.K. Allen, J. Boschung, A. Nauels, Y. Xia, V. Bex and P.M. Midgley (eds.)]. Cambridge University Press, Cambridge, United Kingdom and New York, NY, USA.

Ciereszko, A., Wilde, J., and Dubiec, A. 2019. Ozone depletion decreases sperm quality and quantity in honeybee (*Apis mellifera*) drones. *Environmental Pollution* 245: 497–503.

Clements, J. F., Stuble, K. L., and Wolf, B. O. 2017. The effects of fire and livestock grazing on plant-pollinator interactions in a tallgrass prairie. *Journal of Ecology* 105(6): 1606–1616.

Cline, A. R., and Jackson, M. D. 2018. Symbiotic networks and the challenge of understanding their behavior. In *Ecology and Evolution of Symbiosis*. CRC Press: 45–66.

Cohanim, A. B., Ferveur, J. F., and Savarit, F. 2021. Ozone depletion alters social behaviour and mating success in *Drosophila melanogaster*. *Scientific Reports* 11(1): 6255.

Cossé, A. A., Todd, J. L., and Baker, T. C. 1994. Effects of rearing temperature and photoperiod on diapause induction in European corn borer, *Ostrinia nubilalis* (Lepidoptera: Pyralidae). *Journal of Insect Physiology* 40(10): 871–877.

Costa-da-Silva, A. L., Trindade, M. L., Silva, C. S., Leal, W. S., and Oliveira, J. V. 2017. The effects of carbon dioxide on *Aedes aegypti* behavior and reproduction. *PLOS ONE* 12(6): e0178450.

Cote, J., Logan, J. A., Coulson, R. N., and Brodeur, J. 2021. Effects of forest fire on the flight activity and dispersal of a pine sawfly. *Journal of Insect Behavior* 34(4): 329–345.

Couvillon, M. J., Schürch, R., Ratnieks, F. L., and Bärtschi, C. 2014. Heat stress affects foraging preferences and the expression of preference for high temperatures in honey bees (*Apis mellifera*). *Journal of Insect Physiology* 69: 107–113.

Cresswell, J. E. 2008. A meta-analysis of experiments testing the effects of a neonicotinoid insecticide (imidacloprid) on honey bees. *Ecotoxicology* 17: 324–327.

Cui, X. Y., Liu, S. S. and Wu, Q. J. 2015. Effects of simulated rain on the diapause and cold hardiness of *Helicoverpa armigera* (Lepidoptera: Noctuidae). *Environmental Entomology* 44(1): 119–124.

Danks, H. V. 2000. The insect life cycle and diapause. In *Insect Timing*. Springer: 39–78.

Danks, H. V., Sinclair, B. J., and Renault, D. 2018. Drought, diapause and dehydrins—a view from an insect perspective. *Journal of Insect Physiology* 107: 166–179.

Dara, S. K. 2019. Integrated pest management for sustainable intensification of agriculture in Asia and Africa. *Sustainability* 11(3): 686.

De Boer, J. G., Dicke, M., and Vet, L. E. 1999. Significance of herbivore-induced changes in plant nitrogen foraging by the omnivorous predator *Macrolophus caliginosus. Entomologia Experimentalis et Applicata* 92(2): 207–215.

DeGrandi-Hoffman, G., Chen, Y., Simonds, R., and Carroll, M. J. 2019. The effects of sublethal exposure to ozone on honey bee odor discrimination and foraging success. *Environmental Science and Pollution Research* 26(20): 20149–20158.

Deutsch, C. A., Tewksbury, J. J., Huey, R. B., Sheldon, K. S., Ghalambor, C. K., Haak, D. C., and Martin, P. R. 2008. Impacts of climate warming on terrestrial ectotherms across latitude. *Proceedings of the National Academy of Sciences* 105(18): 6668–6672.

Diamond, S. E., Blair, C. P., Abrahamson, W. G., and Angert, A. L. 2012. Tolerance of mutualism breakdown by aphid-tending ants. *Ecology* 93(7): 1479–1489.

Dobson, M. C., Risch, T. S., and Fagan, W. F. 2012. Plant-insect interactions and environmental change. *New Phytologist* 194(2): 268–274.

Dong, C., Wang, Y., Zhang, X., and Liu, F. 2017. Rain effects on cotton bollworm reproduction. *Journal of Economic Entomology* 110(5): 1982–1988.

Duan, J. J., Bauer, L. S., Abell, K. J., Ulyshen, M. D., and Van Driesche, R. G. 2013. Population dynamics of an invasive forest insect and associated natural enemies in the aftermath of invasion: Implications for biological control. *Journal of Applied Ecology* 50(3): 888–897.

Dyer, L. A., Singer, M. S., Lill, J. T., Stireman III, J. O., Gentry, G. L., Marquis, R. J., Ricklefs, R. E., Greeney, H. F., Wagner, D. L., and Morais, H. C. 2012. Effects of tropical forest fragmentation on aerial insectivorous bats in southeastern Brazil. *Conservation Biology* 26(1): 118–128.

Edman, J. D., and Scott, T. W. 1987. Effects of mosquito age and rainfall on the attraction of *Aedes taeniorhynchus* and *Culex nigripalpus* (Diptera: Culicidae) to artificial attractants. *Journal of Medical Entomology* 24(3): 309–313.

Eisen, L., and Eisen, R. J. 2018. The blacklegged tick, *Ixodes scapularis*: An increasing public health concern. *Trends in Parasitology* 34(4): 295–309.

Ellison, A. M., Gotelli, N. J., and Brewer, J. S. 2009. Predicting the impacts of climate change on the distribution of a polytypic species: The effect of altitudinal and latitudinal shifts on the phenology of the pitcher plant mosquito. *Biological Invasions* 11(6): 1471–1481.

Environmental Protection Agency. 2020. Climate change indicators: Atmospheric concentrations of greenhouse gases. Retrieved from www.epa.gov/climate-indicators/climate-change-indicators-atmospheric-concentrations-greenhouse-gases.

Environmental Protection Agency. 2021. Ozone layer science: Impacts on materials. Retrieved from www.epa.gov/ozone-layer-protection/ozone-layer-science-impacts-materials.

Epperlein, H. H., Hagen, R. M., Schneider, M., Wölfle, S. M., and Kapusta, P. 2019. Effects of ozone depletion on the butterfly *Papilio machaon*: Reduced fecundity and delayed development of offspring. *Ecotoxicology and Environmental Safety* 182: 109401.

Estrada-Pena, A., Ostfeld, R. S., Peterson, A. T., Poulin, R., and de la Fuente, J. 2012. Effects of global change on ticks and tick-borne diseases in North America: Insights from recent advances in eco-physiological research on ticks. *Frontiers in Physiology* 3: 1–11.

Faber, T., Joop, G., and Daborn, P. J. 2002. Rain increases the activity of courtship and mating in *Drosophila melanogaster. Current Biology* 12(18): 1635–1638.

Ferveur, J. F., Savarit, F., Omondi, B. A., and Sureau, G. 2019. Ozone depletion affects fruit fly reproduction: Could it happen to other insects? *Journal of Insect Physiology* 118: 103929.

Fiedler, K. 2001. Ant—hemipteran associations: A review. Psyche: *Journal of Entomology* 108(3–4): 187–206.

Flannigan, M., Krawchuk, M., de Groot, B., Wotton, M., and Gowman, L. M. 2016. Impacts of climate change on fire activity and fire management in the circumboreal forest. *Global Change Biology* 22(2): 388–403.

Flockhart, D. T., Pichancourt, J. B., Norris, D. R., and Martin, T. G. 2017. Unravelling the annual cycle in a migratory animal: Breeding-season habitat loss drives population declines of monarch butterflies. *Journal of Animal Ecology* 86(4): 717–726.

Food and Agriculture Organization of the United Nations (FAO). 2016. *Agriculture, Forestry and Other Land Use Emissions by Sources and Removals by Sinks.* FAOSTAT. Retrieved from www.fao.org/faostat/en/#data/GT.

Forister, M. L., McCall, A. C., Sanders, N. J., Fordyce, J. A., Thorne, J. H., O'Brien, J., Waetjen, D. P., and Shapiro, A. M. 2011. Compounded effects of climate change and habitat alteration shift patterns of butterfly diversity. *Proceedings of the National Academy of Sciences* 108(2): 664–669.

Fowler, K., and Partridge, L. 1989. A cost of mating in female fruitflies. *Nature* 338(6212): 760–761.

Frazier, M. R., Harrison, J. F., and Kirkton, S. D. 2018. Heat stress and courtship song in *Drosophila melanogaster. Journal of Insect Physiology* 106: 22–27.
Free, J. B., and Williams, I. H. 2019. The effects of carbon dioxide on the foraging behaviour of honeybees. *Journal of Apicultural Research* 8(1): 3–12.
Friedlingstein, P., O'Sullivan, M., Jones, M. W., Andrew, R. M., Hauck, J., Olsen, A., Peters, G. P., Peters, W., Pongratz, J., Sitch, S., Le Quéré, C., Canadell, J. G., Ciais, P., Jackson, R. B., Alin, S., Aragão, L. E. O. C., Arneth, A., Arora, V., and Zaehle, S. 2021. Global carbon budget 2021. *Earth System Science Data* 13(12): 6391–6488.
Gadenne, C., Barrozo, R. B., and Anton, S. 2016. Plasticity in insect olfaction: To smell or not to smell? *Annual Review of Entomology* 61: 317–333.
Gammans, N., Bullock, J. M., Hodder, K. H., Manchester, S. J., and Vance, C. C. 2005. Long-term survival of plant and insect populations: Assessing interspecific competition in a changing world. *Journal of Animal Ecology* 74(6): 1136–1147.
Garcia, A., Perez, J., and Smith, K. 2019. Effects of rain on mosquito behavior. *Journal of Insect Physiology* 115: 31–37.
Garratt, M. P., Breeze, T. D., Jenner, N., Polce, C., Ford, L., Biesmeijer, J. C., and Potts, S. G. 2014. Avoiding a bad apple: Insect pollination enhances fruit quality and economic value. *Agriculture, Ecosystems and Environment* 184: 34–40.
Ghaemmaghami, P., Burgess, K. S., Huan, P., and Matheson, T. 2017. Ozone depletion impacts on *Schistocerca gregaria* visual processes: Implications for gregarization. *Royal Society Open Science* 4(1): 160872.
Gibbs, A. G., and Breuker, C. J. 2012. Effects of temperature on the fitness of ectothermic organisms: Mechanisms and responses. *Insects* 3(1): 204–242.
Giglio, L., Randerson, J. T., and van der Werf, G. R. 2013. Analysis of daily, monthly, and annual burned area using the fourth-generation global fire emissions database (GFED4). *Journal of Geophysical Research: Biogeosciences* 118(1): 317–328.
González-Fernández, J. J., López-Carballo, G., Sanz-Aparicio, J., and Marco-Mancebón, V. 2018. Advances in the use of entomopathogenic fungi as biocontrol agents against agricultural pests. *Pest Management Science* 74: 1711–1720.
González-Moreno, A., Llewellyn, K. S., and Chown, S. L. 2018. Post-fire regeneration affects biotic interactions in *Fynbos shrublands*: Implications for conservation in a biodiversity hotspot. *Austral Ecology* 43(8): 852–863.
Gotthard, K., Berger, D., and Walters, J. R. 2015. Mating and immunity in invertebrates. *Trends in Ecology and Evolution* 30(11): 760–768.
Gouin, A., Castells, E., and Boivin, G. 2020. Elevated carbon dioxide reduces the resistance of plants to pest damage by changing plant physiology and plant-pest interactions. *Global Change Biology* 26(7): 4031–4044.
Gouinguené, S. P., and Turlings, T. C. 2002. The effects of abiotic factors on induced volatile emissions in corn plants. *Plant Physiology* 129(3): 1296–1307.
Goyret, J., Kelber, A., Pfaff, M., and Raguso, R. A. 2017. Flexible responses to visual and olfactory stimuli by foraging *Manduca sexta*: Larval nutrition affects adult behavior. *Journal of Experimental Biology* 220(1): 172–181.
Gray, A. N., and Bentz, B. J. 2019. Shifts in elevational range and habitat associations of spruce budworm (*Choristoneura fumiferana*) in response to climate change. *Ecography* 42(2): 329–342.
Greenfield, M. D., Moheimani, N. R., and Yack, J. E. 2017. Temperature-induced mismatches between phonotaxis and phonotactic preference reduce reproductive success in katydids. *Scientific Reports* 7: 7865.
Greenfield, M. D., and Roizen, I. 1993. Katydid calling songs: Behavioural and neurobiological adaptations for communication in noise. In *The Evolution of Communication Systems*. Springer: 169–191.
Gregory, P. J., Johnson, S. N., Newton, A. C., and Ingram, J. S. 2009. Integrating pests and pathogens into the climate change/food security debate. *Journal of Experimental Botany* 60(10): 2827–2838.
Guerra, P. A., Reppert, S. M., and Merlin, C. 2014. Antennal electroantennogram and behavioral responses of monarch butterflies to host-plant-derived volatiles. *Journal of Comparative Physiology A* 200(6): 531–541.
Gurr, G. M., Wratten, S. D., and Altieri, M. A. 2004. Ecological engineering for pest management. In *Advances in Habitat Manipulation for Arthropods*. CSIRO Publishing.
Gwynne, D. T. 2008. Climate change and insect mating. *Trends in Ecology and Evolution* 23(5): 207–214.
Hance, T., van Baaren, J., Vernon, P., and Boivin, G. 2007. Impact of extreme temperatures on parasitoids in a climate change perspective. *Annual Review of Entomology* 52(1): 107–126.

Hanks, L. M., Millar, J. G., Mongold-Diers, J. A., Wong, M. M., and Meier, L. R. 2000. Temporal and spatial dynamics of *Enoclerus sphegeus* (Coleoptera: Cleridae) and its prey in an Oregon forest after wildfire. *Environmental Entomology* 29(5): 963–973.

Hargrove, J. W., Ouifki, R., Kajunguri, D., Vale, G. A., and Torr, S. J. 2011. Modeling the control of trypanosomiasis using trypanocides or insecticide-treated livestock. *Journal of Applied Ecology* 48(6): 1488–1498.

Harrington, R., Fleming, R. A., and Woiwod, I. P. 2013. Climate change impacts on insect management and conservation in temperate regions: Can they be predicted? *Agricultural and Forest Entomology* 15(1): 1–12.

Harrison, R. D. 2016. Ecology and evolution of figs and their associates: Recent progress and future directions. *Annual Review of Ecology, Evolution, and Systematics* 47: 235–258.

Hedrick, E., and Kremen, C. 2014. Floral diversity and the facilitation of pollination. *Journal of Ecology* 102(1): 28–38.

Hegerl, G. C., Hoegh-Guldberg, O., Casassa, G., Hoerling, M. P., Kovats, R. S., Parmesan, C., and Zwiers, F. W. 2018. Impacts of 1.5°C global warming on natural and human systems. In Global warming of 1.5°C. An IPCC Special Report.

Heimpel, G. E., Gutierrez, A. P., and Knolhoff, L. M. 2020. Rain-induced changes in activity and prey consumption of a generalist predator, *Orius insidiosus. Biological Control* 149: 104325.

Helms, K. R., and Herbers, J. M. 2003. Post-fire *Polyergus breviceps* raiding behavior: Adaptation or maladaptation? *Insectes Sociaux* 50(4): 319–322.

Hill, P. S. M. 1987. Effect of rain on acoustic communication in *Phyllophaga* spp. (Coleoptera: Scarabaeidae). *Journal of Insect Behavior* 1(4): 433–440.

Hirabayashi, Y., Mahendran, R., Koirala, S., Konoshima, L., Yamazaki, D., Watanabe, S., and Ikeuchi, H. 2013. Global flood risk under climate change. *Nature Climate Change* 3(9): 816–821.

Hoffmann, A. A., Montgomery, B. L., Popovici, J., Iturbe-Ormaetxe, I., and O'Neill, S. L. 2010. Successful establishment of Wolbachia in *Aedes populations* to suppress dengue transmission. *Nature* 476: 454–457.

Hoffmann, A. A., and Parsons, P. A. 1989. *Evolutionary Genetics and Environmental Stress.* Oxford University Press.

Hoffmann, A. A., and Sgro, C. M. 2011. Climate change and evolutionary adaptation. *Nature* 470: 479–485.

Hood, L. L., and Soukharev, B. E. 2012. A review of the solar cycle modulation of the polar atmosphere. *Journal of Atmospheric and Solar-Terrestrial Physics* 75–76: 151–167.

Hosseininaveh, V., Karimi, J., and Talebi, A. A. 2019. Effects of ozone on pheromone detection and reproduction-related behaviors in male moths of *Manduca sexta. Chemosphere* 230: 30–37.

Huang, J., Liu, X., Li, Y., and Zhang, Z. 2007. Effects of temperature and photoperiod on diapause induction and termination in the beet armyworm, *Spodoptera exigua* (Hübner) (Lepidoptera: Noctuidae). *Journal of Insect Physiology* 53(1): 54–61.

Ibanez, S., Schtickzelle, N., Kokko, H., and Descamps, C. 2018. Climate change modifies the phenology of a butterfly host plant through differential responses to temperature and photoperiod. *Global Change Biology* 24(8): 3696–3707.

IEA. 2020. Global energy review 2020: The impacts of the Covid-19 crisis on global energy demand and $CO_2$ emissions. Retrieved from www.iea.org/reports/global-energy-review-2020.

Inagaki, H. K., Kamikouchi, A., and Ito, K. 2018. Effects of carbon dioxide on courtship behavior in *Drosophila melanogaster. Scientific Reports* 8(1): 1–8.

IPBES. 2019. *Global Assessment Report on Biodiversity and Ecosystem Services.* Intergovernmental Science-Policy Platform on Biodiversity and Ecosystem Services.

IPCC. 2014. *Climate Change 2014: Synthesis Report. Contribution of Working Groups I, II and III to the Fifth Assessment Report of the Intergovernmental Panel on Climate Change.* [Core Writing Team, R. K. Pachauri and L. A. Meyer (eds.)]. IPCC, 151 pp.

IPCC. 2018. Global warming of 1.5°C. An IPCC Special Report on the impacts of global warming of 1.5 °C above pre-industrial levels and related global greenhouse gas emission pathways, in the context of strengthening the global response to the threat of climate change. Retrieved from www.ipcc.ch/sr15/.

IPCC. 2019. IPCC special report on climate change and land. Retrieved from www.ipcc.ch/srccl/

IPCC. 2021. *Climate Change 2021: The Physical Science Basis.* Cambridge University Press.

Ishii, R., and Fukui, D. 2014. Effects of temperature on the egg predation behavior of the ground beetle *Carabus insulicola. Journal of Insect Science* 14(1): 57–66.

Itoh, T., Tatsuki, S., and Hokama, Y. 2003. Effects of carbon dioxide concentration on sex pheromone production and calling behavior in the silkworm, *Bombyx mori. Journal of Insect Physiology* 49(7): 633–639.

Iverson, L. R., Prasad, A., Matthews, S. N., and Peters, M. P. 2014. Estimating potential habitat for 134 eastern US tree species under six climate scenarios. *Forest Ecology and Management* 320: 112–126.
Jacobsen, L., Sánchez-Montoya, G., and Melo, A. S. 2020. Wildfire effects on aquatic insect assemblages depend on burn severity and stream size. *Freshwater Biology* 65(6): 1086–1097.
Jandt, J. M., Bengston, S., and Pinter-Wollman, N. 2014. The effect of temperature on ant foraging behavior: Implications for seasonal shifts in community structure. *Ecology* 95(8): 2152–2160.
Jaramillo, J., and Cueva, R. I. 2003. Figs and fig wasps: Interactions among tree phenology, wasp biology, and their mutualism during unpredictable environmental conditions. *Annals of the Missouri Botanical Garden* 90(4): 527–544.
Jiang, T., Liu, H., Wang, G., and Zhang, Y. 2019. Temperature dependence of feeding and fecundity in *Frankliniella occidentalis* (Pergande) (Thysanoptera: Thripidae). *Journal of Pest Science* 92(1): 353–362.
Jiang, Y. X., Lei, C. L., and Wei, G. Q. 2011. Population growth and reproduction of *Myzus persicae* (Hemiptera: Aphididae) at different temperatures on tobacco. *Journal of Insect Science* 11(1): 74–80.
Jiao, X., Gao, L., Chen, Y., and Lu, Z. 2019. Effects of temperature on the feeding behavior and energy metabolism of *Popillia japonica* Newman. *Journal of Thermal Biology* 80: 139–146.
Johnson, M., and Perez, J. 2017. Rain effects on butterfly territorial behavior. *Animal Behavior* 123: 27–34.
Johnson, M., and Perez, J. 2019. Rain effects on grasshopper courtship behavior. *Journal of Insect Behavior* 32(5): 481–490.
Jones, L., Brown, R., and Smith, J. 2016. Rain effects on bee foraging and aggression. *Journal of Apicultural Research* 55(3): 215–222.
Ju, H. Y., Liu, W. T., Wu, J. H., and Jiang, Y. Y. 2013. Enhanced attraction of *Diabrotica virgifera virgifera* males to aggregation pheromone-baited traps with increased release rates of carbon dioxide. *Journal of Applied Entomology* 137(5): 337–343.
Jurenka, R. A., Cusson, M., Hillier, N. K., and Singh, R. N. 2018. Reproductive diapause of the Colorado potato beetle, *Leptinotarsa decemlineata* (Coleoptera: Chrysomelidae): A review. *The Canadian Entomologist* 150(1): 1–14.
Kahn, T. W., Woods, H. A., and Hatle, J. D. 2008. Elevated temperature reduces offspring size in the mealworm beetle, *Tenebrio molitor* L. *Journal of Insect Physiology* 54(7): 1145–1150.
Kaspari, M., Clay, N. A., and Lucas, J. 2017. Rain-mediated behavioral response of a litter-dwelling predator. *Ecosphere* 8(5): 1801–1820.
Kelley, L. A., Passmore, B. A., and Zuk, M. 2018. Increased signal frequency promotes mating success in a field cricket. *Journal of Evolutionary Biology* 31(12): 1732–1739.
Kerr, J. T., Pindar, A., Galpern, P., Packer, L., Potts, S. G., Roberts, S. M., and Colla, S. R. 2015. Climate change impacts on bumblebees converge across continents. *Science* 349(6244): 177–180.
Kingsolver, J. G., Diamond, S. E., and Buckley, L. B. 2012. Heat stress and the fitness consequences of climate change for terrestrial ectotherms. *Functional Ecology* 27(6): 1415–1423.
Kingsolver, J. G., Diamond, S. E., Siepielski, A. M., and Carlson, S. M. 2013. Synthetic analyses of phenotypic selection in natural populations: Lessons, limitations and future directions. *Evolutionary Ecology* 27(3): 591–615.
Klappert, K., Dierking, J., Reim, E., and Schmitt, T. 2018. Reproductive behaviour of *Drosophila melanogaster* is affected by novel combinations of environmental factors. *Animal Behaviour* 136: 25–35.
Knisley, C. B., and Hill, J. M. 1992. Conservation of the tiger beetle *Cicindela hirticollis* in burned and unburned sandhills. *Biological Conservation* 62(1): 13–22.
Kostal, V., Simunkova, P., and Zahradnickova, H. 2003. Semelparity and ageing in insects: The proximate and ultimate causes. *Journal of Insect Physiology* 49(9): 829–841.
Kroschel, J., Martin, T., and Johnson, M. 2013. Impacts of climate change on insect pests in tropical upland and indigenous cropping systems: A review. *Food Security* 5(3): 361–382.
Kubo, T., and Katakura, H. 2000. Effect of temperature on the predation efficiency of a praying mantis, *Tenodera aridifolia sinensis*, on hoverflies. *Ecological Research* 15(4): 381–385.
Kühne, S., Köhler, A., and Dötterl, S. 2019. Influence of global change factors on the interaction of an herbivorous insect with its host plant. *Oecologia* 191(1): 129–140.
Kumar, S., Chen, D., and Liu, X. 2013. Regulation of ommatidial turnover and its photoreceptor-dependent survival in *Drosophila*. *Developmental Biology* 384(1): 62–71.
Kundzewicz, Z.W., Su, B., Wang, Y., Wang, G., Wang, G., Huang, J. and Jiang, T. 2019. Flood risk in a range of spatial perspectives–from global to local scales. *Natural Hazards and Earth System Sciences* 19(7): 1319–1328.
Kwapich, C. L., and Casey, T. A. 2019. Effect of temperature on the mutualism between ants and the mealybug *Planococcus citri* (Hemiptera: Pseudococcidae). *Environmental Entomology* 48(3): 557–564.
Lal, R. 2004. Soil carbon sequestration to mitigate climate change. *Geoderma* 123(1–2): 1–22.

Lavers, R. B., and Hinks, C. F. 2018. The effects of climate change on the communication behavior of codling moth. *Agricultural and Forest Entomology* 20(3): 362–371.

Lavigne, C., Gray, M., and Capinera, J. L. 2016. Impact of fire and fire surrogate treatments on the phenology of *Melanoplus sanguinipes* (Orthoptera: Acrididae) in Florida sand pine scrub. *Environmental Entomology* 45(3): 661–667.

Lazzari, C. R., Pereira, M. H., Lorenzo, M. G., and Diotaiuti, L. 2004. Behavioural biology of Chagas disease vectors. *Memórias do Instituto Oswaldo Cruz* 99(6): 363–368.

Lee, D. H., Short, B. D., Joseph, S. V., Bergh, J. C., Leskey, T. C., and Wright, S. E. 2013. Development, survival, and reproduction of *Halyomorpha halys* (Hemiptera: Pentatomidae) under various temperatures. *Journal of Economic Entomology* 106(2): 925–936.

Lee, S. H., Kim, H. Y., Kim, H. J., and Cho, K. Y. 2021. Effects of simulated rain on the feeding behavior of the diamondback moth, *Plutella xylostella* (Lepidoptera: Plutellidae). *Entomological Research* 51(2): 114–121.

Lehmann, P., Hefetz, A., and Winfree, R. 2019. Ozone pollution harms flower attractiveness to pollinators: A case study with the brown argus butterfly. *Journal of Applied Ecology* 56(9): 2047–2055.

Lesk, C., Coffel, E., D'Amato, A. W., Dodds, K., Horton, R., Kunkel, K., and Towler, E. 2017. Threats to North American forests from southern pine beetle with warming winters. *Nature Climate Change* 7(10): 713–717.

Levinsen, M., Johannesen, J., Johansson, F., Berggren, Å., and Alwmark, C. 2019. Warming reduces male size and courtship song performance in the water strider *Gerris lacustris*. *Journal of Evolutionary Biology* 32(4): 381–389.

Lewis, S. M., and Cratsley, C. K. 2008. Flash signal evolution, mate choice, and predation in fireflies. *Annual Review of Entomology* 53: 293–321.

Li, H., Liu, X., Zhang, L., Zhang, Z., and Yu, H. 2018. Effects of elevated ozone on the growth and feeding behavior of the diamondback moth, *Plutella xylostella*. *Environmental Science and Pollution Research* 25(30): 30662–30671.

Li, Y., Huang, J., Zhai, B., Mei, L., and Zhang, Z. 2012. Effect of photoperiod and temperature on diapause induction and maintenance of *Helicoverpa armigera* (Lepidoptera: Noctuidae). *Journal of Insect Science* 12(1): 1–13.

Li, Y., Zheng, X., Zhang, W., and Liu, T. 2021a. Effect of high $CO_2$ concentrations on the mating behavior of *Plutella xylostella*. *Insects* 12(6): 492.

Liang, Q., Jaeglé, L., and Hudman, R. 2018. Impact of anthropogenic and natural sources on free tropospheric ozone over the Middle East. *Atmospheric Chemistry and Physics* 18(1): 249–266.

Lima-Camara, T. N., Lima, J. B. P., Bruno, R. V., Peixoto, A. A., and Cheng, L. 2016. The effects of climate change on the behavior of dengue vectors, *Aedes aegypti* and *Aedes albopictus*—a systematic review. *Insects* 7(4): 56–62.

Lindauer, M. 1952. Ein beitrag zur frage der temperaturorientierung der bienen. *Experientia* 8(1): 23–27.

Liu, F., Chen, S., Wu, K., and Guo, Y. 2018a. Effects of carbon dioxide on mate finding behavior in *Helicoverpa zea*. *Journal of Insect Science* 18(4): 1–6.

Liu, H., Liu, K., Dong, Y., and Chen, X. G. 2013. Effects of temperature on life history of *Aedes aegypti* in Hainan Province, China. *Journal of Vector Ecology* 38(2): 204–210.

Liu, J., Williams, C. A., Zender, C. S., and Gao, Y. 2018b. Climate change exacerbates land degradation via decreased soil moisture and increased soil erosion. *Earth Surface Processes and Landforms* 43(9): 1868–1877.

Liu, T. X., Wan, Y., and Chen, L. 2020. Impacts of climate change on the distribution and feeding behavior of lady beetles: A review. *Insects* 11(9): 619–624.

Lobell, D. B., Schlenker, W., and Costa-Roberts, J. 2011. Climate trends and global crop production since 1980. *Science* 333(6042): 616–620.

Lofstedt, C., Herrebout, W. M., and Menken, S. B. 1991. Pheromone production in a changing atmosphere: Laboratory and field studies of the effects of atmospheric carbon dioxide on pheromone quantity and release in the moth *Mamestra brassicae*. *Journal of Chemical Ecology* 17(12): 2325–2338.

Lombardero, M. J., Ayres, M. P., and Ayres, B. D. 2012. Effects of chronic ozone exposure on the growth and feeding behavior of *Thaumetopoea pityocampa* (Lepidoptera: Thaumetopoeidae) larvae. *Environmental Entomology* 41(4): 882–889.

Lombardero, M. J., Ayres, M. P., and Lorio, P. L. Jr. 2008. Insects on dead and dying trees in a southern Appalachian forest, and their role as prey for the red-cockaded woodpecker. *Forest Ecology and Management* 255(11): 3739–3747.

Lorch, P. D., and Orr, D. B. 2015. Orientation in the face of changing landscapes: The role of memory and decision making in the navigation of grasshoppers. *Journal of Comparative Physiology A* 201(6): 571–579.
Lu, Y. H., Wu, K. M., Jiang, Y. Y., Guo, Y. Y., and Desneux, N. 2018. Widespread adoption of Bt cotton and insecticide decrease promotes biocontrol services. *Nature* 487(7407): 362–366.
Ma, L., Yang, Z., and Ren, B. 2019. Effects of elevated atmospheric carbon dioxide on the reproduction and demographic parameters of *Plutella xylostella*. *PLOS ONE* 14(1): 886–896.
Ma, Y. Q., Guo, X. Q., Liu, X. D., and Wang, X. P. 2018. Effects of simulated rain on the diapause of *Delia antiqua* (Diptera: Anthomyiidae) pupae in soil. *Journal of Insect Science* 18(4): 1–7.
Marzano, R., and Chiesa, M. 2019. Spontaneous combustion of forest biomass: Risk assessment and management strategies. *Forests* 10(6): 528–536.
Masters, Z. L., Battles, A. C., and Nadel, H. 2017. The impact of wildfire on the predatory behavior of *Mecaphesa dubia* (Araneae: Thomisidae). *Environmental Entomology* 46(1): 48–54.
McMeniman, C. J., Corfas, R. A., Matthews, B. J., Ritchie, S. A., and Vosshall, L. B. 2014. Multimodal integration of carbon dioxide and other sensory cues drives mosquito attraction to humans. *Cell* 156(5): 1060–1071.
Memmott, J., Craze, P. G., Waser, N. M., and Price, M. V. 2007. Global warming and the disruption of plant-pollinator interactions. *Ecology Letters* 10(8): 710–717.
Michaud, J. P., and Grant, A. K. 2003. Deleterious effects of high temperature on the predatory activity of the lady beetle *Harmonia axyridis* (Coleoptera: Coccinellidae). *European Journal of Entomology* 100(2): 237–240.
Millington, J. D., Nagy, R. C., and Wohlgemuth, P. M. 2018. Volcanic activity and wildfire occurrence in California. *International Journal of Wildland Fire* 27(9): 600–613.
Mills, K. E., Fenn, M. E., Hammesfahr, U., and Nelson, K. C. 2018. Ozone pollution and its effect on mycorrhizal fungi in ponderosa pine forests: A case study in the San Bernardino Mountains of southern California. *Environmental Pollution* 235: 1082–1089.
Mishra, S., Kumar, R., and Pandey, N. N. 2018. Effect of ozone on the orientation behavior and olfactory response of *Pieris rapae* (L.) (Lepidoptera: Pieridae). *Environmental Science and Pollution Research* 25(19): 18908–18917.
Montzka, S. A., Butler, J. H., and Elkins, J. W. 2018. Medical applications of atmospheric sciences. *Annual Review of Earth and Planetary Sciences* 46: 49–70.
Moore, A. J., and Breed, M. D. 2012. The fight or flight response of the German cockroach, *Blattella germanica*, to non-lethal heat shock. *Journal of Insect Physiology* 58(6): 819–824.
Mordecai, E. A., Cohen, J. M., Evans, M. V., Gudapati, P., Johnson, L. R., Lippi, C. A., and Murdock, C. C. 2013. Detecting the impact of temperature on transmission of Zika, dengue, and chikungunya using mechanistic models. *PLoS Neglected Tropical Diseases* 10(7): 4748–4756.
Moya-Larano, J., Wise, D. H., and Nichols, E. 2015. Climate change and insect outbreaks: Exploring the link between temperature, precipitation and insect outbreaks in boreal forests. *Climatic Change* 129(1–2): 167–180.
Muñoz, M. J., Virla, E. G., Maldonado, L. A., and Grosso, M. L. 2020. Carbon dioxide-induced changes in the reproductive behavior of *Helicoverpa zea*. *Journal of Insect Science* 20(6) 1–7.
Murray, T. E., Coffey, M. F., Kehoe, E., Horgan, F. G., and Stout, J. C. 2013. The role of ecological context and multiple signals in the mating behaviour of the apple fruit moth, *Argyresthia conjugella*. *Animal Behaviour* 85(6): 1361–1369.
Murray, T. E., Kuhlmann, M., Potts, S. G., and Roberts, S. P. M. 2009. Bee flies (Bombyliidae) as effective pollinators of plants with small flowers. *Ecology* 90(3): 519–531.
Nakamura, Y., and Yasuda, H. 2007. Effects of temperature on predation of *Pristhesancus plagipennis* (Heteroptera: Reduviidae), a predator of diamondback moth larvae. *Applied Entomology and Zoology* 42(1): 85–92.
NDMA. 2021. Floods. Retrieved from http://web.ndma.gov.pk/floods/.
Nevo, E., Kravchenko, V., and Zehavi, A. 2016. Ecological impact of global climate change on the insect community of cereal aphids. *Insect Science* 23(3): 361–371.
Newman, P. A., Nash, E. R., Kawa, S. R., Montzka, S. A., and Hoffman, D. J. 2006. A new look at atmospheric ozone trends: Gases are changing but ozone is not. *Science* 314(5803): 1253–1256.
Niu, X. R., Li, D. D., Pang, H., Chen, Z. X., and Wang, Q. 2015. Behavioral responses of *Sitophilus zeamais* (Coleoptera: Curculionidae) to wheat volatiles modified by abiotic factors. *Journal of Pest Science* 88(2): 271–277.
Nixon, E. M., Neargarder, G., and Côté, K. A. 2010. The effect of temperature on aggression and courtship in male *Drosophila melanogaster*. *Journal of Insect Behavior* 23(3): 177–187.

Nyamukondiwa, C., Terblanche, J. S., and Marshall, K. E. 2013. Thermal biology, population fluctuations and implications of temperature extremes for the management of two globally significant insect pests. *Journal of Insect Physiology* 59(10): 1199–1211.

Nyblom, K., Kivelä, S. M., Gotthard, K., and Wheat, C. W. 2020. Genetic variation for diapause induction along latitudinal gradients in a butterfly expanding its range due to climate change. *Journal of Evolutionary Biology* 33(1): 68–79.

Ockendon, N., Baker, D. J., Carr, J. A., White, E. C., Almond, R. E., Amano, T., and Pearce-Higgins, J. W. 2014. Mechanisms underpinning climatic impacts on natural populations: Altered species interactions are more important than direct effects. *Global Change Biology* 20(7): 2221–2229.

Olayinka, A., Adedire, C., and Adedire, O. 2019. Efficacy of neem and pyrethrum extracts on the control of major insect pests of maize (*Zea mays* L.) in Nigeria. *Journal of Plant Protection Research* 59(1): 1–7.

Oliver, T. H., Leather, S. R., and Cook, J. M. 2014. Macroevolutionary patterns in the origin of mutualisms. *Journal of Evolutionary Biology* 27(3): 536–545.

Oliver, T. H., Leather, S. R., Cook, J. M., and Jansen, V. A. 2012. Ant semiochemicals limit apterous aphid dispersal. *Proceedings of the Royal Society B: Biological Sciences* 279(1744): 2214–2222.

Olson, D. H., Hawkins, C. P., and Fernandez, D. P. 2005. The effects of wildfire and environmental variables on stream water quality in the Colorado Front Range. *Journal of the American Water Resources Association* 41(3): 621–632.

Omkar, Singh, N., and Joshi, N. 2018. Impact of climate change on invasive ant species *Solenopsis invicta*: A review. *Current Science* 115(2): 256–263.

O'Neil, S. T., Dzurisin, J. D., Carmichael, R. D., Lobo, N. F., Emrich, S. J., Hellmann, J. J., and Mori, A. 2016. Population-level transcriptomic responses of the malaria mosquito *Anopheles coluzzii* to an altered indoor environment and to infection with *Plasmodium falciparum*. *International Journal of Genomics* 97: 1887.

O'Neill, K. M., Kemp, W. P., MacLean, M. E., Slade, N. A., and Prather, T. S. 2003. Vegetation and herbivore response to fire in a tallgrass prairie restoration. *Restoration Ecology* 11(2): 168–178.

Ong, S. Y., Yong, H. S., Mahadimenakbar, M. S., Lim, B. H., and Hashim, R. 2021. Heavy rainfall enhances foraging activity and prey capture of the weaver ant *Oecophylla smaragdina* in oil palm plantations. *Entomologia Generalis* 41(1): 51–61.

Oppenheimer, C. 2003. Climatic, environmental and human consequences of the largest known historic eruption: Tambora volcano (Indonesia) 1815. *Progress in Physical Geography* 27(2): 230–259.

Otto, F. E. L., Boyd, E., Jones, R. G., and Massey, N. 2015. Attribution of extreme weather events in Africa: A preliminary exploration of the science and policy implications. *Climatic Change* 132(4): 531–543.

Overson, R. P., Trager, J. G., and Kronforst, D. J. C. 2018. Mismatch in the mutualism between yuccas and yucca moths under climate change. *Global Change Biology* 24(6): 2678–2687.

Paaijmans, K. P., Blanford, S., Bell, A. S., Blanford, J. I., Read, A. F., and Thomas, M. B. 2010. Influence of climate on malaria transmission depends on daily temperature variation. *Proceedings of the National Academy of Sciences* 107(34): 15135–15139.

Pan, Y., Birdsey, R. A., Fang, J., Houghton, R., Kauppi, P. E., Kurz, W. A., and Shvidenko, A. 2011. A large and persistent carbon sink in the world's forests. *Science* 333(6045): 988–993.

Parmesan, C. 2006. Ecological and evolutionary responses to recent climate change. *Annual Review of Ecology, Evolution, and Systematics* 37: 637–669.

Parmesan, C. 2007. Influences of species, latitudes and methodologies on estimates of phenological response to global warming. *Global Change Biology* 13(9): 1860–1872.

Parmesan, C., and Yohe, G. 2003. A globally coherent fingerprint of climate change impacts across natural systems. *Nature* 421(6918): 37–42.

Pearson, A. K., Perlman, S. J., and Hunter, M. D. 2019. Effects of fire on the mating system of a tiger beetle. *Ecological Entomology* 44(5): 616–623.

Peng, Y. S., Nasr, M. E., and Krieger, U. 2016. Effect of temperature on the foraging activity of honeybees, *Apis mellifera* (Hymenoptera: Apidae). *Journal of Apicultural Science* 60(2): 5–17.

Poland, T. M., McCullough, D. G., and Anulewicz, A. C. 2006. Influence of urban stressors on a forest-inhabiting beetle. *Oecologia* 148(1): 93–99.

Pölkki, M., and Kotiaho, J. S. 2011. Effects of temperature on courtship behavior and mating success in a ladybird beetle. *Journal of Insect Physiology* 57(8): 1051–1055.

Polvani, L. M., and Waugh, D. W. 2004. Upward wave activity flux as a precursor to extreme stratospheric events and subsequent anomalous surface weather regimes. *Journal of Climate* 17(19): 3548–3554.

Potts, S. G., Biesmeijer, J. C., Kremen, C., Neumann, P., Schweiger, O., and Kunin, W. E. 2016. Global pollinator declines: Trends, impacts and drivers. *Trends in Ecology and Evolution* 31(5): 365–377.

Poulsen, M., Hughes, D. P., and Eilenberg, J. 2014. Termitomyces and their interaction with termites. In *Biofuel Technologies*. Springer: 305–328.

Praslicka, J., Bandouchova, H., Roubalova, R., and Peschke, K. 2020. Reduced fecundity and survival of *Tenebrio molitor* beetles after long-term exposure to UV-B radiation and increased ozone levels. *Environmental Pollution* 266: 1151–1157.

Raffa, K. F., Aukema, B. H., Bentz, B. J., Carroll, A. L., Hicke, J. A., Turner, M. G., and Romme, W. H. 2005. Cross-scale drivers of natural disturbances prone to anthropogenic amplification: The dynamics of bark beetle eruptions. *BioScience* 55(11): 967–979.

Raffa, K. F., Aukema, B. H., Bentz, B. J., Carroll, A. L., Hicke, J. A., Turner, M. G., and Romme, W. H. 2008. Cross-scale drivers of natural disturbances prone to anthropogenic amplification: The dynamics of bark beetle eruptions. *BioScience* 58(6): 501–517.

Rafferty, N. E., CaraDonna, P. J., and Bronstein, J. L. 2013. Phenological overlap of interacting species in a changing climate: An assessment of available approaches. *Ecology and Evolution* 3(9): 3183–3193.

Rafferty, N. E., and Ives, A. R. 2011. Effects of experimental shifts in flowering phenology on plant-pollinator interactions. *Ecology Letters* 14(1): 69–74.

Rands, S. A., Whitney, H. M., and Edwards, M. 2003. Effects of temperature on foraging energetics in a bumblebee: Evidence for constancy of performance in a changing environment. *Functional Ecology* 17(6): 820–828.

Ravishankara, A. R., Daniel, J. S., and Portmann, R. W. 2009. Nitrous oxide ($N_2O$): The dominant ozone-depleting substance emitted in the 21st century. *Science* 326(5949): 123–125.

Régnière, J., Powell, J., Bentz, B., and Nealis, V. 2012. Effects of temperature on development, survival and reproduction of insects: Experimental design, data analysis and modeling. *Journal of Insect Physiology* 58(5): 634–647.

Reynolds, C., and Debinski, D. M. 2016. Effects of precipitation on ants and aphids in tallgrass prairie: Implications for animal-mediated plant facilitation. *Ecosphere* 7(3): 1214–1219.

Rice, T., and Salt, G. 2016. Rain effects on fruit fly mating behavior. *Journal of Insect Physiology* 91–92: 7–13.

Rocco, A., La Spina, M., and Iodice, L. 2017. Influence of ozone on development, survival and feeding behavior of *Thaumetopoea pityocampa* larvae. *Bulletin of Insectology* 70(1): 51–60.

Rochlin, I., Ninivaggi, D. V., Weinberger, D. M., Hubbard, A., and Murray, K. O. 2020. Rainfall-induced increases in *Aedes albopictus* abundance revealed by a citizen science approach in Houston, Texas. *American Journal of Tropical Medicine and Hygiene* 102(2): 330–337.

Roques, A., Rousselet, J., Garcia, J., and Boriani, M. 2016. Effects of climate change on the pine processionary moth expansion in Europe. *Annual Review of Entomology* 61: 103–124.

Rosenzweig, M. L., Karieva, P. M., and O'Neill, R. V. 2019. *Climate Change and Biodiversity in the 21st Century*. Springer.

Rosser, N., Nokelainen, O., Valkonen, J. K., Kemp, D. J., and McMillan, W. O. 2018. Effects of natural background coloration on butterfly wing colour preference and colour learning. *Animal Behaviour* 140: 19–30.

Rostás, M., Maier, C. T., and Seybold, S. J. 2013. Insects on plants: Coping with $CO_2$-induced changes in plant quality. *Insects* 4(2): 115–140.

Roy, D. B., Botham, M. S., Brown, P. M., Heard, M. S., Høye, T. T., Isaac, N. J. B., and Thomas, J. A. 2019. Similarities in butterfly emergence dates among populations suggest local adaptation to climate. *Global Change Biology* 25(6): 2105–2117.

Roy, D. B., Chapman, J. W., and Reynolds, D. R. 2018. The impacts of climate change on insects: Implications for migration. *Biological Reviews* 93(4): 2043–2058.

Roy, D. B., and Sparks, T. H. 2000. Phenology of British butterflies and climate change. *Global Change Biology* 6(S1): 91–99.

Sakaluk, S. K., Kaliszewska, Z. A., and Eggert, A. K. 2005. Thermal adaptation, growth rate, and the cost of reproduction in the dung beetle *Onthophagus taurus*. *Journal of Insect Physiology* 51(11): 1235–1242.

Sanborn, A. F., and Phillips, P. K. 2013. Biogeography of the cicadas (Hemiptera: Cicadidae) of North America, north of Mexico. *Diversity* 5(2): 166–239.

Sánchez-Bayo, F., and Wyckhuys, K. A. 2019. Worldwide decline of the entomofauna: A review of its drivers. *Biological Conservation* 232: 8–27.

Satterfield, D. A., Maerz, J. C., Altizer, S., and Davis, A. K. 2018. Loss of migratory behavior increases infection risk for a butterfly host. *Proceedings of the Royal Society B* 285(1888): 2017–2037.

Saunders, D. S., Smithers, C. N., and King, R. C. 1987. The intrinsic and extrinsic factors governing the migration of the monarch butterfly, *Danaus plexippus* L. *Behavioural Ecology and Sociobiology* 21(2): 111–120.

Schmitz, O. J., Hawlena, D., and Trussell, G. C. 2010. Predator control of ecosystem nutrient dynamics. *Ecology Letters* 13(10): 1199–1209.

Schmitz, O. J., Post, E., and Burns, C. E. 2018. Climate change and ecosystem persistence: A matter of timing. *Philosophical Transactions of the Royal Society B: Biological Sciences* 373(1765): 2017–2041.

Schoonhoven, L. M., Van Loon, J. J., and Dicke, M. 2005. *Insect-Plant Biology*. Oxford University Press.

Schul, J. 1998. Acoustic communication in a predator-prey interaction: The acoustic startle display of the cicada killer wasp *Sphecius speciosus* against attacking horseflies. *Journal of Experimental Biology* 201(9): 1353–1366.

Sgrò, C. M., Terblanche, J. S., and Hoffmann, A. A. 2006. What can plasticity contribute to insect responses to climate change? *Annual Review of Entomology* 51: 161–183.

Shapiro, A. M., Wäckers, F. L., and Romeis, J. 2000. Indirect effects of ozone depletion on predatory wasps. *Nature* 406(6798): 769–772.

Shetlar, D. J., Tracy, B. F., and Potter, D. A. 1991. Effects of ozone on the interactions between ants (Hymenoptera: Formicidae) and Hemiptera: Aphididae. *Journal of Economic Entomology* 84(5): 1465–1469.

Shingleton, A., Zhai, Y., and Robinson, G. 2018. Rain effects on aphid reproduction. *Ecology and Evolution* 8(23): 11892–11898.

Shirai, Y., Nakano, R., and Ishii, S. 2002. Effects of carbon dioxide concentration on calling behavior of male Heteroptera. *Journal of Insect Physiology* 48(4): 369–374.

Shirangi, T. R., Dufour, H. D., Williams, T. M., and Carroll, S. B. 2013. Rapid evolution of sex pheromone-producing pathway in *Drosophila*. *PLoS Biology* 11(3): 1001–1009.

Shrestha, A. K., Reddy, G. V. P., and Wu, S. 2019. Effects of simulated rainfall on feeding behavior and performance of the cabbage white butterfly, *Pieris rapae* (L.). *Journal of Insect Science* 19(3): 1–6.

Shu, Y., and Park, Y. 2020. Carbon dioxide modulates aggressive behavior in *Tribolium castaneum*. *Journal of Pest Science* 93(2): 763–771.

Sih, A., Ferrari, M. C. O., and Harris, D. J. 2011. Evolution and behavioural responses to human-induced rapid environmental change. *Evolutionary Applications* 4(2): 367–387.

Silva, R. B., Specht, A., Bueno, A. F., Michelotto, M. D., and Paula-Moraes, S. V. 2016. Effects of simulated rain on *Spodoptera frugiperda* (J. E. Smith) (Lepidoptera: Noctuidae) feeding behavior and mortality. *Neotropical Entomology* 45(1): 98–104.

Simmons, L. W., and Tomkins, J. L. 2017. Sexual selection and climate change. *Journal of Evolutionary Biology* 30(1): 8–19.

Simpson, J. D., Johns, R. C., Pitt, D. G., and Silk, P. J. 2019. Delayed and reduced diapause induction in spruce budworm (*Choristoneura fumiferana*) following chronic ozone exposure. *Environmental Entomology* 48(1): 166–174.

Simpson, S. J., Sword, G. A., and Lo, N. 2006. Polyphenism in insects. *Current Biology* 16(17): 738–749.

Singer, M. C., Kearney, M. R., Behm, J. E., and Walters, R. J. 2017. Feeding behavior mediates direct and indirect effects of climate change on a widespread predator. *Ecology* 98(4): 903–912.

Singer, M. C., and Parmesan, C. 2010. Phenological asynchrony between herbivorous insects and their hosts: Signal of climate change or pre-existing adaptive strategy? *Philosophical Transactions of the Royal Society B: Biological Sciences* 365(1555): 3161–3176.

Singer, M. S., Andersen, A. N., Porter, S. D., and Hoffmann, A. A. 2020. The impact of climate change on an ant-mealybug mutualism. *Journal of Animal Ecology* 89(9): 1947–1958.

Singer, M. S., Mace, K. C., and Bernays, E. A. 2009. Self-medication as adaptive plasticity: Increased ingestion of plant toxins by parasitized caterpillars. *PLOS ONE* 4(3): 4796–4807.

Six, D. L., Gara, R. I., Kibbe, K., Grady, A. N., and Holsten, E. H. 2005. Effects of wildfire and restoration treatments on bark beetle (Coleoptera: Scolytidae) activity in a southwestern ponderosa pine forest. *Environmental Entomology* 34(1): 12–21.

Smith, J., and Brown, S. 2018. Rain effects on ant foraging behavior. *Ecological Entomology* 43(1): 93–99.

Snell-Rood, E. C., van Dyken, J. D., Cruickshank, T., Wade, M. J., and Moczek, A. P. 2010. Toward a population genetic framework of developmental evolution: The costs, limits, and consequences of phenotypic plasticity. *BioEssays* 32(1): 71–81.

Solomon, S., Ivy, D. J., Kinnison, D., Mills, M. J., Neely III, R. R., and Schmidt, A. 2016. Emergence of healing in the Antarctic ozone layer. *Science* 353(6296): 269–274.

Solomon, S., Qin, D., Manning, M., Chen, Z., Marquis, M., Averyt, K. B., and Miller, H. L. 2007. *Climate Change 2007: The Physical Science Basis. Contribution of Working Group I to the Fourth Assessment Report of the Intergovernmental Panel on Climate Change*. Cambridge University Press.

Soroye, P., Newbold, T., and Kerr, J. T. 2020. Climate change contributes to widespread declines among bumble bees across continents. *Science* 367(6478): 685–688.

Staley, J. T., Girling, R. D., Stewart-Jones, A., Poppy, G. M., Leather, S. R., and Wright, D. J. 2017. Organic and conventional fertilizer effects on a tri-trophic system: Parasitoids, aphids and pea plants. *Journal of Applied Ecology* 54(1): 112–121.

Stuble, K. L., Patterson, C. M., Rodriguez-Cabal, M. A., and Fowler, D. A. 2017. Wildfire disturbance and climate change: Implications for plant—ant interactions and riparian vegetation in the western US. *Current Forestry Reports* 3(2): 120–129.

Sullivan, S., and Lewis, S. 2018. Rain effects on firefly mating behavior. *Animal Behavior* 141: 183–190.

Sun, Y., Pu, X., Zhou, L., Zhang, L., Li, J., Li, R., and Guo, Y. 2019. Elevated $CO_2$ enhances attraction of a specialist herbivore to its host plants in a concentration-dependent manner. *Plant, Cell and Environment* 42(2): 673–684.

Sun, Y. F., Deaker, R., and James, T. K. 2012. The effect of carbon dioxide on aphid host plant location by *Aphidius ervi* Haliday (Hymenoptera: Braconidae). *Austral Entomology* 51(1): 21–25.

Tanwar, R. K., Bhattacharya, R., Kranthi, K. R., Kranthi, S., Bhatnagar, V. S., and Gupta, G. P. 2018. Effect of simulated rain on *Helicoverpa armigera* (Hübner) (Lepidoptera: Noctuidae) feeding behaviour. *Journal of Applied Entomology* 142(7): 749–759.

Tariel, J., Giraudeau, M., Parisot, N., and Boulanger, N. 2019. Increasing atmospheric $CO_2$ modifies predator-prey interactions through altered plant quality. *Oecologia* 189(1): 191–203.

Tauber, M. J., Tauber, C. A., and Masaki, S. 2016. Environmental control of diapause in the Colorado potato beetle, *Leptinotarsa decemlineata. Journal of Insect Physiology* 95: 1–11.

Telonis-Scott, M., Biessmann, H., Jacobson, J., McKechnie, S. W., and Shingleton, A. W. 2019. Fire-induced environmental stress affects developmental and reproductive traits in *Drosophila melanogaster. Journal of Evolutionary Biology* 32(1): 1–13.

Tenczar, P., Andrieux, T., Le Conte, Y., and Sandoz, J. C. 2019. Rainfall disrupts honey bee foraging and recruitment, but not learning of floral odours. *Journal of Experimental Biology* 222(11): jeb201855.

Thackeray, S. J., Henrys, P. A., Hemming, D., Bell, J. R., Botham, M. S., Burthe, S., and Wanless, S. 2016. Phenological sensitivity to climate across taxa and trophic levels. *Nature* 535(7611): 241–245.

Thackeray, S. J., Jones, I. D., and Maberly, S. C. 2010a. Long-term change in the phenology of spring phytoplankton: Species-specific responses to nutrient enrichment and climatic change. *Journal of Ecology* 98(2): 555–570.

Thackeray, S. J., Sparks, T. H., Frederiksen, M., Burthe, S., Bacon, P. J., Bell, J. R., and Webb, T. J. 2010b. Trophic level asynchrony in rates of phenological change for marine, freshwater and terrestrial environments. *Global Change Biology* 16(12): 3304–3313.

Thaler, R. H. 2012. *The Winner's Curse: Paradoxes and Anomalies of Economic Life*. Simon and Schuster.

Tian, H., Peng, C., Zhang, L. and Yan, X. 2013. Effects of elevated ozone on the feeding preference and growth of the diamondback moth, *Plutella xylostella. Journal of Environmental Sciences* 25(10): 2004–2010.

Tobin, P. C., Berec, L., Liebhold, A. M., and Rossi, R. E. 2014. Expanding gypsy moth (*Lymantria dispar*) range in the United States: Climate change, forest connectivity, and population bottlenecks. *Landscape Ecology* 29(2): 213–226.

Tregenza, T., and Wedell, N. 2000. Effects of resource variation during early adulthood on male calling behaviour and mating success in the field cricket *Gryllus bimaculatus. Ecological Entomology* 25(3): 277–284.

Trpis, M., Hausermann, W., and Craig Jr, G. B. 2018. The effects of ozone on *Aedes aegypti* mosquitoes and mosquito larvae. *Bulletin of the World Health Organization* 56(6): 959–965.

Tsvetkov, N., Samson-Robert, O., Sood, K., Patel, H., Malena, D., and Gajiwala, P. 2017. Rain effects on honey bee reproductive health. *Scientific Reports* 7(1): 1–10.

Tuan, S. J., Lin, Y. T., Chen, C. C., Huang, T. C., Lee, H. C., and Chang, N. T. 2020. Effects of simulated rain on *Bemisia tabaci* (Hemiptera: Aleyrodidae) mortality and feeding behavior. *Journal of Economic Entomology* 113(2): 576–582.

Van Asch, M., Salis, L., Holopainen, J. K., and Blande, J. D. 2007. Nonlinear effects of photoperiod on diapause induction in the pine beauty moth, *Panolis flammea. Journal of Insect Physiology* 53(11): 1200–1208.

Van Bergen, E., Brakefield, P. M., and Heinen, R. 2018. Rain effects on butterfly fecundity and larval performance in the tropics: Shifts in chemical signalling traits. *Ecological Entomology* 43(1): 109–119.

Van der Sluijs, J. P., Amaral-Rogers, V., Belzunces, L. P., Bijleveld van Lexmond, M. F., Bonmatin, J. M., Chagnon, M., and Goulson, D. 2013. Conclusions of the worldwide integrated assessment on the risks of neonicotinoids and fipronil to biodiversity and ecosystem functioning. *Environmental Science and Pollution Research* 20(5): 957–984.

Van Renterghem, J., Smith, S., Barott, K., and Doherty, J. A. 2019. Auditory communication of *Aedes triseriatus* (Diptera: Culicidae) mosquitoes during the post-fire succession period. *Journal of Insect Behavior* 32(4): 343–358.

Venette, R. C., Kriticos, D. J., Magarey, R. D., Koch, F. H., Baker, R. H., and Worner, S. P. 2002. Pest risk maps for invasive alien species: A roadmap for improvement. *BioScience* 52(9): 785–792.

Visser, M. E., te Marvelde, L., and Lof, M. E. 2006. Adaptive phenological mismatches of birds and their food in a warming world. *Journal of Ornithology* 147: 83–94.

Von Frisch, K., and Lindauer, M. 1954. Über die "Sprache" der Bienen I. Über die durch Rundtänze und Schwänzeltänze mitgeteilte Entfernungs- und Richtungsangaben. *Zeitschrift für vergleichende Physiologie* 36(1): 1–23.

Wada-Katsumata, A., Silverman, J., Schal, C., and Roelofs, W. 2019. Effects of ozone on acoustic signal detection and responses to danger cues in the field cricket, *Gryllus assimilis*. *Environmental Science and Pollution Research* 26(20): 20121–20127.

Wagner, D. L., and Kay, S. A. 2002. Evaluating the impact of prescribed fire on ant predation in a New Mexico grassland. *Journal of Arid Environments* 52(2): 167–175.

Walker, T. J., Moore, P. A., and Wagner, W. E. Jr. 1997. The effect of rain on acoustic communication in the field cricket *Gryllus pennsylvanicus*. *Journal of Comparative Physiology A* 181(1): 97–104.

Wang, S., Gotelli, N. J., DeSalle, R., and Zhang, Y. 2016. Trait-based community assembly and succession of the infant gut microbiome. *Nature Communications* 7: 11–19.

Wang, X., Parisien, M.-A., Flannigan, M. D., Parks, S. A., Anderson, K. R., and Little, J. M. 2017. Sensitivity of burned area in Canadian forest to climate change, fire weather conditions, and management. *Global Change Biology* 23(1): 51–65.

Wang, Y. H., Cheng, J. A., and Zhu, Z. R. 2015. Effects of elevated atmospheric $CO_2$ concentration on the mating behavior of *Harmonia axyridis*. *Entomologia Experimentalis et Applicata* 157(2): 183–191.

Ward, P. S., and Novotny, V. 2006. The response of a rainforest insect community to climate warming. *Journal of Animal Ecology* 75(3): 607–616.

Wehner, R., and Müller, M. 2006. Rain and navigation in desert ants: The role of the wet-dry transition zone. *Journal of Comparative Physiology A* 192(10): 1089–1102.

Wermelinger, B., Signer, C., Moser, D., Schuepp, H., and Zschokke, S. 2011. Rain reduces predation by *Chrysoperla carnea* (Neuroptera: Chrysopidae) on the green apple aphid, *Aphis pomi* (Hemiptera: Aphididae). *Biological Control* 56(2): 124–129.

Wheeler, A. G., Hanks, L. M., and Lewis, E. E. 1995. Temperature-dependent predation by the spined soldier bug (Heteroptera: Pentatomidae) on Colorado potato beetle larvae (Coleoptera: Chrysomelidae). *Environmental Entomology* 24(3): 753–757.

Wiley, C., and O'Connor, M. 2013. Habitat degradation, climate change, and insect invasions: Drivers of non-native beetle outbreaks in a montane landscape. *Biological Invasions* 15(9): 1901–1912.

Willetts, E., Guadagno, L., and Ikkala, N. 2010. *Addressing Climate Change: Issues and Solutions from Around the World*. IUCN.

Williams, N. M., Crone, E. E., Roulston, T. H., Minckley, R. L., and Packer, L. 2010. Ecological and life-history traits predict bee species responses to environmental disturbances. *Biological Conservation* 143(10): 2280–2291.

Willink, B., Brockerhoff, A. M., and Rojas, M. C. 2019. Effects of precipitation on the orientation behavior of two species of nocturnal beetles (Coleoptera: Carabidae). *Environmental Entomology* 48(6): 1276–1283.

Winder, L., and Schäfer, M. A. 2011. Climate change uncouples trophic interactions in an aquatic ecosystem. *Ecology Letters* 14(5): 484–489.

WMO. 2021. Scientific assessment of ozone depletion: 2021. Global ozone research and monitoring project—report No. 60. Retrieved from www.wmo.int/globalozoneassessment/2021/

Woods, H. A., Dillon, M. E., Pincebourde, S., and Klok, C. J. 2012. Changes in temperature sensitivity and energy use in a butterfly across a latitudinal gradient. *Journal of Insect Physiology* 58(7): 959–967.

Yan, L., Cai, Z., Yang, L., and Hu, C. 2016. Effects of elevated atmospheric carbon dioxide on insect-plant interactions: A meta-analysis. *PLOS ONE* 11(10): 164–174.

Yan, Z., Zhang, F., Zhu, X., Wang, G., and Lei, C. 2017. Influence of carbon dioxide on olfactory-mediated responses of *Plutella xylostella* to host and non-host plants. *Journal of Pest Science* 90(2): 581–590.

Zanotto, F. P., Moura, M. O., and Lopes, J. R. S. 2000. The influence of rain on the hatching of *Lymantria dispar* (Lepidoptera: Lymantriidae) eggs in the laboratory. *Environmental Entomology* 29(2): 361–366.

Zavala, J. A., Casteel, C. L., DeLucia, E. H., and Berenbaum, M. R. 2013. Anthropogenic increase in carbon dioxide compromises plant defense against invasive insects. *Proceedings of the National Academy of Sciences* 110(17): 7039–7044.

Zhan, S., Wang, X., Dong, Y., Zhang, H., Lai, S., Huang, J., and Yan, G. 2018. Rainfall, abundance, and temperature effects on the larval ecology of the dominant Culex mosquito in western China. *Parasites and Vectors* 11(1): 240–248.

Zhang, L., Lu, M., and Ma, K. 2019. Effects of temperature on feeding and development of *Brevicoryne brassicae* (Hemiptera: Aphididae) on Chinese cabbage. *Journal of Economic Entomology* 112(4): 1874–1880.

Zhang, Q., Li, Y., Li, Q., Xie, X., Zhang, Y., and Liu, Y. 2020. Effects of ozone depletion on the reproductive system of the greater wax moth, *Galleria mellonella. Environmental Science and Pollution Research* 27(7): 7312–7320.

Zhang, X., Alexander, L., Hegerl, G. C., Jones, P., Tank, A. K., Peterson, T. C., and Trewin, B. 2011. Indices for monitoring changes in extremes based on daily temperature and precipitation data. Wiley Interdisciplinary Reviews: *Climate Change* 2(6): 851–870.

Zhao, H., Jia, M., Zhang, Y., Wang, R., Dai, P., and Wang, Q. 2020a. Ozone pollution reduces the ability of *Apis mellifera* olfactory learning. *Environmental Pollution* 264: 114–126.

Zhao, X., Luo, L., Wang, X. and Lü, J. 2020b. Effects of photoperiod and temperature on diapause induction and termination in the mulberry pyralid *Glyphodes pyloalis. Entomologia Experimentalis et Applicata* 168(10): 856–864.

Zhao, Z. H., Wang, D. X., Huang, J. H., and Li, Y. H. 2017. Effects of CO2 concentration on range shifts: A case study using *Helicoverpa armigera. Ecological Entomology* 42(3): 342–350.

Zhou, L., Zhang, Y., Chen, J., and Gao, Z. 2017. Microbial communities in the fig wasp pollinators (*Blastophaga psenes*) and their host figs (*Ficus carica* L.). *Microbial Ecology* 73(4): 855–867.

Zipkin, E. F., Ries, L., Reeves, R., Regetz, J., and Oberhauser, K. S. 2012. Tracking climate impacts on the migratory monarch butterfly. *Global Change Biology* 18(10): 3039–3049.

Ziska, L. H., McConnell, L. L., and Sipes, S. 2016. Anthropogenic climate change and the nature of insect outbreaks. *Global Change Biology* 22(12): 505–517.

# 8 Insect Population Dynamics and Climate Change

*Vyshnavi Sunil, Waqar Majeed, Sanhita Chowdhury, Aqsa Riaz, Farah Rauf Shakoori, Muhammad Tahir, and Vinod Kumar Dubey*

## 8.1 INTRODUCTION

Climate change and global warming pose significant challenges to agriculture worldwide. Fluctuating precipitation patterns, increased carbon dioxide ($CO_2$) levels in the atmosphere, and rising temperatures significantly impact agricultural productivity and insect pests (Morshed et al. 2023). A fluctuating climate is considered responsible for rapid expansion of pests, more severe pest infestations, crop yield loss, a higher chance of insidious pest species, the proliferation of vector-borne diseases, a decline in the success of biological agents, less effectiveness with current integrated pest management (IPM) strategies, transition in pest population structures, pest biology, and physiology, like the average number of generations per year (Kaur et al. 2023). Most pests are anticipated to react to climate change because they are poikilothermic species. They are a significant problem because of the substantial damage they pose to agricultural production and livestock populations (Getahun 2020).

The overall effect of climate change is to increase the number of epidemics and zone shifts. Although it is becoming easier to understand the relationship between climate and the dynamics of insect pests, the exact mechanisms behind these interactions are still poorly understood. The temperature has a significant direct influence on the population dynamics of insect pests through their life histories and morphological characteristics, as well as an indirect effect through its impact on perennial plants and natural predators (Johnson and Haynes 2023; Costes et al. 2013). There is a widespread agreement among scientists that the climate of the earth is fluctuating due to increasing greenhouse gases resulting from human activities. The increasing temperature may favour mobile organisms with relatively short generation times. They can move and escape from vulnerable conditions and develop adaptations more rapidly than non-motile species with prolonged generation times. Therefore, to cope with global warming, insect outbreaks have been reported that are considered responsible for the devastating damage to crops (Kurz et al. 2008; Lamichhane et al. 2015). Yet climate change goes beyond an increase in temperature. Over the past century, the atmospheric $CO_2$ level increased from 280 to 400 ppm, and it may reach 550 ppm by 2050. Severe droughts have become more frequent and intense, and this trend is expected to continue with grave consequences (Hartmann et al. 2018). The same tendencies and forecasts apply to storm damage, with increasing winter precipitation weakening tree root anchoring in wetter soils as an example of how it affects crops. All factors mentioned may increase the risk of insect and pest infestations in forests (Jactel et al. 2019; Hlásny and Turčáni 2009).

Projected climate change might also encourage pest population explosion, increase epidemic frequencies, and allow many pest species to be distributed geographically. It will likely exacerbate the current and future challenges posed by phytophagous insect pests, resulting in increased economic losses and decreased food security (Thackeray et al. 2016). Fluctuating climate and severe weather conditions significantly affect crop productivity and various pest species. As adaptive species, insect pests respond differently to diverse sources of climate fluctuations, as the temperature

DOI: 10.1201/9781003382089-8

is considered the most crucial ecological factor influencing insect population hierarchy. Worldwide climate change will likely affect how insects socialize with their hosts and natural predators, and will alter their geographical distribution, winter dormancy survival rates, sociocultural rates, and the risk of invasive insect species and crop diseases. When the pest problem worsens owing to climate change, future pest management strategies are critically needed (Skendžić et al. 2021).

## 8.2 CLIMATE UNDER CHANGE

Large-scale climatic fluctuations are frequently reflected in population dynamics. Many insects, for instance, have relatively short annual cycles and have higher population densities in the summertime (Yamamura et al. 2006), whereas other animal populations react to long-term climatic fluctuations (Srygley et al. 2023). Global warming and climate change are among the topics that are currently being addressed with the most considerable significance to agriculture everywhere (Gopalakrishnan et al. 2019). Climate variables, such as increasing temperatures, increasing $CO_2$ levels in the atmosphere, and altering precipitation patterns, significantly impact crop productivity and agricultural insect pest activities. A changing climate can have a variety of effects on insect pests. It may affect their spatial range, cause their overwintering subsistence to increase, increase the generation span, change plant-pest synchronicity, alter species communication, cause growth of the risk of migration patterns of pest invasion, significantly raise the onset of insect-transmitted crop diseases, and cause biological pest control, mainly through natural enemies, to become less productive. As a result, crop economic damage increases, and human food security is jeopardized (Singh 2022; Rao and Jeyaveeran 2014).

### 8.2.1 Effect of Increased Temperature

Insects are susceptible to fluctuations in temperature, and an increase of 10°C tends to cause their metabolic rate to almost double (Dukes et al. 2009). Most of the research has elaborated that enhanced temperature can increase the utilization of food for insects, their growth, and their movement. The insect's reproduction, density, survival, stage of development, and distribution might be affected by all these factors (Bale et al. 2002). While some species can flourish in warmer climates and reproduce quickly, others cannot adapt and frequently face difficulty stabilizing their numbers. Temperature changes alter the dynamics of pest populations because this significantly impacts their motility, biochemical reactions, and metamorphosis (Shrestha 2019). According to the distribution and behaviour of current insects, it may be anticipated that more significant grazing should go alongside rising temperatures. According to insects' behaviour and their distribution pattern, it is concluded that a temperature change can change the growth of insects and increase herbivory (Delucia et al. 2008; Pureswaran et al. 2018). This means there is a decrease in the growth rate of insects in tropical zones at the current temperature, which is nearly the optimum, while the growth rate of temperate zones insects is increasing, similarly, by simulating the changes in the population of insects that would benefit the cultivation of the world's three major cereal crops under various climate change scenarios. In the temperate zone, the increase in temperature can increase the density of pests for wheat. They predicted inconsistent responses to the rise of pest populations for maize planted in different areas and a drop in the pest populations for rice grown in tropical regions (Deutsch et al. 2018).

Global warming is expected to increase in the next few years, which will change the occurrence patterns of insects. According to climate predictions, the world's mean temperature will increase by 1.5°C to 2°C by the end of 2100 (Barbarossa et al. 2021; Malhi et al. 2021). In global warming scenarios, the intensity of pest confliction is anticipated to enhance as ambient temperatures typically rise towards ideal levels for the development and growth of several species of insect pests, possibly decreasing thermal limitations on insect dynamics. Due to the limited biological requirements, adaptive responses to problems, and variable climate effects on their phenology and life history, climate warming may not always increase pest populations and corresponding losses of

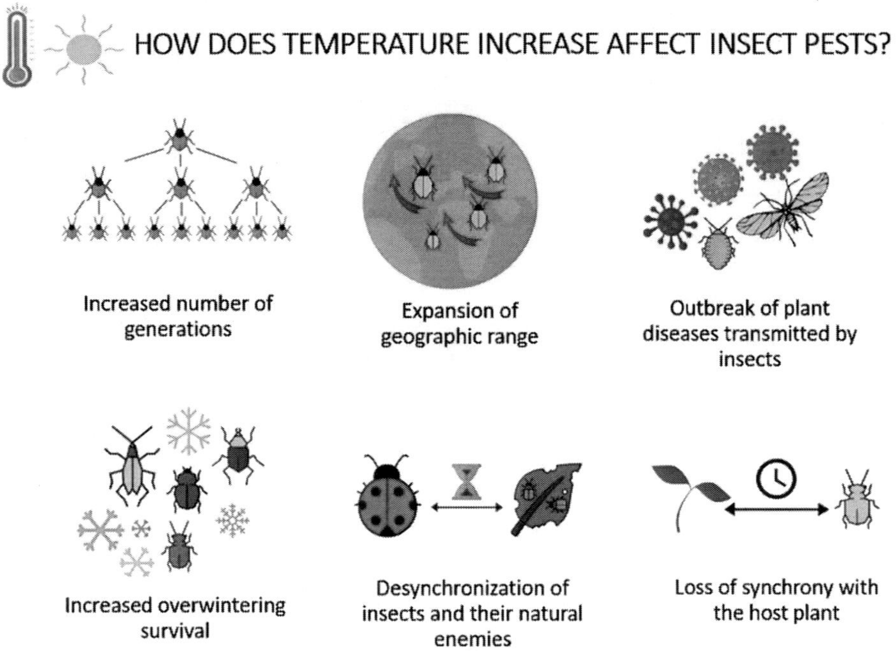


**FIGURE 8.1** Impact of temperature increase on insect pests.

**Source: Adapted with permission from Skendžić et al. (2021).**

commercial crops (Deutsch et al. 2008, 2018). Figure 8.1 illutrated that how temperature effects the environemnta, which ultimately influences the life cycle and survial and diseases for insects.

### 8.2.2 Effect of Increased $CO_2$ Concentration

The distribution, productivity, and density of insects that consume their feed from plants are impacted by enhanced $CO_2$ levels in the atmosphere. Such enhancements also affect insect pests' growth, fertility, consumption rates, and relative abundance (Fuhrer 2003) (Figure 8.2). An infamous gas that has always been linked to climate change is $CO_2$. Climate change research has always focused on rising $CO_2$ levels. $CO_2$ affects the nutrition level of plants, which contributes to an increase in the insect population. Because plants can absorb carbon more quickly at higher $CO_2$ levels, they become essentially composed of carbon and grow more swiftly. Increases in $CO_2$ are frequently interpreted as increases in the photosynthetic rate, and it is generally accepted that plants become more drought tolerant as $CO_2$ concentrations rise. This enables a situation where insects have access to hosts even under challenging circumstances. As a result, it creates an environment where pest frequency is high. Most critically, the lower nutritional level in plants is the primary cause of an increasing level of damage with higher $CO_2$ levels. Plants exposed to more $CO_2$ are likely to have lower nitrogen levels, which affects the host plants' ability to absorb nutrients. Because of the accumulation of non-structural carbohydrates in plant tissues, an enhanced $CO_2$ level in the atmosphere typically results in an enhancement of the carbon-nitrogen ratio, which lowers the nutritional quality for insects with low protein requirements by neutralizing the nutritional quality by 15% to 25% in the body tissue (Pathak et al. 2012). To meet their increased nitrogen needs, insects rapidly increase their feed consumption. For the Japanese beetle, *Popillia japonica*, a higher concentration of carbohydrates like fructose, sucrose, and glucose in soybean leaf grown under a higher $CO_2$ level is considered a favourable factor. According to reports, aphid fecundity is also linked to greater $CO_2$

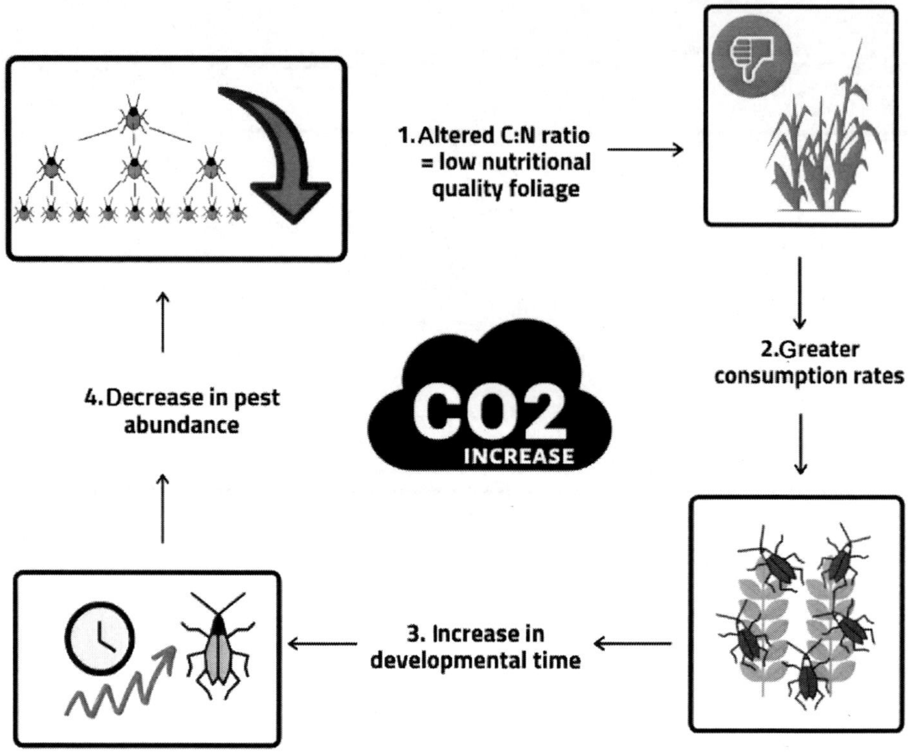


**FIGURE 8.2** Impact of $CO_2$ increase on insect pests.

**Source: Adapted with permission from Skendžić et al. (2021).**

levels. Certain insects, including the pine sawfly, have demonstrated enhanced levels of nitrogen use efficiency when raised under higher $CO_2$ concentrations (Shrestha 2019).

High $CO_2$ concentrations are anticipated to affect plant physiology by boosting photosynthetic activity, enhancing growth, and increasing plant density, as discussed in the preceding section. So, change in the quantity and quality of plants indirectly affects insect density. Changes in leaf chemistry are common in plants growing in high $CO_2$ conditions, which can impact plant nutrition and the attractiveness of plants to leaf-eating insects. Moreover, these plants frequently build starches and sugars in their leaves, changing the carbon-to-nitrogen ratio and lowering the palatability. As $N_2$ has a significant role in insect growth, some pest groups consume more plants when $CO_2$ concentrations increase. Insects can cause more plant damage by destroying more plant tissue to get the same amount of food. Leaf feeders like chewers, caterpillars, and miners frequently increase their consumption rates when nitrogen levels decline, as anticipated from $CO_2$ fertilization and compensatory eating (Skendžić et al. 2021).

### 8.2.3 Impact of Changeable Precipitation Pattern

Variations in rainfall frequency are crucial climate change indicators. As seen in most events, the rainfall rate has decreased while the severity has increased. This precipitation pattern has favoured the incidence of droughts and floods. Overlapping rainfall patterns directly affect the insects that burrow in the soil. In other terms, heavy rains may cause protracted water stagnation and floods. This occurrence threatens insect survival and interferes with their diapause. Moreover, heavy rains and floods can potentially sweep away bug eggs and their larvae (Shrestha 2019). Tiny pests like whiteflies, jassids, mites, and others can be washed away when it rains heavily. Insect populations

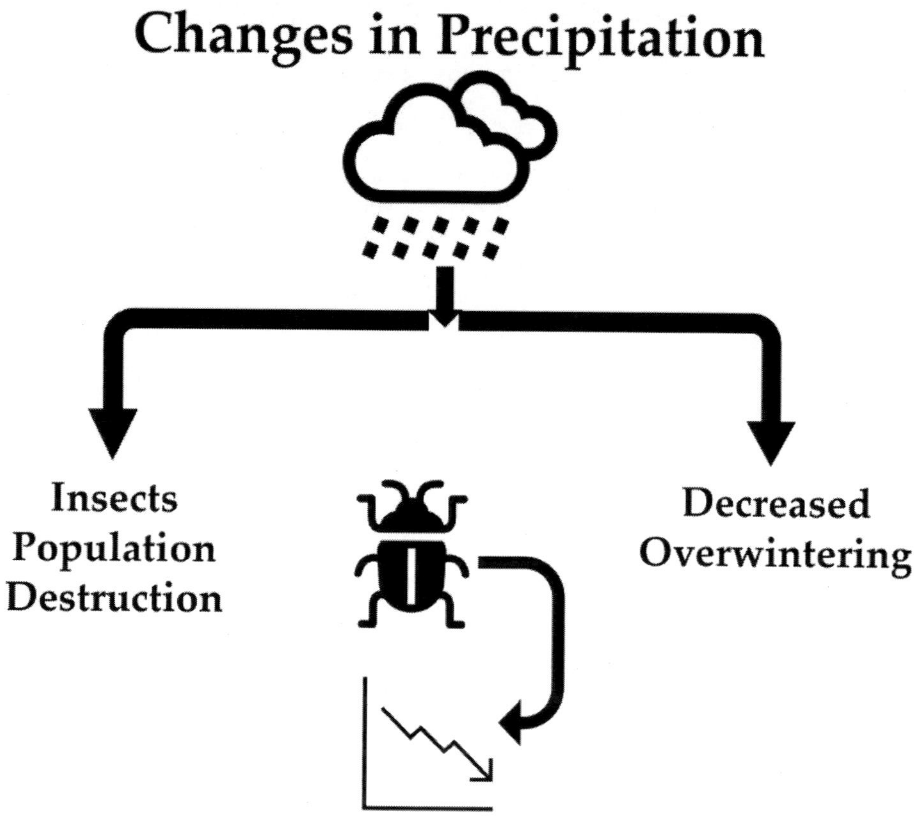


**FIGURE 8.3** Impact of precipitation on insect population dynamics. Modified from Skendžić et al. (2021).

are impacted majorly by irregular rainfall (Shahzad et al. 2021). For instance, Staley et al. (2007) investigated how soil-dwelling wireworms (*Agriotes lineatus* L.) responded to drought and increased summer rainfall in grassland plots because they cause significant damage to crops like potatoes, corn, sugar beetroot, and others planted on grassland plots. The figure illustarion provided explains the influence of precipitation fluctuation on the insects and their ultimate decline (Figure 8.3).

The enhanced rainfall in summer, conflicting with atmospheric and desiccation conditions, resulted in the rapid development of wireworms in the outer part of the soil (Gregory et al. 2009; Staley et al. 2007). Drought affects herbivorous insects in several ways: (a) desiccated climates may offer favourable environmental conditions for their growth; (b) drought-stressed plants may attract insect species (when water columns in the xylem separate or cavitate during the transpiration phase, for example, harmful bark beetles [Scolytidae] can trace the ultrasonic acoustic emission produced); and (c) plants under stress from drought are more susceptible to insect attack because the production of bioactive molecules alongside a decreased defensive system (Yihdego et al. 2019).

### 8.2.4 Global Climate Change: The Thermal Context

According to MacDonald's detailed review of global climate change, three aspects are crucial to our understanding of how higher temperatures will impact terrestrial invertebrates, particularly mites and insects, regarding physiology and ecology (MacDonald 2010):

1. Throughout this century, the earth will continue to experience a period of climate warming, which started around 1860.

**TABLE 8.1**
**Impact of Climate Change on Agricultural Insect Pests**

| Response of Insect Pests | Species Involved | Impact of Climate Change | References |
|---|---|---|---|
| Shifting distribution | Blue oat mites, *Penthaleus* spp. | Distributions of the three *Penthaleus* species in Australia are correlated with different climatic variables, suitable climate space likely to decrease in the future. Cryptic species respond differently. | (Hill et al. 2012) |
| | Fruit fly, *Dacus* spp. | Results from distribution models revealed general patterns of pole ward movement for the group. For the individual species, distribution shifts appear to also be eastwards and at finer scales, varying amounts of species turnover were apparent. These changes in response across different scales present regional management challenges for these species under climate change. | (Hill et al. 2016) |
| Altering phenology | Apple codling moth, *Cydia pomonella* | Under future conditions of increased temperature in Switzerland, the risk of an additional third generation will increase from 0%–2% to 100%, and there will be a two-week shift in earlier overwintering adult flight. The shifts in phenology and voltinism will require change to plant-protection strategies. | (Stoeckli et al. 2012) |
| Adjusting to persist in situ | Red legged earth mite, *Halotydeus destructor* | Greater thermal tolerance in population of Australia than in Africa | (Hill et al. 2013) |

2. Temperature increases are likely greatest at higher latitudes like those in the high Arctic.
3. Climate change is likely to be gradual, but both the rate and future levels of $CO_2$ are unprecedented. Climate change is expected to increase temperatures in the next 50–100 years, despite the current temperatures being within organisms' range in previous climate cycles. Understanding how environmental change impacts insect population dynamics, directly and indirectly, is essential for forecasting the effects of rising temperatures. Insects may be indirectly impacted by global warming depending on the quantity and quality of vegetation change (Jamieson et al. 2012a). The synchronization of phenology between insects and their host plants or natural enemies may be disrupted by variations in the reaction pattern to temperature fluctuations (Cornelissen 2011). Unlike resistance to lower lethal temperatures, insect resistance to higher lethal temperatures varies substantially less between species. Hence, a crucial determinant in determining insects' geographic distribution and abundance is frequently a high temperature (Hoffmann et al. 2013). Contrary to popular belief, heat stress can affect most insects between 28°C and 32°C. Heat stress may cause unavoidable harmful effects on all arthropods (Kiritani 2006).

## 8.3 DIRECT INFLUENCE OF CLIMATE CHANGE ON INSECT POPULATION DYNAMICS

### 8.3.1 Response of Insects to Increased Temperature and $CO_2$

Rising temperatures and $CO_2$ have complex physiological effects that can profoundly affect agricultural crop and insect interactions (Skendžić et al. 2021). $CO_2$ is an essential compound for photosynthesis. It enters the plant through stomatal openings and is converted into starch and sugars by solar energy during photosynthesis. Increased $CO_2$ can improve light use efficiency and photosynthesis, stimulating and increasing plant growth and development, and their response differs among C3 and C4 plants (Leakey et al. 2019). According to the carbon nutrient balance hypothesis, under elevated $CO_2$ levels, all carbon-rich secondary metabolites should increase, but this is not always the case. For instance, *Plantago lanceolata* has iridoid glycosides in their foliage that are unaffected under increased $CO_2$ concentrations (Marak et al. 2002). An experimental increase in $CO_2$ showed no variation in the volatile terpenoids of loblolly pine (Hunter 2001), big sagebrush (Takahashi and Huntly 2010), and peppermint. An increase in $CO_2$ level increases the C/N ratio, enhancing the secondary metabolites in the host plants, resulting in reduced insect feeding. For instance, in Aspen, $CO_2$ induced an increase in phenolic glycosides and reduced gypsy moth performance (Osier and Lindroth 2001). Variable responses of elevated $CO_2$ during different seasons are due to the ageing of leaves and herbivores. In oaks, increases in the C/N ratio of foliage are more pronounced in older leaves than younger ones. Old larvae can better compensate for reduced foliage quality than young larvae (Hunter 2001).

The effect of change in plant phenotype due to increased $CO_2$ levels in insects is not always negative. Some phloem feeder performs better when fed on plants grown at elevated $CO_2$ levels (Klaiber et al. 2013). Soybean crops exposed to elevated levels of $CO_2$ under natural field conditions resulted in Japanese beetle *Popillia japonica* foliar damage more than doubling. In addition, Mexican bean beetles *Epilachna varivestis* and Japanese beetles *P. japonica* preferred leaves high in $CO_2$ (Hamilton et al. 2005).

Physiologically, insects are very sensitive to changes in temperature, and with an increase in temperature, their metabolic rate tends to double (Dukes et al. 2009). A mass outbreak of *Cephalcia arvensis* Panzer, an oligophagous hymenopteran species associated with spruce, was observed in the Southern Alps between 1985 and 1992. Because of limited dispersal ability and low female fertility, as well as long diapause followed by high mortality in the population, this species does not tend to form mass outbreaks. The larval outbreak was caused during June–July due to high temperatures. But low humidity and high temperature lead to the short development of *C. arvensis* larvae, helping them to avoid diapause and leading to faster pupation (Battisti 2008). Temperature changes have a greater impact on insects aboveground than on those that live in soil since the soil is a thermally insulating medium that can reduce the effects of temperature changes (Bale et al. 2002).

### 8.3.2 Impact on Physiological and Behavioural Processes

Climate change may, directly and indirectly, affect insect physiology and behaviour. Depending on the temperature, an insect may undergo one to five additional life cycles following a 2°C temperature increase. Changing the leaf's chemistry is one example of indirect impact (Prasannakumar et al. 2012). Due to increasing $CO_2$ concentrations, insects eat more, so they need more food to maintain the same nutritional level. Also, aphids are less responsive to alarm pheromones, parasitism is reduced, and *Bacillus thuringiensis*–based transgenics are less efficient. In comparison to ambient $CO_2$ concentrations (121 hoppers/hill), increased $CO_2$ accelerated brown plant hopper (BPH) multiplication, resulting in a significant increase in their population (435 hoppers/hill). The signals insects use to communicate are thermally sensitive because they mate within critical temperature ranges. Several aspects of songs are affected by temperature (e.g. the pulse duration, frequency, and inter-pulse length), but the pulse rate is most significantly affected. Regarding the type of call and time characteristics of courting songs, female crickets are more selective than male crickets (Gray et al. 2016). There is a possibility that increasing temperatures might cause species to mismatch their signal preferences (Sharma and Singla 2021).

Environmental factors, such as photoperiod and temperature, profoundly affect diapause initiation, maintenance, and termination. There are several ways in which temperature influences diapause. First, they often stimulate diapause by interacting with the photoperiod. Low temperatures promote diapause in species that do not undergo diapause in long days. In contrast, a high temperature tends to suppress diapause in species undergoing short-day or aestivate diapause. Secondly, for maintaining diapause in insects, the temperature is important. Rapid diapause usually occurs in a temperature range significantly lower than non-diapause development. A low thermal threshold during diapause ensures that diapause is maintained and development is prevented by warm conditions that may occur during or before the winter season (Schebeck et al. 2017). Finally, the temperature may also act as an active diapause-terminating stimulus.

### 8.3.3 Change in Insect Pest Diversity

Insects' ability to maintain homeostasis is limited, so the temperature is considered the most significant environmental factor influencing their behaviour, distribution, development, survival, and reproduction (Bale et al. 2002; González-Tokman et al. 2020). There is a direct correlation between insect diversity and ecosystem health since insects are excellent indicators of environmental change and play a significant role in food chains. Different insect species may decline because of climate change, and those unable to adapt to the changes may disappear in the future (Thomas et al. 2004). There may be differences in the aggregate responses to multiple pest species at the regional level, resulting in contrasting or masked patterns that reflect overall trends (Hill et al. 2016). The altitude-dependent range expansion of corn earworms *Heliothis zea* and *Helicoverpa armigera* caused by global warming may result in heavy yield losses in maize and pose a major pest management challenge (Shrestha 2019).

Many methods exist for observing the effects of climate change. Real insights into climate change can be gained by analyzing climatic factors such as temperature, precipitation (amount, frequency, and timing), humidity, wind (velocity, timing), gaseous concentration, etc. Also, analyzing the results of changing these components can serve as a signal for climate change. Agribusiness will unavoidably see significant effects from the complexity of climate variables (Rauff and Bello 2015). The growth, development, and multiplication of organisms like insects, fungi, bacteria, and viruses, among others, are influenced by climate variables such as temperature, humidity, precipitation, and others. The population of pests is also anticipated to vary because of climate change (Khan et al. 2009). With this, host-plant tolerance to diseases and pests is anticipated to vary because of climate change. Faster disease cycles and altered insect pest physiologies can overcome the resistance (Reddy 2013).

The pests' population can rise or fall because of changing climate conditions. Animals that can survive climate extremes adapt to the altered atmosphere, maintain their number, and eat crop vegetation (Shahzad et al. 2021). Temperature changes have a considerable impact on insect species in temperate regions compared to those in tropical and temperate climates. Pest insects have been seen to alter their elevations and invade higher altitudes in temperate zones of the world. This has created additional difficulties for agriculture's sustainability (Juroszek and Von Tiedemann 2013).

### 8.3.4 Impact on Distribution and Abundance

As a result of climate change, insect pests and natural enemies' communities are likely to have decreased abundances of decomposers, predators, and herbivores, which can negatively affect ecosystem structure and services. The incidence and intensity of the pest population are remarkably affected by climate change (Sharma 2014, 2016). In recent years, abnormal weather patterns and insecticide misuse have resulted in widespread outbreaks of rice plant hoppers, *Nilaparvata lugens* (Stal.) and *Sogatella furcifera* (Horvath), leading to crop failures over 33,000 ha. Higher densities of *Maruca vitrata* (Fab.) and *H. armigera* (Hub.) due to climate change and their shifting to temperate areas will have drastic effects on crops, including legumes (Czepak et al. 2013; Tay et al. 2013). Due to climate change, *H. armigera* is emerging as an invasive pest in North America and Brazil. Similarly, there is an increased emergence of *Spodoptera litura* due to higher $CO_2$ levels (Nagrare et al. 2009).

The effect of climate change on pest distribution is more determined by lower temperatures than higher temperatures (Boullis et al. 2015). Thus, insect species endemic to warmer climates can invade temperate areas occasionally but cannot complete their development due to inadequate thermal units during the summer. Alternatively, one or more generations may be completed during a good summer but then caught during winter when they are not cold-hardy or unable to migrate (Danks 2007). It has been shown that when temperatures rise, insect pests will shift from tropical and subtropical regions to temperate regions at higher altitudes, along with shifts in the cultivation areas of their host plants (Sharma 2010, 2014; Parmesan and Yohe 2003). Consequently, tropical insect species may become more abundant (Bale et al. 2002; Grimbacher and Stork 2009; Diffenbaugh et al. 2008), and outbreaks of insect pests can wipe out entire crops (Kannan and James 2009).

Crop growth simulation models have been extensively used in studying global climate change. But most simulation models do not usually consider pests, weeds, and diseases (Moore and Allard 2016). Forecasting the future spread and development of pests and diseases due to global climate change is crucial to crop production worldwide. New pests and niches develop due to increasing temperature. Due to their inability to regulate their body temperature, insects are directly affected by the temperature they are exposed to. Climate change may have some beneficial effects on crops and forest health, but many of its effects will be detrimental. Forest communities persist under altered climatic conditions due to their ability to cope with changing conditions. However, pest outbreaks, reduced forest growth, and tree mortality can result from the changeover of pests into new territories without natural enemies keeping them under check. Polyphagous insects that exhibit phenotypic and genotypic plasticity across various environments, but are restricted to a narrow niche, are more likely to be adversely affected than specialist insects (Bale et al. 2002).

A long-term study of pest populations and behaviour, particularly in climate change-sensitive regions, may offer some insight into climate change-related biological responses. When inadequate and inappropriate biological and climatic data limit the development of predictive population models, CLIMEX can be a more practical method of predicting distributions (Sutherst 2003). To generate an ecoclimatic index, CLIMEX uses a global climatic database and parameters describing each species' moisture and temperature response. The life cycle of pests with temperate origins usually includes some physiological mechanism, such as diapause, that allows their survival over cold winters. Moreover, local populations of widely distributed species may develop more rapidly in cooler parts of their range, and their developmental thresholds may be lower (Aitken et al. 2008).

### 8.3.5 Enhanced Overwintering Survival

Insects are affected by climate change primarily through processes that occur in summer (reproduction, development) or through distribution changes rather than their winter survival. In some climatic zones, predicted temperatures of 1°C to 5°C during the next 50–100 years are expected to increase winter survival for species that regularly suffer from cold stress. There is, however, no guarantee that this effect will be universal.

Insect cold-hardiness can be divided into two strategies: freeze tolerance and freeze avoidance by supercooling. In freeze-tolerant species, nucleating agents (proteins) are synthesized; in freeze-avoiding species, all potential nucleators, such as gut contents, are removed. As more studies on freeze-tolerant species have been conducted, there has been a consistent pattern of temperatures at which nucleators initiate freezing at the super cooling point, meaning typically above 10°C. Nevertheless, the lethal temperature varies significantly from −10°C for the sub-Antarctic beetle *Hydromedion sparsutum* (Müller) (J S Bale et al. 2000) to −25°C for *Syrphus ribesii* Linn. (Soley and Sipes 2021) and −40°C or lower in the North American Gallfly, *Eurosta solidaginis* Fitch (Robich and Layne 2013). A decrease in temperature or an increase in exposure time usually affects insect survival under experimental conditions.

Winter temperatures can influence population dynamics and ecological and evolutionary processes, especially in ectotherms (Burraco et al. 2020). The winter weather in temperate zones is changing fast and modifying cold stress regimes from their historical patterns. Even though insect population models typically assume that responses to low-temperature stress are static, organismal responses to repeated stress events remain poorly understood (Marshall and Sinclair 2015). As the duration and intensity of cold exposure grow, insects exposed to low temperatures may suffer from chilling injury, which results in mortality unrelated to freezing (Renault et al. 2002). Hence, low-temperature stress is ecologically important and offers a manageable framework for comprehending the effects of recurrent stress. Numerous systems have been used to study the effects of repeated cold exposure on insects. Still, these have typically only varied the number of exposures and frequently lacked adequate controls, confounding repeated cold with other factors like cumulative cold exposure and the age of the experimental animals (Hoffmann et al. 2003). The rewarming phase allows the repair of cold-induced damage between low-temperature exposures; in general, repeated cold exposure usually increases investment in polyol cryoprotectants like glycerol. The rewarming effect is particularly significant in less ecologically significant fluctuating thermal regimes where prolonged cold exposure alternates with brief benevolent periods (McClanahan 2020).

### 8.3.6 Early Emergence

Many species, however, experience shorter diapause durations at higher temperatures. For instance, in *S. crassipalpis*, diapause lasts 118 days at 17°C, 70 days at 25°C, and 57 days at 28°C (Bale and Hayward 2010). According to (Hahn and Denlinger 2007), one hypothesis is that diapause causes nutrient reserves to be depleted sooner due to a higher metabolic rate. Minimum energy is required for entering diapause than continuing in it. Low temperatures have been found to reduce the energy costs associated with diapause, contributing to the preservation of nutrient reserves needed for post-diapause processes. In larvae, energy reserves are required to sustain pupation, metamorphosis, the construction of adult tissues, and post-diapause feeding and reproduction (Hahn and Denlinger 2007). Consequently, survival, fecundity, and diapause duration may be adversely affected if climate warming disrupts diapause metabolism.

Adati *et al.* examined the influence of temperature on the growth and survival of the legume pod borer, *M. vitrata*, under *in vivo* settings. The development times for eggs, larvae, and pupae are sped up as the temperature rises from 14.4°C to 29.3°C. For development in the egg, larval, and pupal stages, the thermal constant and lower thermal threshold were 51.1°C, 234.7°C, and 116.5°C days and 10.5°C, 10.0°C, and 10.9°C, respectively (Adati et al. 2004).

### 8.3.7 Late Emergence

The presence of high temperatures during early diapause can extend the dormancy period in some cases (Benoit 2010). Temperature again plays a role in diapause duration, which is shortest at the optimal temperature and longer at higher or lower temperatures. If a chilling cue is not received, the northern grasshopper *Melanoplus borealis* (Fieber) may not emerge (Fielding 2008). With little free water available and low relative humidity, the winter and summer months are particularly difficult for maintaining water balance for species like *M. destructors* that endure the cold and heat as pupae that cannot migrate to more hospitable climates (Benoit et al. 2010). The literature on insect diapause provides in-depth analyses of diapause's ecology, clock, hormone, energetics, and cold tolerance mechanisms (Denlinger and Armbruster 2014). Over the past ten years, significant progress has been made in our comprehension of the molecular and physiological mechanisms underlying mosquito diapause and the crucial role diapause plays in the extension of mosquito range and response to global climate change (Yaro et al. 2012).

### 8.3.8 Loss of Synchrony with the Environment

Whether phenologies of interacting species respond simultaneously to climate change is a key factor in climate change research. There is little understanding of how climate change will affect the

**TABLE 8.2**
**Outbreak of Insect Pests in Relation to Changing Climate Scenario**

| Insect Pest | Order/Family | Host Plant | Region/Location | Probable Reason | Impact of Pest Outbreak | Reference |
|---|---|---|---|---|---|---|
| Sugarcane woolly aphid, *Ceratovacuna lanigera* Zehntner | Hemiptera: Aphididae | Sugarcane | Sugarcane belt of Karnataka and Maharashtra states during 2002–2003. | Recent abnormal weather patterns and insecticide misuse | 30% yield losses, and reduced cane recovery | (Joshi and Viraktamath 2004; Srikanth et al. 2012) |
| Rice brown plant hoppers, *Nilaparvata lugens* (Stal) and white backed plant hopper, *Sogatella furcifera* (Horvath) | Hemiptera: Delphacidae | Rice | North India | Recent abnormal weather patterns and insecticide misuse | Crop failure over more than 33,000 ha. paddy area | ICAR 2008 IRRI 2009 |
| Mealybug, *Phenacoccus solenopsis* Tinsley | Hemiptera: Pseudococcidae | Cotton, vegetables, and ornamentals | Cotton-growing belt of the country | Recent abnormal weather patterns, insecticide misuse, and changed cropping environment (introduction of Bt cotton) | Heavy yield (30%–40%) loss to the cotton, and increased cost of crop protection due to overuse of pesticides | (Dhawan et al. 2009) |
| Papaya mealybug, *Paracoccus marginatus* | Hemiptera: Pseudococcidae | Papaya | Tamil Nadu, Karnataka, and Maharashtra | Recent abnormal weather patterns and insecticide misuse | Significant yield loss to the papaya growers | (Tanwar et al. 2010) |
| Litchi stink bug, *Tessaratoma javanica* (Thunberg) | Hemiptera: Tessaratomidae | Litchi | Jharkhand | Drought conditions and resultant water stress in plants | Severe infestation on inflorescence and developing fruits resulted in up to 80% yield loss to the litchi | (Choudhary et al. 2013) |

phenological synchrony between insect herbivores and their host plants (Schwartzberg et al. 2014; Ayres and Lombardero 2000). According to Hegland et al. (2009), some plant-pollinator interactions remain cohesive, while others have become poorly matched under recent climate change

(Memmott et al. 2007). Researchers have suggested that herbivorous insects may be at a greater risk of emerging before they can eat their food plants early in the season, based on an early study of the winter moth (*Operophtera brumata*). It has been observed that egg hatching has advanced significantly in the winter moth *O. brumata* (Linn.) since budburst in its host plant, *Quercus robur* (Linn.), has not yet begun (van Asch and Visser 2007; Visser and Holleman 2001). During warming experiments with forest tent caterpillars (*Malacosoma disstria*), the budburst of the host trees preceded egg hatch by one week (Uelmen et al. 2016).

### 8.3.9 Establishment of Migratory Species

As a result of climate change, several insect pest species are expanding their ranges in pole ward waves during favourable seasons or migrating in mass to temporary locations where they can cause enormous damage like migratory locusts (Zhang et al. 2019). Climate change, especially precipitation and temperature, affect the pest population in host plants (Peng et al. 2020). As the temperature increased in China, Zeng et al. (2020) found that noctuid *Agrotis ipsilon* expanded its migration range.

Although travelling vast distances exposes migrants to additional risks from climate change, it may also be their only chance for survival. For instance, migrant insects transfer their adults between breeding sites more frequently than resident insects do, and among migrant species, those with greater dispersal showed more favourable recent population trends (Robinson et al. 2009). Like seabirds and other animals, insects with greater dispersal are better equipped to adapt to rising temperatures (Şekercioğlu et al. 2012).

Some of the largest distance migrations are made possible by birds, insects, and numerous marine animals riding winds or currents to aid their migrations (Chapman et al. 2011). Changes to these currents or winds may facilitate travel, but they are more likely to result in higher energy demands and/or the delivery of people to undesirable locations or times (J. D. Ford et al. 2006). Increased sea temperatures will change the location, strength, and duration of these fronts, potentially modifying migratory routes and timings and ultimately population structure. In marine ecosystems, the presence of thermal front boundaries and stratified water restricts migration pathways (Kiyofuji et al. 2019).

Many migratory animals need an ecologically connected network of habitats in addition to sufficient habitat at either end of their voyage. The ability of populations to adjust changing conditions will be significantly impacted by the degree of connection across places, which varies between species (Pearson and Dawson 2003). The density-dependent selection of habitat for migratory species and mortality of their populations indicates that changes in any area they use are likely to affect all populations of a species (Robinson et al. 2009).

### 8.3.10 Impact on Invasive Pests

A major threat to biodiversity is invasive alien species. According to the Convention on Biological Diversity (CBD), invasive alien species alter agricultural, forestry, and aquatic ecosystems' structure, diversity, and function, resulting in high costs for agriculture, forestry, and aquatic ecosystems (Mooney and Hobbs 2000; Timoney 2003). By altering phenological events such as flowering times, global warming may exacerbate ecological consequences like the introduction of new pests (Willis et al. 2008; Parmesan and Yohe 2003; Fitter and Fitter 2002), especially in temperate plants (Jamieson et al. 2012b), where several tropical plants are capable of enduring phenological changes.

Climate and landscape play a major role in determining the dispersal limits of invasive alien species, as well as the seasonal conditions of their development, growth, and survival. These habitats might have been previously unsuitable for dispersal due to geographical barriers like sea, ocean, or mountain which might have changed and favoured their migration due to global climate change (Vanhanen 2008). Due to their wide tolerance range and bioclimatic range, alien insects can find more suitable habitats than native insects (Walther et al. 2009). They also exhibit plasticity which helps them to develop under unfavourable climatic conditions (Fox et al. 2019; Snell-Rood et al. 2018). The increase in temperature can positively influence the population number of introduced species whose development and survival were previously limited by low temperatures. Climate change can result in alien species adapting and spreading faster than native species (Capdevila-Arguelles and Zilletti 2008). In European forest ecosystems, there are *Parectopa robiniella* (Clem.) and *Phyllonorycter robiniella* (Clem.), moths belonging to the Gracillaridae family that arrived in Europe from North America (Šefrová 2003). The black locust, *Robinia pseudoacacia*, which was introduced to Europe at the beginning of the 17th century, develops leaf mines. Although the host plant has been around for centuries in Europe, both insect species were recorded for the first time in the southern regions of the continent in the second half of the 20th century, which then expanded to the Northern region, whose cause of expansion was said to be due to global warming. To prevent the spread of invasive species in new areas, effective monitoring and management are essential (Hellmann et al. 2008; Heeb et al. 2019).

### 8.3.11 Increased Number of Generations

The changing life cycle of insects is one of the effects of increased temperatures. Voltinism in insect populations assumes that increasing surface temperatures would allow multivoltine species to produce more offspring. According to a model based on nitrogen phenology, the concentrations of secondary metabolites increase in insect herbivores. When temperatures rise above 2°C, the oviposition period, which currently coincides with diapause-inducing photoperiods, may shift dramatically, increasing insect voltinism (Tobin et al. 2008).

Thermal development tolerance has been used to link climate change to phenological shifts in various species traits and climate variables. Growing degree days (GDD) are one method of measuring this tolerance. GDD measure heat accumulation calculated annually by accumulating the total daily sum of degrees between a minimum and maximum temperature threshold (Dmin and Dmax). Aphids and some lepidopteran species, such as the large cabbage white butterfly (*Pieris brassicae* Linn.), that are multivoltine can develop faster at higher temperatures, assuming all other parameters remain the same, allowing them to breed more generations readily within a year (Bale et al. 2002). For example, as aphids develop at higher temperatures, they spend less time in the larval and nymphal stages when they are heavily threatened by predators allowing them to become adults earlier (Menendes et al. 2008).

## 8.4 THE INDIRECT INFLUENCE OF CLIMATE CHANGE ON INSECT POPULATION DYNAMICS

Ecosystems with few species are more sensitive to climatic extremes (Harvey et al. 2004), while ecosystems with low productivity take longer to recover from perturbations (Hooper et al. 2005). Fecundity and growth stage duration may be higher in insects from higher temperature environments due to their increased fitness. For example, the egg duration in BPH was 10.4 days at 25°C and 7.9 days at 27°C and 28°C (Rashmi and Chakravarthy 2020). It is predicted that the winter mortality of adults of *Nezara viridula* and *Halyomorpha halys* will be reduced by 15% for each 1°C rise and *Chilo suppressalis* two generations per year after 2°C warming (Kiritani 2006).

More frequent and severe pest outbreaks have been recorded due to global warming (Jat et al. 2021). The population of pests becomes unstable as $CO_2$ concentration rises, causing severe outbreaks. Moreover, this causes changes in pest population density, seasonal incidence, abundance, and geographic distribution (Selonen et al. 2021). Increased prevalence of *H. armigera* and *M. vitrata*, populations on legume crops due to climate change, causes severe, sudden outbreaks. Extreme temperature ranges, among other abiotic climate changes, significantly impact where phytophagous herbivores are found (Buckley 2013). A rising average temperature will increase the likelihood of new pests and pest niches emerging. A pigeon pea crop's *H. armigera* and *M. vitrata* pod infection varies depending on the temperature at different planting dates. Temperature changes directly impact insect physiology and development, and they have a similar impact on the nutritional makeup of host plants (Holopainen et al. 2013). Gandhi has documented the direct and indirect effects of temperature and $CO_2$ on the growth of insects (Gandhi and Herms 2010).

### 8.4.1 Impacts of Climate Change on Natural Enemies of Insects and Interspecific Competition

The temperature response of parasitoids is crucial to the success of biological control systems against insect pests. When temperatures were raised to 32°C, *Cyrtorhinus lividipennis*, the egg predator of BPH, *Nialaparva lugens* experienced increased attack rates and decreased handling times. The attack rate and handling time drastically decreased at 35°C, showing that the predator activity increased to a critical temperature and decreased afterwards. Climate change can affect crop distribution. Due to these changes in crop distribution, herbivores may migrate to areas where their predators and parasitoids are not tracking them, resulting in spatial desynchronization (Hullé et al. 2010). Plants that respond to elevated $CO_2$ levels, temperature extremity, and reduced precipitation provide various nutrients for herbivores, indirectly affecting parasitoids and predators (Thomson et al. 2010). Elevated temperature and $CO_2$ reduced nutritional titer of alfalfa plants host of *S. exigua*. This shortened the development time of *S. exigua* larvae, and their natural enemy, the *Cotesia marginiventris* population, declined (Dyer et al. 2013).

### 8.4.2 Impacts of Climate Change on Insect Host Plants

Several researchers have reviewed the impact of climate change on interactions between crop plants and insect pests (Rao et al. 2006; Petzoldt and Seaman 2010; Fand et al. 2012). Insect larvae compensate for decreased nutritional value by increasing consumption and development time, which exposes them to predators and other stresses due to reduced host-plant quality (Johns and Hughes 2002). A mismatch in phenology has occurred in Britain over the past couple of decades between the return flights of the migratory red admiral butterfly (*Vanessa atalanta*) and their host plants, stinging nettle (*Urtica dioica*) (Sparks et al. 2005).

The nutritional makeup of the plant was altered by increased $CO_2$ concentration and temperature, which might have made the foliage less appetizing or more nutritious for insects (Trumble and Butler 2009). Lower leaf nitrogen, higher carbon, higher relative proportions of carbon to nitrogen, and higher polyphenol content have all been seen in groundnut plants cultivated under elevated $CO_2$ levels (550 ppm and 700 ppm). Compared to ambient $CO_2$ levels, this leads to a longer larval lifespan, larger larval weights, and increased intake of groundnut plants (Baraza et al. 2007).

### 8.4.3 Effect on Pest Control Measures

Increased temperatures and $CO_2$ emissions are widely associated with conditional resistance, which reduces pesticide sensitivity. As a result of environmental changes, phenotypic plasticity or epigenetic changes may be involved in this type of conditional resistance. By altering pesticide metabolism and translocation, elevated temperatures and $CO_2$ enrichment can reduce pesticide efficacy directly or indirectly by increasing pesticide detoxification in host plants, thus reducing pesticide availability for the target pest (Matzrafi 2019). Several aspects of chemical control can be affected by climate, including residue persistence, toxic action, and treatment dosage and frequency. The number of days for chemical sprays can change if weather factors such as strong winds and rain make application impossible. Biological and cultural controls may also be affected by climate. The effects of climate change may therefore alter the effectiveness of various control methods and the emphasis placed on a particular control method (Howden et al. 2007).

The overwintering range of the diamondback moth (*Plutella xylostella*) has expanded by 2.4 million km$^2$ over the past 50 years due to climate change, according to experiment-based field-tested models. Based on the analysis of global data sets, pesticide resistance levels are associated with overwintering ranges: overwintering sites had 158 times higher pesticide resistance than sites with seasonal occurrence (Ma et al. 2021).

Many other pesticides with conditional resistance are also characterized by nontarget site resistance (e.g. deltamethrin, avermectin, glyphosate, and paraquat). Changes in environmental conditions would reduce the efficacy of pesticides with predominate nontarget resistance mechanisms. A model predicting the risk of conditional pesticide resistance under changing environmental conditions can be developed by understanding how pesticide resistance evolves (Matzrafi 2019). When diflubenzuron (an insect growth regulator) was used at higher temperatures, it caused rapid mortality and was more effective at 35°C (Amarasekare and Edelson 2004). This might be because higher temperatures increase the speed of insect moulting (Potter et al. 2011).

### 8.4.4 Effects of Pest–Host Plant–Natural Enemy Relationships

Temperature strongly impacts herbivores (Potter et al. 2011). Southerly distributed insects could benefit from increasing temperatures. Physiological measurements and published mortality records indicate that southern pine beetle mortality is almost 100% when air temperatures are below 16°C (Trân et al. 2007). Due to winter temperatures, southern pine beetles thrive in such conditions ($P > 0.50$), which indicates that this species occurs as far north as possible. As a result of climate warming, outbreaks may increase in the northern and alpine regions of the insect species' current distribution (Robinet and Roques 2010). Natural enemies belong to the third trophic level (Furlong and Zalucki 2017). When the other species in trophic levels are affected by climate change, it can alter the dynamics between pests and natural enemies, potentially negatively affecting biological control (Welch and Harwood 2014).

The ability of insects and host plants to cope with climate change depends on their mobility. Climatically favourable locations of the crop can be identified easily by mobile pests like European corn borer, *Ostrinia nubilalis* (Hübner) (Lepidoptera: Pyralidae), bollworm and diamondback moth *P. xylostella* Linn. (Lepidoptera: Plutellidae) (Trnka et al. 2007). Conversely, smaller organisms such as coccids, mealybugs, or eriophyid mites, *Aceria* spp. may have difficulty locating new hosts as crops are planted in new locations. Researchers found a lower mortality rate from parasitoids in newly colonized areas for the Brown Argus butterfly (*Aricia agestis*). These species have shifted northward due to climate change (Menendes et al. 2008). The results showed no difference between the native and expanded ranges of *A. agestis* except in parasitoid species composition. Most species attacking *A. agestis* in its newly colonized range were generalists attacking *Polyommatus icarus*, a native butterfly. In contrast to generalist parasitoids that have switched hosts, specialist parasitoids are likely to be more efficient predators.

As herbivore quality declines, predators and parasitoids will decline in fitness. Several groups of predators are affected by host quality, including spiders (Pekár and Toft 2015; Hvam and Toft 2005), predatory bugs (Butler and O'Neil 2007), and carabid beetles (Tamburini et al. 2016). In addition to host size, age, stage of development, and diet, parasitoid larval growth and development are frequently influenced by host size, age, stage of development, and diet (Harvey et al. 2004). In koinobiont parasitoids, hosts that continue to feed after parasite activity can affect parasitoids' fitness. A study demonstrated that if a natural enemy develops at a lower temperature than its prey (e.g., the aphid) and develops faster when the temperature rises, too early and a warm spring, this can result in its early emergence and high mortality rates due to a lack of food (Hance et al. 2007) which if continued for a longer period can lead to a huge decline in natural enemy population.

In some cases, a decrease in prey size does not necessarily translate into a reduction in natural enemies' effectiveness. Predators might consume more prey, improving pest control (Coll and Hughes 2008; Chen et al. 2005). Due to temperature influences, phytophagous insects are also susceptible to enemy attacks. Under higher temperature conditions, alarm pheromones, which signal predator or parasitoids attack, are less effective against aphids (Beale et al. 2006).

In grape berry moth *Endopiza viteana* (Clem.) (Lepidoptera: Noctuidae) (Tobin et al. 2008) and the European maize borer *O. nubilalis* (Hübner) (Lepidoptera: Crambidae) (Trnka et al. 2007), an increase in global temperature has been predicted to shift to higher voltinism which can alter the size and demographic rate of pest populations and reduce predator-prey interactions, drastically affecting the biological control. As a result of mid-season temperature increases, aphid populations regularly crash (Karley et al. 2004), while they benefit from low predator foraging efficiency in cooler temperatures. Early infestation of aphids due to warming temperatures in spring can drastically affect biological control (Hullé et al. 2010).

## 8.5 MANAGEMENT OF INSECTS UNDER CLIMATE CHANGE

Agricultural pests must be controlled through modified cropping practices and adaptive management strategies in response to a changing climate. Pest outbreaks can be reduced by planting different crop varieties at different times of the year and increasing biodiversity at field margins (Thomson et al. 2010; Andrew and Hill 2017). Crops grown under variable conditions like resistance to salinity, drought, water stress etc., can be adopted. Drought-tolerant cassava, *M. esculenta* Crantz, displays higher rates of mealybug infestations, and the mealybugs feeding on it exhibit lower parasitism rates (Calatayud et al. 2002).

Maintaining biodiversity and revegetating areas support a huge diversity of natural enemies. As a result of the high abundance and diversity of predators and parasitoids in vegetated areas, pest control is highly effective in those areas (Schmidt et al. 2004; Gurr et al. 2003; Olson and Wäckers 2007). Regarding long-term conservation biological control and sustainable crop production, complex landscapes characterized by highly connected crop-to-non-crop mosaics may be most effective (Clough et al. 2007). A vineyard using straw and compost as mulches reduced water loss and increased the abundance of predatory Hemiptera, parasitic Hymenoptera, and carabids (Thomson and Hoffmann 2007). In addition to increasing spider abundance, mulch increases habitat complexity, and reduced tillage increases beetle abundance (von Beeren et al. 2012).

Several studies have examined the dynamics of pest populations and the causes of outbreaks (Wermelinger 2004). These studies have identified several inherent characteristics of pest species and conditions that tend to lead to outbreaks. These data can be used to assess the likely risks different insect species pose under varying conditions, including climate change. More functional diversity is needed to increase system resilience and reduce losses due to insect pests in agroecosystems prone to climate change (Newton et al. 2009). Rabi/cold season crops can experience delayed and shorter growing seasons due to the late onset and short duration of winter. Hence, it is important to breed varieties that can tolerate adverse climatic conditions and pest and disease outbreaks (Pareek et al. 2017).

In accordance with some research findings, neonicotinoids insecticides induce salicylic acid–associated plant defense responses that enhance plant vigour and tolerance to abiotic stress without affecting insecticidal activity (Horii et al. 2007; Chiriboga 2009; Ford et al. 2010). The role of these insecticides in enhancing plant stress tolerance can be studied, identified, and used for pest management in the future. Rescheduling a crop can be done to manage pests during climate change. Since the impact of climate change can vary, mere physical and cultural control measures without proper change in crop calendars in accordance with climate change are ineffective for pest management.

Geographic information systems (GIS) can be used for area-wide pest management, predicting insect pest outbreaks based on geographic features. The effect of climate change on pest distribution, abundance, population dynamics, etc. can be studied using this method by predicting and mapping trends of potential changes in geographical distribution. Using phenology models in phytosanitary risk assessments in agroecosystems is important to assess, understand, and predict insect populations under a wide range of environmental conditions and management practices (Kroschel et al. 2013; Jarvis and Baker 2001a, 2001b). Phenology models can be coupled with GIS to prepare risk maps. Most of the currently available computer-aided modelling tools, such as ILCYM (Sporleder et al. 2012), HABITAT (Venette et al. 2010), ECAMON (Trnka et al. 2007), DYMEX (Kriticos et al. 2003), and CLIMEX (Peacock and Worner 2006; Senaratne et al. 2006) are coupled with basic GIS platforms which allow for spatial simulations of estimated pest risk indices. Using the risk maps, one can analyze the future distribution and damage potential of many invasive insects and diseases in economically important crops under projected climate changes.

In a mechanical model, environmental variables from a given area are combined with information on the species' environmental tolerances to make predictions. Predicting probable future pest activity will enable us to be more prepared to combat outbreaks of serious insect pests well in advance (Fand et al. 2014). In contrast to correlative models, mechanism-based species distribution models investigate how the environment influences physiological performance within a region. A process of elimination then predicts future species distributions by eliminating regions that hamper physiological performance so much that they inhibit survival, growth, or reproduction (Kearney et al. 2010).

For improved assessments well in advance, a decision support system involving mechanisms for collecting and disseminating insect pest data under diverse environmental conditions must be developed (Pareek et al. 2017). Considering the changing pest scenario resulting from climate change, it is recommended to focus future research on more general forms of resistance against abiotically stressed insects or diseases. As a result of climate change, the current extinction rate is 100 to 1,000 times higher than before, and nearly 45 to 275 species are going extinct daily (Sharma 2010). Climate change will have a significant impact on agriculture's sustainability and productivity, especially in developing countries like India, so it is crucial to educate farmers, extension workers, and

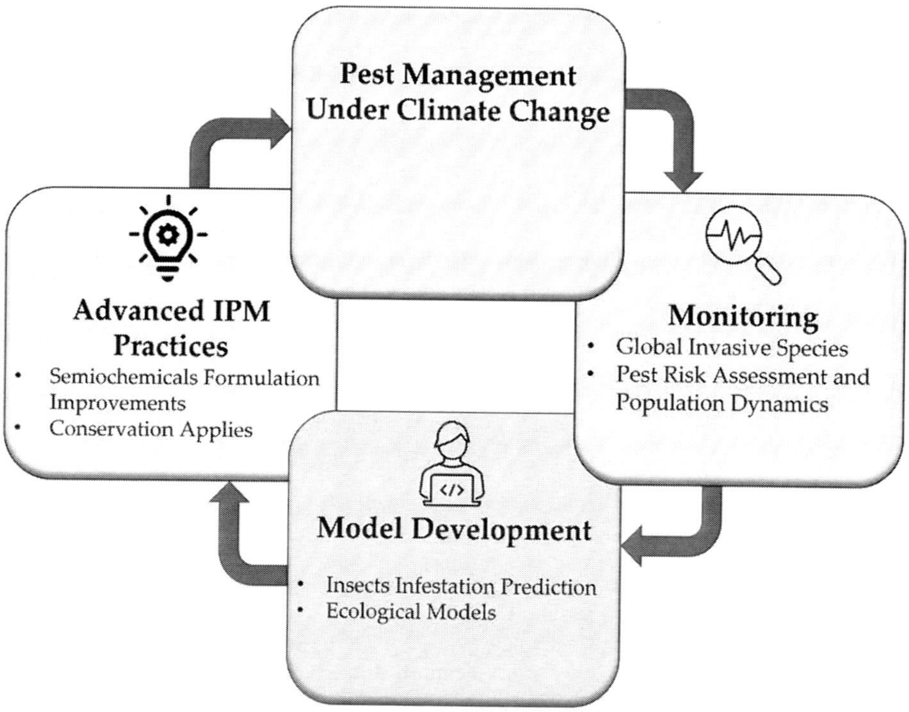


**FIGURE 8.4** Insect pest management strategies according to new environmental conditions. Modified from Skendžić et al. (2021).

other stakeholders involved in supply chain management about climate change–related changes in pest and disease incidences of major crops in their regions, as well as the different adaptation strategies needed to cope with it. Climate change and its uncertainties must be considered in pest management strategies. Pest management personnel could use this strategy to inform adaptation measures for pest control under new environmental conditions (Figure 8.4).

## Acknowledgement

The authors acknowledge the article permission agency for allowing the reuse/modification of images.

## REFERENCES

Adati, Tarô, Satoshi Nakamura, Manuele Tamò, and Kei Kawazu. 2004. "Effect of Temperature on Development and Survival of the Legume Pod Borer, *Maruca vitrata* (Fabricius) (Lepidoptera: Pyralidae) Reared on a Semi-Synthetic Diet." *Applied Entomology and Zoology* 39 (1): 139–45.

Aitken, Sally N, Sam Yeaman, Jason A Holliday, Tongli Wang, and Sierra Curtis-McLane. 2008. "Adaptation, Migration or Extirpation: Climate Change Outcomes for Tree Populations." *Evolutionary Applications* 1 (1): 95–111.

Amarasekare, Kaushalya G, and J V Edelson. 2004. "Effect of Temperature on Efficacy of Insecticides to Differential Grasshopper (Orthoptera: Acrididae)." *Journal of Economic Entomology* 97 (5): 1595–1602.

Andrew, Nigel R, and Sarah J Hill. 2017. "Effect of Climate Change on Insect Pest Management." In *Environmental Pest Management: Challenges for Agronomists, Ecologists, Economists and Policymakers*, 1st ed., 195–223. John Wiley & Sons Ltd.

Asch, Margriet van, and Marcel E Visser. 2007. "Phenology of Forest Caterpillars and Their Host Trees: The Importance of Synchrony." *Annual Review of Entomology* 52: 37–55.

Ayres, Matthew P, and María J Lombardero. 2000. "Assessing the Consequences of Global Change for Forest Disturbance from Herbivores and Pathogens." *Science of the Total Environment* 262 (3): 263–86.

Bale, J S, W Block, and M R Worland. 2000. "Thermal Tolerance and Acclimation Response of Larvae of the Sub-Antarctic Beetle *Hydromedion sparsutum* (Coleoptera: Perimylopidae)." *Polar Biology* 23: 77–84.

Bale, J S, and S A L Hayward. 2010. "Insect Overwintering in a Changing Climate." *Journal of Experimental Biology* 213 (6): 980–94.

Bale, Jeffery S, Gregory J Masters, Ian D Hodkinson, Caroline Awmack, T Martijn Bezemer, Valerie K Brown, Jennifer Butterfield, Alan Buse, John C. Coulson, John Farrar, John E. G. Good, Richard Harrington, Susane Hartley, T. Hefin Jones, Richard L. Lindroth, Malcolm C. Press, Ilias Symrnioudis, Allan D. Watt, John B. Whittaker. 2002. "Herbivory in Global Climate Change Research: Direct Effects of Rising Temperature on Insect Herbivores." *Global Change Biology* 8 (1): 1–16. https://doi.org/10.1046/j.1365-2486.2002.00451.x.

Baraza, Elena, Regino Zamora, José A Hódar, and José M Gómez. 2007. "Plant-Herbivore Interaction: Beyond a Binary Vision." In *Functional Plant Ecology*, 481–514. CRC Press.

Barbarossa, Valerio, Joyce Bosmans, Niko Wanders, Henry King, Marc F P Bierkens, Mark A J Huijbregts, and Aafke M Schipper. 2021. "Threats of Global Warming to the World's Freshwater Fishes." *Nature Communications* 12 (1): 1701.

Battisti, Andrea. 2008. "Forests and Climate Change-Lessons from Insects." *IFOREST-Biogeosciences and Forestry* 1 (1): 1.

Beale, Michael H, Michael A Birkett, Toby J A Bruce, Keith Chamberlain, Linda M Field, Alison K Huttly, Janet L Martin, Rachel Parker, Andrew L Phillips, and John A Pickett. 2006. "Aphid Alarm Pheromone Produced by Transgenic Plants Affects Aphid and Parasitoid Behavior." *Proceedings of the National Academy of Sciences* 103 (27): 10509–13.

Beeren, Christoph von, Rosli Hashim, and Volker Witte. 2012. "The Social Integration of a Myrmecophilous Spider Does Not Depend Exclusively on Chemical Mimicry." *Journal of Chemical Ecology* 38 (3): 262–71. https://doi.org/10.1007/s10886-012-0083-0.

Benoit, Joshua B. 2010. "Water Management by Dormant Insects: Comparisons between Dehydration Resistance during Summer Aestivation and Winter Diapause." In *Aestivation: Molecular and Physiological Aspects*, Springer, Berlin, Heidelberg. 209–229.

Benoit, Joshua B, Philip K Morton, Susan E Cambron, Kevin R Patrick, and Brandon J Schemerhorn. 2010. "Aestivation and Diapause Syndromes Reduce the Water Balance Requirements for Pupae of the Hessian Fly, *Mayetiola destructor*." *Entomologia Experimentalis et Applicata* 136 (1): 89–96.

Boullis, Antoine, Frederic Francis, and François J Verheggen. 2015. "Climate Change and Tritrophic Interactions: Will Modifications to Greenhouse Gas Emissions Increase the Vulnerability of Herbivorous Insects to Natural Enemies?" *Environmental Entomology* 44 (2): 277–86.

Buckley, Lauren B. 2013. "Get Real: Putting Models of Climate Change and Species Interactions in Practice." *Annals of the New York Academy of Sciences* 1297 (1): 126–38.

Burraco, Pablo, Germán Orizaola, Pat Monaghan, and Neil B Metcalfe. 2020. "Climate Change and Ageing in Ectotherms." *Global Change Biology* 26 (10): 5371–81.

Butler, Casey D, and Robert J O'Neil. 2007. "Life History Characteristics of *Orius insidiosus* (Say) Fed Diets of Soybean Aphid, *Aphis glycines* Matsumura and Soybean Thrips, *Neohydatothrips variabilis* (Beach)." *Biological Control* 40 (3): 339–46.

Calatayud, P A, M A Polania, C D Seligmann, and A C Bellotti. 2002. "Influence of Water-Stressed Cassava on *Phenacoccus herreni* and Three Associated Parasitoids." *Entomologia Experimentalis et Applicata* 102 (2): 163–75. https://doi.org/10.1046/j.1570-7458.2002.00936.x.

Capdevila-Arguelles, L, and B Zilletti. 2008. "A Perspective on Climate Change and Invasive Alien Species." *Council of Europe. T-PVS/Inf* 5.

Chapman, Jason W, Raymond H G Klaassen, V Alistair Drake, Sabrina Fossette, Graeme C Hays, Julian D Metcalfe, Andrew M Reynolds, Don R Reynolds, and Thomas Alerstam. 2011. "Animal Orientation Strategies for Movement in Flows." *Current Biology* 21 (20): R861–70.

Chen, Fajun, Feng Ge, and Megha N Parajulee. 2005. "Impact of Elevated $CO_2$ on Tri-Trophic Interaction of *Gossypium hirsutum*, *Aphis gossypii*, and *Leis axyridis*." *Environmental Entomology* 34 (1): 37–46.

Chiriboga, Christian Alejandro. 2009. "Effects of Imidacloprid (Merit® 2F) on Physiology of Woody Plants and Performance of Two-spotted Spider Mite, Fall Webworm, and Imported Willow Leaf Beetle." *Entomology*. Graduate student poster competition, Ohio Agricultural Research and Development Center, The Ohio State University.

Choudhary, Jaipal S, Chandra S Prabhakar, Moanaro, Bikash Das, and Shivendra Kumar. 2013. "Litchi Stink Bug (*Tessaratoma javanica*) Outbreak in Jharkhand, India, on Litchi." *Phytoparasitica* 41 (1): 73–77. https://doi.org/10.1007/s12600-012-0265-x.

Clough, Yann, Andreas Kruess, and Teja Tscharntke. 2007. "Local and Landscape Factors in Differently Managed Arable Fields Affect the Insect Herbivore Community of a Non-Crop Plant Species." *Journal of Applied Ecology* 44 (1): 22–28. https://doi.org/10.1111/j.1365-2664.2006.01239.x.

Coll, Moshe, and Lesley Hughes. 2008. "Effects of Elevated $CO_2$ on an Insect Omnivore: A Test for Nutritional Effects Mediated by Host Plants and Prey." *Agriculture, Ecosystems & Environment* 123 (4): 271–79.

Cornelissen, Tatiana. 2011. "Climate Change and Its Effects on Terrestrial Insects and Herbivory Patterns." *Neotropical Entomology* 40: 155–63.

Costes, E, P E Lauri, S Simon, and B Andrieu. 2013. "Plant Architecture, Its Diversity and Manipulation in Agronomic Conditions, in Relation with Pest and Pathogen Attacks." *European Journal of Plant Pathology* 135 (3): 455–70. https://doi.org/10.1007/s10658-012-0158-3.

Czepak, Cecília, Karina Cordeiro Albernaz, Lúcia Madalena Vivan, Humberto Oliveira Guimarães, and Tiago Carvalhais. 2013. "First Reported Occurrence of *Helicoverpa armigera* (Hübner) (Lepidoptera: Noctuidae) in Brazil." *Pesquisa Agropecuária Tropical* 43: 110–13.

Danks, H V. 2007. "The Elements of Seasonal Adaptations in Insects." *Canadian Entomologist* 139 (1): 1–44.

Delucia, Evan H, Clare L Casteel, Paul D Nabity, and Bridget F O'neill. 2008. "Insects Take a Bigger Bite Out of Plants in a Warmer, Higher Carbon Dioxide World." 105 (6): 1781–1782. https://doi.org/10.1073/pnas.0712056105.

Denlinger, David L, and Peter A Armbruster. 2014. "Mosquito Diapause." *Annual Review of Entomology* 59: 73–93.

Deutsch, Curtis A., Joshua J. Tewksbury, Raymond B. Huey, Kimberly S. Sheldon, Cameron K. Ghalambor, David C. Haak, and Paul R. Martin. 2008. "Impacts of Climate Warming on Terrestrial Ectotherms across Latitude." *Proceedings of the National Academy of Sciences of the United States of America* 105 (18): 6668–72. https://doi.org/10.1073/pnas.0709472105.

Deutsch, Curtis A., Joshua J. Tewksbury, Michelle Tigchelaar, David S. Battisti, Scott C. Merrill, Raymond B. Huey, and Rosamond L. Naylor. 2018. "Increase in Crop Losses to Insect Pests in a Warming Climate." *Science* 361 (6405): 916–19. https://doi.org/10.1126/science.aat3466.

Dhawan, A K, Kamaldeep Singh, Anand Aneja, and Sarika Saini. 2009. "Distribution of Mealy Bug, *Phenacoccus solenopsis* Tinsley in Cotton with Relation to Weather Factors in South-Western Districts of Punjab." *Journal of Entomological Research* 33 (1): 59–63.

Diffenbaugh, Noah S, Christian H Krupke, Michael A White, and Corinne E Alexander. 2008. "Global Warming Presents New Challenges for Maize Pest Management." *Environmental Research Letters* 3 (4): 044007.

Dukes, Jeffrey S, Jennifer Pontius, David Orwig, Jeffrey R Garnas, Vikki L Rodgers, Nicholas Brazee, Barry Cooke, Kathleen A Theoharides, Erik E Stange, Robin Harrington, Joan Ehrenfeld, Jessica Gurevitch, Manuel Lerdau, Kristina Stinson, Robert Wick, and Matthew Ayres. 2009. "Responses of Insect Pests, Pathogens, and Invasive Plant Species to Climate Change in the Forests of Northeastern North America: What Can We Predict?" *Canadian Journal of Forest Research* 39 (2): 231–48. https://doi.org/10.1139/X08-171.

Dyer, Lee A, Lora A Richards, Stephanie A Short, and Craig D Dodson. 2013. "Effects of $CO_2$ and Temperature on Tritrophic Interactions." *PLOS ONE* 8 (4): e62528.

Fand, Babasaheb B, Jaipal S Choudhary, Mahesh Kumar, and Santanu K Bal. 2014. "Phenology Modelling and GIS Applications in Pest Management: A Tool for Studying and Understanding Insect-Pest Dynamics in the Context of Global Climate Change." In *Approaches to Plant Stress and Their Management*, Springer, New Delhi. 107–24.

Fand, Babasaheb B, Ankush L Kamble, and Mahesh Kumar. 2012. "Will Climate Change Pose Serious Threat to Crop Pest Management: A Critical Review." *International Journal of Scientific and Research Publications* 2 (11): 1–14.

Fielding, Dennis J. 2008. "Diapause Traits of *Melanoplus sanguinipes* and *Melanoplus borealis* (Orthoptera: Acrididae)." *Annals of the Entomological Society of America* 101 (2): 439–48.

Fitter, A H, and R S R Fitter. 2002. "Rapid Changes in Flowering Time in British Plants." *Science* 296 (5573): 1689–91.

Ford, James D, Barry Smit, Johanna Wandel, and John MacDonald. 2006. "Vulnerability to Climate Change in Igloolik, Nunavut: What We Can Learn from the Past and Present." *Polar Record* 42 (2): 127–38.

Ford, Kevin A, John E Casida, Divya Chandran, Alexander G Gulevich, Rachel A Okrent, Kathleen A Durkin, Richmond Sarpong, Eric M Bunnelle, and Mary C Wildermuth. 2010. "Neonicotinoid Insecticides Induce Salicylate-Associated Plant Defense Responses." *Proceedings of the National Academy of Sciences* 107 (41): 17527–32.

Fox, Rebecca J, Jennifer M Donelson, Celia Schunter, Timothy Ravasi, and Juan D Gaitán-Espitia. 2019. "Beyond Buying Time: The Role of Plasticity in Phenotypic Adaptation to Rapid Environmental Change." In *Philosophical Transactions of the Royal Society B*. The Royal Society.

Fuhrer, Jürg. 2003. "Agroecosystem Responses to Combinations of Elevated $CO_2$, Ozone, and Global Climate Change." *Agriculture, Ecosystems & Environment* 97 (1–3): 1–20.

Furlong, Michael J, and Myron P Zalucki. 2017. "Climate Change and Biological Control: The Consequences of Increasing Temperatures on Host-Parasitoid Interactions." *Current Opinion in Insect Science* 20: 39–44.

Gandhi, Kamal J K, and Daniel A Herms. 2010. "Direct and Indirect Effects of Alien Insect Herbivores on Ecological Processes and Interactions in Forests of Eastern North America." *Biological Invasions* 12: 389–405.

Getahun, Daniel. 2020. "Predictions of Climate Change on Agricultural Insect Pests Vis-à-Vis Food Crop Productivity: A Critical Review." *Ethiopian Journal of Science and Sustainable Development* 7 (1): 18–26.

González-Tokman, Daniel, Alex Córdoba-Aguilar, Wesley Dáttilo, Andrés Lira-Noriega, Rosa A Sánchez-Guillén, and Fabricio Villalobos. 2020. "Insect Responses to Heat: Physiological Mechanisms, Evolution and Ecological Implications in a Warming World." *Biological Reviews* 95 (3): 802–21.

Gopalakrishnan, Tharani, Md Kamrul Hasan, A T M Sanaul Haque, Sadeeka Layomi Jayasinghe, and Lalit Kumar. 2019. "Sustainability of Coastal Agriculture under Climate Change." *Sustainability* 11 (24): 7200.

Gray, D A, E Gabel, T Blankers, and R M Hennig. 2016. "Multivariate Female Preference Tests Reveal Latent Perceptual Biases." *Proceedings of the Royal Society B: Biological Sciences* 283 (1842): 20161972.

Gregory, P J, S N Johnson, A C Newton, and J S I Ingram. 2009. "Integrating Pests and Pathogens into the Climate Change/Food Security Debate." *Journal of Experimental Botany* 60 (10): 2827–38. https://doi.org/10.1093/jxb/erp080.

Grimbacher, Peter S, and Nigel E Stork. 2009. "Seasonality of a Diverse Beetle Assemblage Inhabiting Lowland Tropical Rain Forest in Australia." *Biotropica* 41 (3): 328–37.

Gurr, Geoff M, Stephen D Wratten, and John Michael Luna. 2003. "Multi-Function Agricultural Biodiversity: Pest Management and Other Benefits." *Basic and Applied Ecology* 4 (2): 107–16.

Hahn, Daniel A, and David L Denlinger. 2007. "Meeting the Energetic Demands of Insect Diapause: Nutrient Storage and Utilization." *Journal of Insect Physiology* 53 (8): 760–73.

Hamilton, Jason G, Orla Dermody, Mihai Aldea, Arthur R Zangerl, Alistair Rogers, May R Berenbaum, and Evan H Delucia. 2005. "Anthropogenic Changes in Tropospheric Composition Increase Susceptibility of Soybean to Insect Herbivory." *Environmental Entomology* 34 (2): 479–85.

Hance, Thierry, Joan van Baaren, Philippe Vernon, and Guy Boivin. 2007. "Impact of Extreme Temperatures on Parasitoids in a Climate Change Perspective." *Annual Review of Entomology* 52: 107–26.

Hartmann, Henrik, Catarina F Moura, William R L Anderegg, Nadine K Ruehr, Yann Salmon, Craig D Allen, Stefan K Arndt, David D Breshears, Hendrik Davi, and David Galbraith. 2018. "Research Frontiers for Improving Our Understanding of Drought-induced Tree and Forest Mortality." *New Phytologist* 218 (1): 15–28.

Harvey, Jeffrey A, T Martijn Bezemer, Jelmer A Elzinga, and Michael R Strand. 2004. "Development of the Solitary Endoparasitoid Microplitis Demolitor: Host Quality Does Not Increase with Host Age and Size." *Ecological Entomology* 29 (1): 35–43.

Heeb, Luca, Emma Jenner, and Matthew J W Cock. 2019. "Climate-Smart Pest Management: Building Resilience of Farms and Landscapes to Changing Pest Threats." *Journal of Pest Science* 92 (3): 951–69.

Hegland, Stein Joar, Anders Nielsen, Amparo Lázaro, Anne-Line Bjerknes, and Ørjan Totland. 2009. "How Does Climate Warming Affect Plant-pollinator Interactions?" *Ecology Letters* 12 (2): 184–95.

Hellmann, Jessica J, James E Byers, Britta G Bierwagen, and Jeffrey S Dukes. 2008. "Five Potential Consequences of Climate Change for Invasive Species." *Conservation Biology* 22 (3): 534–43.

Hill, Matthew P, C Bertelsmeier, Susana Clusella-Trullas, J Garnas, Mark P Robertson, and J S Terblanche. 2016. "Predicted Decrease in Global Climate Suitability Masks Regional Complexity of Invasive Fruit Fly Species Response to Climate Change." *Biological Invasions* 18: 1105–19.

Hill, Matthew P, Steven L Chown, and Ary A Hoffmann. 2013. "A Predicted Niche Shift Corresponds with Increased Thermal Resistance in an Invasive Mite, *Halotydeus destructor*." *Global Ecology and Biogeography* 22 (8): 942–51. https://doi.org/10.1111/geb.12059.

Hill, Matthew P, Ary A Hoffmann, Stuart A McColl, and Paul A Umina. 2012. "Distribution of Cryptic Blue Oat Mite Species in Australia: Current and Future Climate Conditions." *Agricultural and Forest Entomology* 14 (2): 127–37. https://doi.org/10.1111/j.1461-9563.2011.00544.x.

Hlásny, T, and M Turčáni. 2009. "Insect Pests as Climate Change Driven Disturbances in Forest Ecosystems." *Bioclimatology and Natural Hazards*, Springer, Dordrecht. 165–77.

Mooney, Harold A, and Richard J Hobbs. 2000. *Invasive Species in a Changing World.* Island Press.

Hoffmann, Ary A., Steven L. Chown, and Susana Clusella-Trullas. 2013. "Upper Thermal Limits in Terrestrial Ectotherms: How Constrained Are They?" *Functional Ecology* 27 (4): 934–49. https://doi.org/10.1111/j.1365-2435.2012.02036.x.

Hoffmann, Ary A, Jesper G Sørensen, and Volker Loeschcke. 2003. "Adaptation of Drosophila to Temperature Extremes: Bringing Together Quantitative and Molecular Approaches." *Journal of Thermal Biology* 28 (3): 175–216. https://doi.org/10.1016/S0306-4565(02)00057-8.

Holopainen, Jarmo K, Sari J Himanen, and Guy M Poppy. 2013. "Climate Change and Its Effects on the Chemical Ecology of Insect Parasitoids." *Chemical Ecology of Insect Parasitoids*, John Wiley and Sons. 168–190.

Hooper, D U, F S Chapin, J J Ewel, A Hector, P Inchausti, S Lavorel, J H Lawton, D M Lodge, M Loreau, S Naeem, B Schmid, H Setälä, A J Symstad, J Vandermeer, and D A Wardle. 2005. "Effects of Biodiversity on Ecosystem Functioning: A Consensus of Current Knowledge." *Ecological Monographs* 75 (1): 3–35. https://doi.org/10.1890/04-0922.

Horii, A, P McCue, and K Shetty. 2007. "Enhancement of Seed Vigour Following Insecticide and Phenolic Elicitor Treatment." *Bioresource Technology* 98 (3): 623–32.

Howden, S Mark, Jean-François Soussana, Francesco N Tubiello, Netra Chhetri, Michael Dunlop, and Holger Meinke. 2007. "Adapting Agriculture to Climate Change." *Proceedings of the National Academy of Sciences* 104 (50): 19691–96. https://doi.org/10.1073/pnas.0701890104.

Hullé, Maurice, Armelle Cœur d'Acier, Stéphanie Bankhead-Dronnet, and Richard Harrington. 2010. "Aphids in the Face of Global Changes." *Comptes Rendus Biologies* 333 (6–7): 497–503. https://doi.org/10.1016/j.crvi.2010.03.005.

Hunter, Mark D. 2001. "Effects of Elevated Atmospheric Carbon Dioxide on Insect-Plant Interactions." *Agricultural and Forest Entomology* 3 (3): 153–59.

Hvam, Aino, and Søren Toft. 2005. "Effects of Prey Quality on the Life History of a Harvestman." *The Journal of Arachnology* 33 (2): 582–90.

Indian Council of Agricultural Research. 2008. *ICAR Annual Report.* p. 120.

International Rice Research Institute (IRRI). 2009. *IRRI Annual Report.* pp. 1–37.

Jactel, Hervé, Julia Koricheva, and Bastien Castagneyrol. 2019. "Responses of Forest Insect Pests to Climate Change: Not so Simple." *Current Opinion in Insect Science* 35: 103–8.

Jamieson, Mary A, Amy M Trowbridge, Kenneth F Raffa, and Richard L Lindroth. 2012. "Consequences of Climate Warming and Altered Precipitation Patterns for Plant-Insect and Multitrophic Interactions." *Plant Physiology* 160 (4): 1719–27.

Jarvis, Claire H, and Richard H A Baker. 2001a. "Risk Assessment for Nonindigenous Pests: 1. Mapping the Outputs of Phenology Models to Assess the Likelihood of Establishment." *Diversity and Distributions* 7 (5): 223–35.

Jarvis, Claire H, and Richard H A Baker. 2001b. "Risk Assessment for Nonindigenous Pests: 2. Accounting for Interyear Climate Variability." *Diversity and Distributions* 7 (5): 237–48.

Jat, B L, P Pagaria, G Mali, and T Khan. 2021. "Impact of Major Environmental Conditions on Pest Population Dynamics and Plant Resistance Mechanisms in Legumes." *Indian Journal of Plant Protection* 49 (4): 215–26.

Johns, Caitlin V, and Lesley Hughes. 2002. "Interactive Effects of Elevated $CO_2$ and Temperature on the Leaf-miner *Dialectica scalariella* Zeller (Lepidoptera: Gracillariidae) in Paterson's Curse, *Echium plantagineum* (Boraginaceae)." *Global Change Biology* 8 (2): 142–52.

Johnson, Derek M, and Kyle J Haynes. 2023. "Spatiotemporal Dynamics of Forest Insect Populations under Climate Change." *Current Opinion in Insect Science*, 101020.

Joshi, Sunil, and C A Viraktamath. 2004. "The Sugarcane Woolly Aphid, *Ceratovacuna lanigera* Zehntner (Hemiptera: Aphididae): Its Biology, Pest Status and Control." *Current Science*, 307–16.

Juroszek, Peter, and A Von Tiedemann. 2013. "Plant Pathogens, Insect Pests and Weeds in a Changing Global Climate: A Review of Approaches, Challenges, Research Gaps, Key Studies and Concepts." *The Journal of Agricultural Science* 151 (2): 163–88.

Kannan, Ragupathy, and Douglas A James. 2009. "Effects of Climate Change on Global Biodiversity: A Review of Key Literature." *Tropical Ecology* 50 (1): 31.

Karley, A J, W E Parker, J W Pitchford, and A E Douglas. 2004. "The Mid-season Crash in Aphid Populations: Why and How Does It Occur?" *Ecological Entomology* 29 (4): 383–88.

Kaur, Balwinder, Jagmohan Singh, Karansher S Sandhu, Shivreet Kaur, Gagandeep Kaur, Hinal Kharva, Sajjan Grover, Heena Puri, Sandeep Kaur, and Ruchika Kashyap. 2023. "Potential Effects of Future Climate Changes in Pest Scenario." In *Enhancing Resilience of Dryland Agriculture Under Changing Climate: Interdisciplinary and Convergence Approaches*, 459–73. Springer.

Kearney, Michael R, Brendan A Wintle, and Warren P Porter. 2010. "Correlative and Mechanistic Models of Species Distribution Provide Congruent Forecasts under Climate Change." *Conservation Letters* 3 (3): 203–13.

Khan, Shakeel A, Sanjeev Kumar, M Z Hussain, and N Kalra. 2009. "Climate Change, Climate Variability and Indian Agriculture: Impacts Vulnerability and Adaptation Strategies." In *Climate Change and Crops*, 19–38. Springer.

Kiritani, Keizi. 2006. "Predicting Impacts of Global Warming on Population Dynamics and Distribution of Arthropods in Japan." *Population Ecology* 48 (1): 5–12.

Kiyofuji, H, Y Aoki, J Kinoshita, S Okamoto, M Masujima, T Matsumoto, K Fujioka, R Ogata, T Nakao, and N Sugimoto. 2019. "Northward Migration Dynamics of Skipjack Tuna (*Katsuwonus pelamis*) Associated with the Lower Thermal Limit in the Western Pacific Ocean." *Progress in Oceanography* 175: 55–67.

Klaiber, J, Silvia Dorn, and Adriana J Najar-Rodriguez. 2013. "Acclimation to Elevated $CO_2$ Increases Constitutive Glucosinolate Levels of Brassica Plants and Affects the Performance of Specialized Herbivores from Contrasting Feeding Guilds." *Journal of Chemical Ecology* 39: 653–65.

Kriticos, Darren J, Joel R Brown, Gunter F Maywald, Ian D Radford, D Mike Nicholas, Robert W Sutherst, and Steve W Adkins. 2003. "SPAnDX: A Process-Based Population Dynamics Model to Explore Management and Climate Change Impacts on an Invasive Alien Plant, *Acacia nilotica*." *Ecological Modelling* 163 (3): 187–208.

Kroschel, Jürgen, M Sporleder, Henri E Z Tonnang, H Juarez, Pablo Carhuapoma, J C Gonzales, and R Simon. 2013. "Predicting Climate-Change-Caused Changes in Global Temperature on Potato Tuber Moth *Phthorimaea operculella* (Zeller) Distribution and Abundance Using Phenology Modeling and GIS Mapping." *Agricultural and Forest Meteorology* 170: 228–41.

Kurz, Werner A, C C Dymond, Graham Stinson, G J Rampley, E T Neilson, A L Carroll, Tim Ebata, and Les Safranyik. 2008. "Mountain Pine Beetle and Forest Carbon Feedback to Climate Change." *Nature* 452 (7190): 987–90.

Lamichhane, Jay Ram, Marco Barzman, Kees Booij, Piet Boonekamp, Nicolas Desneux, Laurent Huber, Per Kudsk, Stephen R H Langrell, Alain Ratnadass, and Pierre Ricci. 2015. "Robust Cropping Systems to Tackle Pests under Climate Change. A Review." *Agronomy for Sustainable Development* 35: 443–59.

Leakey, Andrew D B, John N Ferguson, Charles P Pignon, Alex Wu, Zhenong Jin, Graeme L Hammer, and David B Lobell. 2019. "Water Use Efficiency as a Constraint and Target for Improving the Resilience and Productivity of C3 and C4 Crops." *Annual Review of Plant Biology* 70: 781–808.

Ma, Chun-Sen, Wei Zhang, Yu Peng, Fei Zhao, Xiang-Qian Chang, Kun Xing, Liang Zhu, Gang Ma, He-Ping Yang, and Volker H W Rudolf. 2021. "Climate Warming Promotes Pesticide Resistance through Expanding Overwintering Range of a Global Pest." *Nature Communications* 12 (1): 5351.

MacDonald, G M. 2010. "Global Warming and the Arctic: A New World beyond the Reach of the Grinnellian Niche?" *Journal of Experimental Biology* 213 (6): 855–61.

Malhi, Gurdeep Singh, Manpreet Kaur, and Prashant Kaushik. 2021. "Impact of Climate Change on Agriculture and Its Mitigation Strategies: A Review." *Sustainability* 13 (3): 1318.

Marak, Hamida B, Arjen Biere, and Jos M M Van Damme. 2002. "Systemic, Genotype-Specific Induction of Two Herbivore-Deterrent Iridoid Glycosides in *Plantago lanceolata* L. in Response to Fungal Infection by *Diaporthe adunca* (Rob.) Niessel." *Journal of Chemical Ecology* 28: 2429–48.

Marshall, Katie E, and Brent J Sinclair. 2015. "The Relative Importance of Number, Duration and Intensity of Cold Stress Events in Determining Survival and Energetics of an Overwintering Insect." *Functional Ecology* 29 (3): 357–66.

Matzrafi, Maor. 2019. "Climate Change Exacerbates Pest Damage through Reduced Pesticide Efficacy." *Pest Management Science* 75 (1): 9–13.

McClanahan, Tim R. 2020. "Coral Community Life Histories and Population Dynamics Driven by Seascape Bathymetry and Temperature Variability." In *Advances in Marine Biology*, 87, 291–330. Elsevier.

Memmott, Jane, Paul G Craze, Nickolas M Waser, and Mary V Price. 2007. "Global Warming and the Disruption of Plant-Pollinator Interactions." *Ecology Letters* 10 (8): 710–17.

Menendes, Rosa, Adela Gonzales-Megias, Owen T Lewis, Mark R Shaw, and Chris D Thomas. 2008. "Escape from Natural Enemies during Climate-Driven Range Expansion: A Case Study." *Ecological Entomology* 33 (3): 413–21. https://doi.org/10.1111/j.1365-2311.2008.00985.x.

Moore, Beverly, and Gillian Allard. 2016. *Climate Change Impacts on Forest Health*. FAO.

Morshed, Md Niaz, Md Abdullah Al Mamun, Sheikh Arafat Islam Nihad, Md Mahfuzur Rahman, Naznin Sultana, and Md Mezanur Rahman. 2023. "Effect of Weather Variables on Seasonal Abundance of Rice Insects in Southeast Coastal Region of Bangladesh." *Journal of Agriculture and Food Research* 11: 100513.

Nagrare, V S, S Kranthi, V K Biradar, N N Zade, V Sangode, G Kakde, R M Shukla, D Shivare, B M Khadi, and K R Kranthi. 2009. "Widespread Infestation of the Exotic Mealybug Species, *Phenacoccus solenopsis* (Tinsley) (Hemiptera: Pseudococcidae), on Cotton in India." *Bulletin of Entomological Research* 99 (5): 537–41. https://doi.org/10.1017/S0007485308006573.

Newton, A C, G S Begg, and J S Swanston. 2009. "Deployment of Diversity for Enhanced Crop Function." *Annals of Applied Biology* 154 (3): 309–22.

Olson, Dawn M, and Felix L Wäckers. 2007. "Management of Field Margins to Maximize Multiple Ecological Services." *Journal of Applied Ecology* 44 (1): 13–21.

Osier, Tod L, and Richard L Lindroth. 2001. "Effects of Genotype, Nutrient Availability, and Defoliation on Aspen Phytochemistry and Insect Performance." *Journal of Chemical Ecology* 27: 1289–1313.

Pareek, Abhishek, B M Meena, Sitaram Sharma, M L Tetarwal, R K Kalyan, and B L Meena. 2017. "Impact of Climate Change on Insect Pests and Their Management Strategies." *Climate Change and Sustainable Agriculture*, New India Publishing. 253–86.

Parmesan, Camille, and Gary Yohe. 2003. "A Globally Coherent Fingerprint of Climate Change Impacts across Natural Systems." *Nature* 421 (6918): 37–42.

Pathak, Himanshu, Pramod K Aggarwal, and S D Singh. 2012. *Climate Change Impact, Adaptation and Mitigation in Agriculture: Methodology for Assessment and Applications*. Indian Agricultural Research Institute.

Peacock, Lora, and Sue Worner. 2006. "Using Analogous Climates and Global Insect Distribution Data to Identify Potential Sources of New Invasive Insect Pests in New Zealand." *New Zealand Journal of Zoology* 33 (2): 141–45.

Pearson, Richard G, and Terence P Dawson. 2003. "Predicting the Impacts of Climate Change on the Distribution of Species: Are Bioclimate Envelope Models Useful?" *Global Ecology and Biogeography* 12 (5): 361–71.

Pekár, Stano, and Søren Toft. 2015. "Trophic Specialisation in a Predatory Group: The Case of Prey-specialised Spiders (Araneae)." *Biological Reviews* 90 (3): 744–61.

Peng, Wanxi, Nyuk Ling Ma, Dangquan Zhang, Quan Zhou, Xiaochen Yue, Shing Ching Khoo, Han Yang, Ruirui Guan, Huiling Chen, and Xiaofan Zhang. 2020. "A Review of Historical and Recent Locust Outbreaks: Links to Global Warming, Food Security and Mitigation Strategies." *Environmental Research* 191: 110046.

Petzoldt, C, and A Seaman. 2010. *Climate Change Effects on Insects and Pathogens*. New York State Agricultural Extension Station.

Potter, Kristen A, Goggy Davidowitz, and H Arthur Woods. 2011. "Cross-stage Consequences of Egg Temperature in the Insect *Manduca sexta*." *Functional Ecology* 25 (3): 548–56.

Prasannakumar, N R, Subhash Chander, and Madan Pal. 2012. "Assessment of Impact of Climate Change with Reference to Elevated $CO_2$ on Rice Brown Planthopper, *Nilaparvata lugens* (Stal.) and Crop Yield." *Current Science*, 1201–5.

Pureswaran, Deepa S., Alain Roques, and Andrea Battisti. 2018. "Forest Insects and Climate Change." *Current Forestry Reports* 4 (2): 35–50. https://doi.org/10.1007/s40725-018-0075-6.

Rao, Jaydeepsingh, and Berliner Jeyaveeran. 2014. "Climate Change: Impact on Crop Pests." www.researchgate.net/publication/275347570.

Rao, M Sreenivasa, M A Khan, K Srinivas, M Vanaja, GGSN Rao, and Y S Ramakrishna. 2006. "Effects of Elevated Carbon Dioxide and Temperature on Insect-Plant Interactions—A Review." *Agricultural Reviews* 27 (3): 200–207.

Rashmi, M A, and A K Chakravarthy. 2020. "Role of Climate Change on Insect Pests of Commercial Flowers." In *Advances in Pest Management in Commercial Flowers*, 137–50. Apple Academic Press.

Rauff, Kazeem O, and Rasaq Bello. 2015. "A Review of Crop Growth Simulation Models as Tools for Agricultural Meteorology." *Agricultural Sciences* 6 (9): 1098.

Reddy, P Parvatha. 2013. "Impact of Climate Change on Insect Pests, Pathogens and Nematodes." *Pest Management in Horticultural Ecosystems* 19 (2): 225–33.

Renault, D, C Salin, G Vannier, and P Vernon. 2002. "Survival at Low Temperatures in Insects: What Is the Ecological Significance of the Supercooling Point?" *CryoLetters* 23 (4): 217–28.

Robich, Sarah R, and Jack R Layne Jr. 2013. "The Effect of Prior Chilling Exposure on Supercooling Ability and Developmental Success in Third-Instars of *Eurosta solidaginis* (Diptera: Tephritidae)." *Annals of the Entomological Society of America* 106 (6): 840–45.

Robinet, Christelle, and Alain Roques. 2010. "Direct Impacts of Recent Climate Warming on Insect Populations." *Integrative Zoology* 5 (2): 132–42.

Robinson, Robert A, Humphrey Q P Crick, Jennifer A Learmonth, Ilya M D Maclean, Chris D Thomas, Franz Bairlein, Mads C Forchhammer, Charles M Francis, Jennifer A Gill, and Brendan J Godley. 2009. "Travelling through a Warming World: Climate Change and Migratory Species." *Endangered Species Research* 7 (2): 87–99.

Schebeck, Martin, E Matthew Hansen, Axel Schopf, Gregory J Ragland, Christian Stauffer, and Barbara J Bentz. 2017. "Diapause and Overwintering of Two Spruce Bark Beetle Species." *Physiological Entomology* 42 (3): 200–210.

Schmidt, Martin H., Ulrich Thewes, Carsten Thies, and Teja Tscharntke. 2004. "Aphid Suppression by Natural Enemies in Mulched Cereals." *Entomologia Experimentalis et Applicata* 113 (2): 87–93. https://doi.org/10.1111/j.0013-8703.2004.00205.x.

Schwartzberg, Ezra G, Mary A Jamieson, Kenneth F Raffa, Peter B Reich, Rebecca A Montgomery, and Richard L Lindroth. 2014. "Simulated Climate Warming Alters Phenological Synchrony between an Outbreak Insect Herbivore and Host Trees." *Oecologia* 175: 1041–49.

Šefrová, H. 2003. "Invasions of Lithocolletinae Species in Europe-Causes, Kinds, Limits and Ecological Impact (Lepidoptera, Gracillariidae)." *Ekologia (Bratislava)* 22 (2): 132–42.

Şekercioğlu, Çağan H, Richard B Primack, and Janice Wormworth. 2012. "The Effects of Climate Change on Tropical Birds." *Biological Conservation* 148 (1): 1–18.

Selonen, Vesa, Samuli Helle, Toni Laaksonen, Markus P Ahola, Esa Lehikoinen, and Tapio Eeva. 2021. "Identifying the Paths of Climate Effects on Population Dynamics: Dynamic and Multilevel Structural Equation Model around the Annual Cycle." *Oecologia* 195: 525–38.

Senaratne, K A D Wilmot, William A Palmer, and Robert W Sutherst. 2006. "Use of CLIMEX Modelling to Identify Prospective Areas for Exploration to Find New Biological Control Agents for Prickly Acacia." *Australian Journal of Entomology* 45 (4): 298–302.

Shahzad, Aqeel, Sana Ullah, Afzal Ahmed Dar, Muhammad Fahad Sardar, Tariq Mehmood, Muhammad Aammar Tufail, Awais Shakoor, and Muhammad Haris. 2021. "Nexus on Climate Change: Agriculture and Possible Solution to Cope Future Climate Change Stresses." *Environmental Science and Pollution Research* 28 (12): 14211–32. https://doi.org/10.1007/s11356-021-12649-8.

Sharma, H C. 2010. "Effect of Climate Change on IPM in Grain Legumes." In *5th International Food Legumes Research Conference (IFLRC V), and the 7th European Conference on Grain Legumes (AEP VII)*, 26–30.

Sharma, Hari C. 2014. "Climate Change Effects on Insects: Implications for Crop Protection and Food Security." *Journal of Crop Improvement* 28 (2): 229–59.

Sharma, Hari C. 2016. "Climate Change Vis-a-Vis Pest Management." *Conference on National Priorities in Plant Health Management February* 4–5, 2016, Tirupati. 1–9.

Sharma, Tarun, and Ayushi Singla. 2021. "Potential Impacts of Climate Change on Insect Behavior." *Journal of Life Sciences* 11 (3): 29–35.

Shrestha, Saroj. 2019. "Effects of Climate Change in Agricultural Insect Pest." *Acta Scientific Agriculture* 3 (12): 74–80.

Singh, T V K. 2022. "09. Climate Change and Its Impact on Agricultural Pests." *Sustainable Agricultural Innovations for Resilient Agri-Food Systems*, Proceedings of the Indian Ecological Society International Conference 2022, SKUAST, Jammu, India. 26–31.

Skendžić, Sandra, Monika Zovko, Ivana Pajač Živković, Vinko Lešić, and Darija Lemić. 2021. "The Impact of Climate Change on Agricultural Insect Pests." *Insects* 12 (5): 440. https://doi.org/10.3390/insects12050440.

Snell-Rood, Emilie C, Megan E Kobiela, Kristin L Sikkink, and Alexander M Shephard. 2018. "Mechanisms of Plastic Rescue in Novel Environments." *Annual Review of Ecology, Evolution, and Systematics* 49: 331–54.

Soley, Nathan M, and Sedonia D Sipes. 2021. "Reproductive Biology and Pollinators of the Invasive Shrub Autumn Olive (*Elaeagnus umbellata* Thunberg)." *Plant Species Biology* 36 (2): 170–86.

Sparks, T H, D B Roy, and R L H Dennis. 2005. "The Influence of Temperature on Migration of Lepidoptera into Britain." *Global Change Biology* 11 (3): 507–14.

Sporleder, M, H E Z Tonnang, H Juarez, P Carhuapoma, J C Gonzales, R Simon, and J Kroschel. 2012. *ILCYM—Insect Life Cycle Modeling (Version 3.0): A Software Package for Developing Temperature-Based Insect Phenology Models with Applications to Regional and Global Analysis of Insect Population and Mapping* (User Manual). International Potato Centre, 124.

Srikanth, J, B Singaravelu, and N K Kurup. 2012. "Natural Control of Woolly Aphid by *Encarsia flavoscutellum* Prevents Yield and Quality Loss in Sugarcane." *Journal of Sugarcane Research* 2 (1): 64–68.

Srygley, Robert B, Robert Dudley, Edgar J Hernandez, Franz Kainz, Andre J Riveros, and Charlie P Ellington. 2023. "Quantifying the Aerodynamic Power Required for Flight and Testing for Adaptive Wind Drift in Passion-Vine Butterflies *Heliconius sara* (Lepidoptera: Nymphalidae)." *Insects* 14 (2): 112.

Staley, Joanna T, Christopher J Hodgson, Simon R Mortimer, Michael D Morecroft, Gregory J Masters, Valerie K Brown, and Michele E Taylor. 2007. "Effects of Summer Rainfall Manipulations on the Abundance and Vertical Distribution of Herbivorous Soil Macro-Invertebrates." *European Journal of Soil Biology* 43 (3): 189–98. https://doi.org/10.1016/j.ejsobi.2007.02.010.

Stoeckli, Sibylle, Martin Hirschi, Christoph Spirig, Pierluigi Calanca, Mathias W Rotach, and Jörg Samietz. 2012. "Impact of Climate Change on Voltinism and Prospective Diapause Induction of a Global Pest Insect—*Cydia pomonella* (L.)." *PLOS ONE* 7 (4): e35723. https://doi.org/10.1371/journal.pone.0035723.

Sutherst, Robert W. 2003. "Guest Editorial: Prediction of Species Geographical Ranges." *Journal of Biogeography* 30 (6): 805–16.

Takahashi, Masaru, and Nancy Huntly. 2010. "Herbivorous Insects Reduce Growth and Reproduction of Big Sagebrush (*Artemisia tridentata*)." *Arthropod-Plant Interactions* 4: 257–66.

Tamburini, Giovanni, Serena De Simone, Maurizia Sigura, Francesco Boscutti, and Lorenzo Marini. 2016. "Conservation Tillage Mitigates the Negative Effect of Landscape Simplification on Biological Control." *Journal of Applied Ecology* 53 (1): 233–41.

Tanwar, R K, P Jeyakumar, and S Vennila. 2010. *Papaya Mealybug and Its Management Strategies*. National Centre for Integrated Pest Management, New Delhi.

Tay, Wee Tek, Miguel F Soria, Thomas Walsh, Danielle Thomazoni, Pierre Silvie, Gajanan T Behere, Craig Anderson, and Sharon Downes. 2013. "A Brave New World for an Old World Pest: *Helicoverpa armigera* (Lepidoptera: Noctuidae) in Brazil." *PLOS ONE* 8 (11): e80134.

Thackeray, Stephen J, Peter A Henrys, Deborah Hemming, James R Bell, Marc S Botham, Sarah Burthe, Pierre Helaouet, David G Johns, Ian D Jones, and David I Leech. 2016. "Phenological Sensitivity to Climate across Taxa and Trophic Levels." *Nature* 535 (7611): 241–45.

Thomas, Chris D, Alison Cameron, Rhys E Green, Michel Bakkenes, Linda J Beaumont, Yvonne C Collingham, Barend F N Erasmus, Marinez Ferreira De Siqueira, Alan Grainger, and Lee Hannah. 2004. "Extinction Risk from Climate Change." *Nature* 427 (6970): 145–48.

Thomson, Linda J, and Ary A Hoffmann. 2007. "Effects of Ground Cover (Straw and Compost) on the Abundance of Natural Enemies and Soil Macro Invertebrates in Vineyards." *Agricultural and Forest Entomology* 9 (3): 173–79.

Thomson, Linda J, Sarina Macfadyen, and Ary A Hoffmann. 2010. "Predicting the Effects of Climate Change on Natural Enemies of Agricultural Pests." *Biological Control* 52 (3): 296–306.

Timoney, K P. 2003. "The Changing Disturbance Regime of the Boreal Forest of the Canadian Prairie Provinces." *The Forestry Chronicle* 79 (3): 502–16.

Tobin, Patrick C, Sudha Nagarkatti, Greg Loeb, and Michael C Saunders. 2008. "Historical and Projected Interactions between Climate Change and Insect Voltinism in a Multivoltine Species." *Global Change Biology* 14 (5): 951–57.

Trân, J Khai, Tiina Ylioja, Ronald F Billings, Jacques Régnière, and Matthew P Ayres. 2007. "Impact of Minimum Winter Temperatures on the Population Dynamics of *Dendroctonus frontalis*." *Ecological Applications* 17 (3): 882–99.

Trnka, M, F Muška, D Semerádová, M Dubrovský, E Kocmánková, and Z Žalud. 2007. "European Corn Borer Life Stage Model: Regional Estimates of Pest Development and Spatial Distribution under Present and Future Climate." *Ecological Modelling* 207 (2–4): 61–84.

Trumble, John, and Casey Butler. 2009. "Climate Change Will Exacerbate California's Insect Pest Problems." *California Agriculture* 63 (2): 73–78.

Uelmen, Johnny A Jr., Richard L Lindroth, Patrick C Tobin, Peter B Reich, Ezra G Schwartzberg, and Kenneth F Raffa. 2016. "Effects of Winter Temperatures, Spring Degree-Day Accumulation, and Insect Population Source on Phenological Synchrony between Forest Tent Caterpillar and Host Trees." *Forest Ecology and Management* 362: 241–50.

Vanhanen, Henri. 2008. "Invasive Insects in Europe-the Role of Climate Change and Global Trade." *Dissertationes Forestales* 57: 33.

Venette, Robert C, Darren J Kriticos, Roger D Magarey, Frank H Koch, Richard H A Baker, Susan P Worner, Nadilia N Gómez Raboteaux, Daniel W McKenney, Erhard J Dobesberger, and Denys Yemshanov. 2010. "Pest Risk Maps for Invasive Alien Species: A Roadmap for Improvement." *BioScience* 60 (5): 349–62.

Visser, Marcel E, and Leonard J M Holleman. 2001. "Warmer Springs Disrupt the Synchrony of Oak and Winter Moth Phenology." *Proceedings of the Royal Society of London. Series B: Biological Sciences* 268 (1464): 289–94.

Walther, Gian-Reto, Alain Roques, Philip E Hulme, Martin T Sykes, Petr Pyšek, Ingolf Kühn, Martin Zobel, Sven Bacher, Zoltán Botta-Dukát, and Harald Bugmann. 2009. "Alien Species in a Warmer World: Risks and Opportunities." *Trends in Ecology & Evolution* 24 (12): 686–93.

Welch, Kelton D, and James D Harwood. 2014. "Temporal Dynamics of Natural Enemy-Pest Interactions in a Changing Environment." *Biological Control* 75: 18–27.

Wermelinger, Beat. 2004. "Ecology and Management of the Spruce Bark Beetle *Ips typographus*—A Review of Recent Research." *Forest Ecology and Management* 202 (1–3): 67–82.

Willis, Charles G, Brad Ruhfel, Richard B Primack, Abraham J Miller-Rushing, and Charles C Davis. 2008. "Phylogenetic Patterns of Species Loss in Thoreau's Woods Are Driven by Climate Change." *Proceedings of the National Academy of Sciences* 105 (44): 17029–33.

Yamamura, Kohji, Masayuki Yokozawa, Motoki Nishimori, Yasuo Ueda, and Tomoyuki Yokosuka. 2006. "How to Analyze Long-Term Insect Population Dynamics under Climate Change: 50-Year Data of Three Insect Pests in Paddy Fields." *Population Ecology* 48 (1): 31–48. https://doi.org/10.1007/s10144-005-0239-7.

Yaro, Alpha S, Adama I Traoré, Diana L Huestis, Abdoulaye Adamou, Seydou Timbiné, Yaya Kassogué, Moussa Diallo, Adama Dao, Sékou F Traoré, and Tovi Lehmann. 2012. "Dry Season Reproductive Depression of *Anopheles gambiae* in the Sahel." *Journal of Insect Physiology* 58 (8): 1050–59.

Yihdego, Yohannes, Hilmi S. Salem, and Hadi Hamaaziz Muhammed. 2019. "Agricultural Pest Management Policies during Drought: Case Studies in Australia and the State of Palestine." *Natural Hazards Review* 20 (1). https://doi.org/10.1061/(ASCE)NH.1527-6996.0000312.

Zeng, Juan, Yongqiang Liu, Haowen Zhang, Jie Liu, Yuying Jiang, Kris A G Wyckhuys, and Kongming Wu. 2020. "Global Warming Modifies Long-Distance Migration of an Agricultural Insect Pest." *Journal of Pest Science* 93: 569–81.

Zhang, Long, Michel Lecoq, Alexandre Latchininsky, and David Hunter. 2019. "Locust and Grasshopper Management." *Annual Review of Entomology* 64 (1): 15–34. https://doi.org/10.1146/annurev-ento-011118-112500.

# 9 Climate Change and Biological Control in Agricultural Systems

## *Principles and Practices*

*Nafiu Bala Sanda*

## 9.1 INTRODUCTION

Agriculture is the act of cultivating crops and rearing animals, also encompassing fisheries and aquaculture, as sources of food, feed and other products for human use. It is an old occupation, whereby food and raw materials are produced for household use and the excess for commercial and industrial use. A country can be said to be food secure if it can produce what it requires to feed its growing population. One of the major challenges of agricultural systems is climate change which is a long-term shift in temperatures and weather patterns through variations in the solar cycles (Devlet, 2021). Certain factors such as drought, rainfall, relative humidity, soil fertility and pests and diseases are some of the major obstacles to arable crop production, which results in low yields. Thus, the decline in yield generally discourages farmers and may result in change in livelihood to farmers, especially in rural settings (Motes, 2010).

In addition, pests and diseases are factors responsible for the decrease in yield of crops. However, insect population dynamics, geographical distribution and disease severity are reported to be influenced by climate change (Lykogianni et al., 2021). Pesticides are the main control measure for these pests, which are harmful to the environment and the farmers. Other control measures include the use of other organisms to control pests and diseases, termed "biological control". Climate change is known to influence natural control agents/enemies and insect pests by changing pest activity, feeding, insect phenology, natural enemy foraging behaviour among others (Furlong & Zalucki, 2017). This is because climate change is causing extreme weather conditions and shifts in season, in addition to degrading the soil quality for agricultural production. This event makes the crop grow weaker in poor soils and in turn makes the agricultural pests stronger. Change in climate further speeds up the way agricultural pests spread around the world. In addition, increases in relative humidity allow these pests to adapt to new environments and to remain active for longer periods of time (Gerard et al., 2010). Longer periods of warmness could allow some insects to reproduce rapidly, resulting in larger populations feeding on agricultural crops. Therefore, there is a need to educate farmers on how to adapt to these changes in feeding behaviour, population distribution and reproduction that may increase pest damages to crops.

## 9.2 EFFECTS OF CLIMATE CHANGE ON AGRICULTURAL PESTS

The word "pest" is from a Latin word *pestis* which means a "plague". It is defined as an organism which is potentially harmful to plants, livestock or humans. The most serious limiting factors to economically efficient crop production are pests and diseases. Food crops make important contributions to the income of the people residing in cultivable lands of the world. Losses from pests and diseases keep yields low (Chidawanyika et al., 2019). However, accurate and reliable estimates of crop losses caused by pests and diseases are hard to estimate. Some available estimates of average annual losses in several crops grown in different countries as a result of

DOI: 10.1201/9781003382089-9

ravages by pests, diseases and weeds at production and storage stages range from 25% to 43% annually. Crop losses vary from small to huge amounts, especially resulting from sporadic outbreaks. Examples of pests and disease-causing organisms or pathogens include insects, mites, rodents, bats, birds, mollusks, nematodes, weeds, parasitic plants, fungi, bacteria, mycoplasma, viruses and sometimes humans (Crowder & Harwood, 2014). Climate change is a significant and long-term variation in the average surface temperature and weather patterns over long periods that results from human activities mainly due to the burning of fossil fuels, cutting down forests (deforestation), and increasing livestock farming, which produces greenhouse gases (Schmitz & Barton, 2014). Several studies reported how climate change affects arthropod distribution through physiological and phenological response, mainly focusing on various fitness parameters of an organism (Schmitz & Barton, 2014; Schuldt et al., 2017; Dong et al., 2018; Mbande et al., 2019a, 2019b).

Climate change is expected to affect insect pests and disease processes, such as survival rate, population size, geographical distribution of insect pests and diseases (Eigenbrode et al., 2015). Generally, all important immature stages of insect pests, weeds and pathogens are directly affected by temperature, relative humidity, carbon dioxide ($CO_2$) level, light quality or quantity, etc. (Juroszek et al., 2020). Specifically, climate change is the unpredictable or erratic nature of temperature and rainfall that results in crop loss and subsequent loss in agricultural productivity (Murrell & Barton, 2017). Warm weather is relatively favourable for insect population growth; this is because they are cold-blooded animals. It encourages high fecundity and shortened life cycles of insect pests, thereby increasing the number of cycles per season and individuals per unit area due to increases in consumption rates and decreases in time to pupation (Furlong & Zalucki, 2017). It was reported that with an increase of 2°C temperature, the number might rise by an additional five life cycles per season (FAO, 2020). Thus, damages caused by insect pests are expected to worsen as climate change progresses. Williams et al. (2003) reported that elevated temperature was shown to increase the performance of gypsy moth by increasing its survival rate and decreasing development time (IPPC, 2021). The reverse is when the level of temperature and precipitation is too high, it can slow down the growth rate and reproduction of certain insect pests by destroying their eggs and young larvae within the host plant. Therefore, pest and disease proliferation are favoured by these climatic factors to some certain level, after which developmental processes will cease in the situation of extreme hotness or coldness of an environment (Selvaraj et al., 2013). Similarly, changes in precipitation patterns were reported to impact the insect pest's predators, parasites and diseases, resulting in complex dynamics. High humidity also favoured rate of infection and incidence of fungal pathogen of insect by climate change that lengthens periods of high humidity and reduces drier conditions (Furlong et al., 2021; Reiners and Petzoldt, 2000).

Atmospheric $CO_2$ can indirectly affect plant pests by changing host-plant architecture, weakening their defense system, and may affect the frequency and intensity of pest outbreaks on crops (Sun et al., 2011). The increased $CO_2$ as a result of industrialization and climate change may influence pests indirectly through the change in plant physiology, chemistry and nutritional content. In addition, the increase in atmospheric $CO_2$ will lead to higher insect feeding rates and greater amounts of crop damage, because this may result in some crops being less nutritious (Mah et al., 2020; Norris et al., 2002). The effects of increased $CO_2$ level on insects were found to be mainly through the $CO_2$-induced changes occurring in their host plants, which then may possibly affect the intensity and frequency of pest outbreaks on crops. The insect finds it difficult to digest leaves eaten from less nutritious plants, because they have to eat more to compensate, causing greater amounts of crop damage, and thereby affecting their performance and behaviour (Kozlov et al., 2015). The effect of elevated $CO_2$ was also reported to cause leaf toughness, leaf thickness and specific leaf weight due to an increase in mechanical defense. Thus, levels of mechanical defense are negatively correlated with herbivore performance (Hamilton et al., 2005).

## 9.3 PRINCIPLES OF BIOLOGICAL CONTROL AND HOST PARASITISM

Biological control is part of integrated pest management which involves the utilization of environmentally friendly, cost-effective control methods for pests of agricultural crops (such as plant diseases, insects, weeds and mites). It is the use of a living organism to control another living organism in an ecosystem. The organism used for control is called the biological control agent or biocontrol agent (Tylianakis & Binzer, 2014; Reeves, 2017). Biocontrol agents include parasites, pathogens and predators (Sanda et al., 2020, 2023; Paper et al., 2022). Some of the advantages of biocontrol agents are that they are target-specific, environmentally friendly and cost-effective options, and they can establish themselves (self-perpetuating or self-sustaining) in new places for longer periods of time. They give control where other options are not physically or economically possible or pose unacceptable risks to the environment. Their main disadvantage is they are a slow and time-consuming method of pest control compared to chemical pesticides (Magan & Medina, 2018). Three biocontrol strategies are applied in integrated pest control programs: conservation biocontrol, classical biocontrol (augmentation) and importation biocontrol (Gilardi et al., 2017). Augmentation is the release of supplemental biocontrol agents, thereby improving or boosting the naturally occurring population. Relatively few natural enemies may be released at a critical time of the season (inoculative release) or millions may be released (inundative release). Classical biocontrol is the intentional introduction of an exotic (non-native), usually coevolved biological control agent by government authorities with the aims of long-term pest control and permanent establishment (Van Driesche, 2008; Sanda and Sunusi, 2014; Sanda et al., 2014a, 2014b). The other method of biological control of agricultural pests is the conservation of existing biocontrol agents in an ecosystem through vegetation manipulation. In this scenario, the biocontrol agents have already established or adapted to the habitat and to the target pest, and their conservation is simple and cost effective (Naranjo and Ellsworth, 2009; Godfray et al., 2010; Sanda et al., 2014a).

## 9.4 EFFECTS OF CLIMATE CHANGE ON BIOLOGICAL CONTROL IN AGRICULTURAL SYSTEMS

Climate change greatly influences pests such as insect pests, weeds and diseases in an ecosystem. As discussed earlier, this section focusses on major climate change variables and their effects on the natural control agents/enemies and their relationship with pests during parasitism. These variables are mainly the rise in surface temperature, elevated $CO_2$ level/concentration and erratic nature of rainfall or precipitation. Climate change drivers could directly or indirectly affect biological control through physiology or behaviour of biological control agents, species distribution, modification of interspecific population dynamics and intraspecific interactions within an agroecosystem (Selvaraj et al., 2013).

### 9.4.1 Climate Change and Predator-Prey Relationships

Differences exist between arthropods/invertebrate species in their response to temperature in particular and other climatic variables in general (Grace et al., 2015; Luedeling et al., 2011; Fuhrer, 2003). The rise in temperature significantly causes variation in overlap between natural control agents/enemies and pests in space and time (Both et al., 2009; Welch and Harwood, 2014). According to Barton and Schmitz (2009), invertebrate behaviour was likely to change under warmer temperature, as in the case of "sit and wait" spiders. These spiders transfer to cooler niches in the ecosystem while active hunting species cover larger territories with a resulting higher chance of negative predator-predator interactions, and this increased hunting activity may have flow-on effects. Similarly, elevated temperature intensified the behavioural shift by grasshoppers to feed preparedly on herbs rather than grasses in the presence of hunting spiders in a field for food (Barton and Schmitz, 2009).

It was reported that the potential change in climate is expected to cause significant change in the level of synchrony between life cycles of different species, which could influence the efficacy of the biocontrol agent in a local scale (Selvaraj et al., 2013). For example, the increase in temperature was reported to have a negative effect on the survival, reproduction and longevity of the ladybird beetle, thereby inhibiting the population of the lady beetle, decreasing its total consumption of prey and weakening the effect of biological control on aphids (Boggs, 2016). Contrarily, a similar night temperature was reported to have no significant effect on the vital traits of the lady beetle, which promotes its population growth of the lady beetle, consequently stabilizing its total prey consumption, maintaining the function of biocontrol to aphids (Thurman et al., 2017).

In China, the effect of elevated $CO_2$ was tested to examine insect responses to their natural control agents (Sun et al., 2011). In the same article, the development of cotton bollworm *Helicoverpa armigera* was prolonged under an increased $CO_2$ level by reducing the foliar nitrogen level of the host plants (Harmon et al., 2009). Similarly, lacewings reared in high $CO_2$ conditions were reported to have developed faster and have lower body mass, mainly because their aphids prey raised in high $CO_2$ had less available soluble protein and hence reduced the predation rate by the predator (lacewing) (Laws, 2017). Therefore, an increase in feeding was needed by the predator due to low food quality as seen in aphids which makes them less at risk of predation (Miller et al., 2014; Gao et al., 2010). Similarly, aphids fed on plants raised in elevated $CO_2$ conditions had decreased responses to alarm pheromone than those fed on plants raised at ambient $CO_2$ conditions (Hentley et al., 2014).

### 9.4.2 Climate Change and Host-Parasitoid Interactions

The ability of a genotype to produce different phenotypes by changing its physiology, morphology and development in response to changes in its environment throughout its ontogeny is termed "phenotypic plasticity" (Sisodia & Singh, 2002; Pigliucci, 2005). Therefore, in response to temperature of the environment during development, a single genotype may express different adult body size. This is consistent with the "temperature-size rule" of Atkinson (1994), which states that as a result of their low metabolic rate, individuals developing at cool temperatures have lower growth rate but still achieve larger body size (by allocating more material to body growth), than the ones developing at higher temperatures (Angilletta et al., 2004; Gillooly et al., 2001; Nespolo et al., 2007). Hence, low temperature results in a slower growth rate, longer development time and larger adult size in parasitoids such as *Trichogramma* species (Sibly & Atkinson, 1994). Therefore, adult size is considered as a key factor in the selection of parasitoids and is found to be positively related to other fitness traits such as longevity and fecundity (Boivin & Lagacé, 1999). It was found in several studies that being large has consequences on reproductive success in females which is influenced by temperature (Roitberg et al., 2001). According to Sanda and Dong (2018), *Trichogramma dendrolimi* adults developed much faster (8.16±0.01days) at the highest temperature of 34°C and took 14.15±0.03 days to emerge at 18°C. Similarly, *T. dendrolimi* adults are much bigger at 18°C and relatively smaller at 34°C, used in experiments (Sanda and Dong, 2018; Sanda et al., 2018). Other parameters tested included mean development time of *T. dendrolimi* and longevity of adult wasps at different temperatures, as shown in Table 9.1.

In the same vein, the biocontrol of cereal beetles *Oulema melanopus* L. was disrupted by the *Tetrastichus julis* Walker parasitoid wasp. Here, the beetles' development was severely affected by the warmer temperature than the parasitoid wasp, thereby creating phonological misalignment between the parasitoid and prey. Thus, change in species ranges as results of rising temperature can also impact biocontrol. This is because pests are able to expand their ranges more rapidly than their biocontrol agents; therefore, they might escape predation and reach outbreak levels in their new ranges (Crowder & Harwood, 2014).

During climate change, variation in prey-predatory interaction can be weakened, strengthened, or have no impact on biocontrol efficacies. At the time of extended heat waves, the biocontrol of *Myzus persicae* (Green peach aphids) was increased by two parasitoid species; therefore, these

**TABLE 9.1**
**Mean Development Time, Tibia Length of *Trichogramma dendrolimi* and Longevity of Adult Wasps at Different Temperatures**

| Temperature (°C) | Mean Development Time (days) (±SD) | Mean Tibia Length (μm) (±SD) | Mean Longevity (days) (±SD) |
|---|---|---|---|
| 18 | 14.15 ± 0.03[e] | 161.01 ± 3.75[d] | 4.73 ± 0.71[a] |
| 22 | 11.05 ± 0.04[d] | 108.78 ± 2.51[a,b] | 4.73 ± 0.97[a] |
| 26 | 10.16 ± 0.02[c] | 107.06 ± 2.29[a] | 7.43 ± 1.85[b] |
| 30 | 8.71 ± 0.05[b] | 111.77 ± 0.86[b,c] | 4.23 ± 0.32[a] |
| 34 | 8.16 ± 0.01[a] | 115.34 ± 0.74[c] | 3.80 ± 0.36[a] |

*Source:* Data from Sanda et al., 2018.
[a] *Means followed by the same letter in a column are not significantly different at P < .05. (ANOVA followed by least significant difference test).*

parasitoids competed less under heat waves and acted more complementarily, strengthening the biocontrol of this pest). In the same vein, high temperature considerably affects host susceptibility to parasitoids and may improve survival of parasitized hosts (Thomas & Blanford, 2003). Similarly, lethal and sublethal damage to parasitoids was found to be associated with exposure to extreme temperature. This was seen in *Wolbachia*, a bacterial endosymbiont that facilitates parthenogenesis, which was removed by short exposure to high temperature. Other climatic factors such as availability of water can affect natural enemies in the biological control of pests. Mealybug parasitism by parasitic wasps was reduced due to water stress associated with drought conditions (Selvaraj et al., 2013). Similarly, when these wasps are raised under water-stressed plants, the mealybug's immune response is improved, resulting in an increase in encapsulation rate of 30%–50% in three different encyrtid mealybug parasitoids species (Calatayud et al., 2002). Under elevated $CO_2$, cotton aphid *Aphis gossypii* survival was shown to increase significantly, but larval development of ladybird beetle took a longer period of time, which will make aphids a more serious pest in the near future (Gao et al., 2009).

### 9.4.3 Climate Change and Host-Pathogen Interactions

Temperature, which is a climatic factor, was reported to affect the development, infectivity, reproduction, time of death and storage of entomopathogenic nematodes (EPNs) (Chen et al., 2003; Bilgrami & Gaugler, 2007). It also affects host searching (Bilgrami & Gaugler, 2007), survival and pathogenicity (Chen et al., 2003; Aydn & Susurluk, 2005), migration activities, sex ratio and insect death rate (Lalramliana & Yadav, 2016). High temperatures could enable EPNs to penetrate into the target host or decrease their efficacy due to desiccation (Gillespie et al., 1997), while a lower temperature could decrease their activity and ability to infect a host. Tolerance of EPNs to temperature differs between species and strains (Lalramliana & Yadav, 2016). Temperature ranges for *Steinernematids* and *Heterorhabditis* spp. are 10°C–39°C and 7°C–35°C, respectively. Therefore, the optimum range for infectivities of these nematodes is between 23°C and 28°C (Georgis et al., 2006) for *Steinernematids* and 25°C for *Heterorhabditis*. Thus, EPNs extracted from particular regions have considerable variations in their abilities to initially infect a host, kill the host and reproduce within it, and they are influenced by temperatures specific to their original habitats (Georgis et al., 2006). According to Sanda et al. (2020), it was reported that the use of two EPNs of *S. carpocapsae* and *H. bacteriophora* for the biocontrol of *Octodonta nipae* third instar larvae caused higher larval mortalities at 30°C as shown in Figure 9.1A and B. For *Ceratitis capitata* larvae, the highest mortality of 86.7% *S. carpocapsae* and 80.0% for *Heterorhabditis* spp. at a higher temperature of 31°C. Similarly in Figure 9.2, the infective juveniles' (IJs) penetrations of

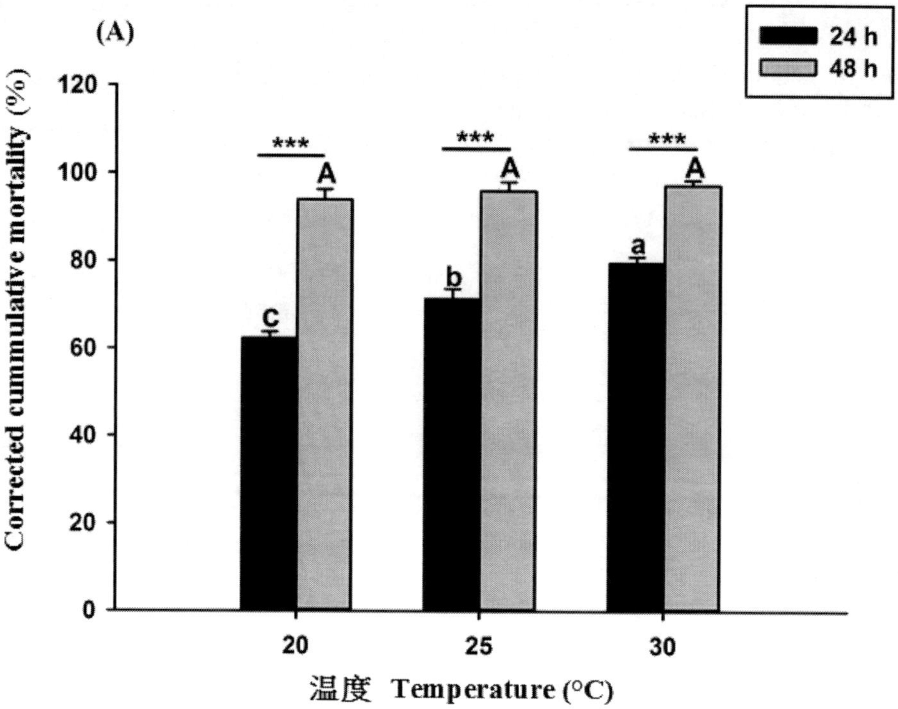


**FIGURE 9.1A** Cumulative mortality of third instar larvae of *Octodonta nipae* at different temperature of *Steinernema carpocapsae* at 24 and 48 hours after treatment. Means with the same letter above bars are not significantly different using one-way ANOVA followed by least significant differences. Means shown by different letters are significantly different at 0.05%.

**Source: Sanda et al. (2020).**

*S. carpocapsae* and *H. bacteriophora* were also highest at 30°C between the temperature conditions (Sanda et al., 2020). According to Lalramliana and Yadav (2016), maximum IJs penetrations into *Conogethes punctiferalis* larvae were recorded for *Steinernema* spp. and *Heterorhabditis* spp. at 30°C. Similarly, Mahar et al. (2005) reported higher penetrations of *H. indica* and *H. bacteriophora* at 30°C in *Phaedon cochleariae* and contrarily higher penetration of *S. carpocapsae* and *S. feltiae* at 25°C (Mahar et al., 2005).

In addition, temperature was shown to have a direct effect on the pathogenicity of entomopathogenic fungi like *Beauveria bassiana* and *Metarhizium anisopliae* which can either be highly pathogenic, a situation causing extensive and rapid mortality of pests in days, or virtually benign, with the same hosts surviving for weeks or months (Thomas & Blanford, 2003). In some climate change models, the depletion of the ozone layer increases UV radiation, and this can have significant impact on some entomopathogens such as bacteria and fungi because they are more sensitive to damage by UV radiation than weeds and insects (Reeves, 2017).

The increase in temperature and $CO_2$ level change the leaf quality of the plant which results in extra feeding by insect pests, resulting in a greater intake of entomopathogens such as *B. thuringiensis* (Bt). Surprisingly, the change in C:N ratio caused by the increased $CO_2$ level significantly reduces the efficacy of the Bt toxins' expression level in transgenic cotton leaves. This might be a result of inadequate production of toxin protein, *Bt*-resistant insect populations, environmental effect on transgene expression and development of resistance due to inadequate management (Adamczyk et al., 2001).

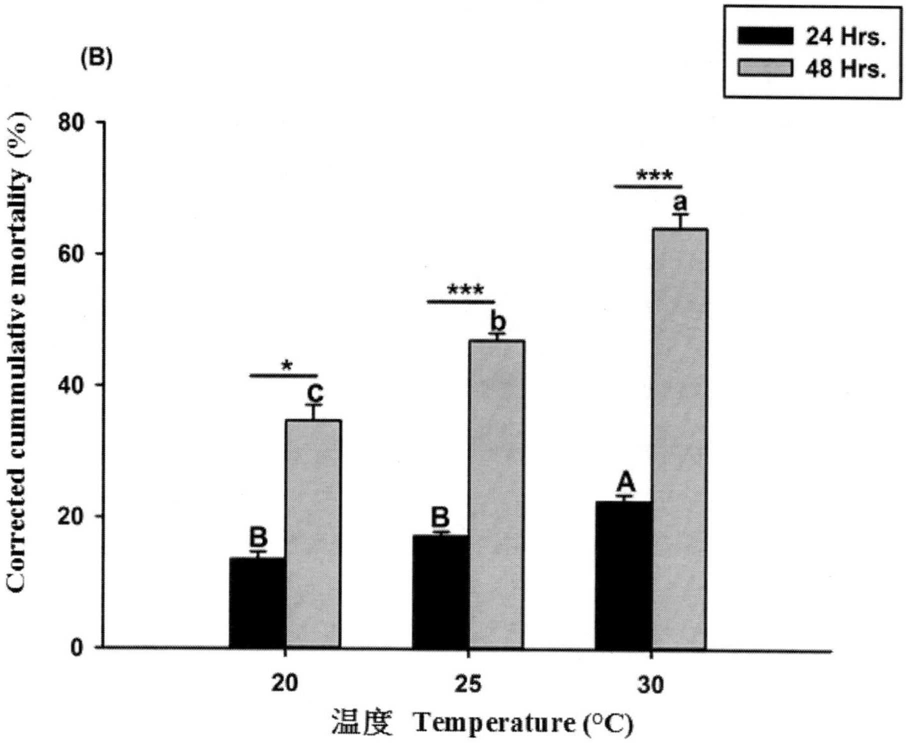


**FIGURE 9.1B** Cumulative mortality of third instar larvae of *Octodonta nipae* at different temperature of *Heterorhabditis bacteriophora* at 24 and 48 hours after treatment. Means with the same letter above bars are not significantly different using one-way ANOVA followed by least significant differences. Means shown by different letters are significantly different at 0.05%.

**Source: Sanda et al. (2020).**

### 9.4.4 Climate Change and Weed Biocontrol

The efficacies of biological control agents on invasive weed species such as effectiveness and risk assessment were reported to be affected by climate change (Diez et al., 2012). Several studies reported that climate change has both negative and positive impacts on the performance of biological control agents of invasive plant species (Müller-Schärer and Schaffner, 2008). In field surveys of Lu et al. (2017), global warming allowed the *Agasicles hygrophila* beetle to expand its range and benefit biocontrol of noxious weed *Alternanthera philoxeroides* (Mart.) Griseb in regions that are too cold for the beetle. In the same article, the phenology of the nontarget indigenous plant was shifted from annual to perennial due to elevated temperature and the increase in overwintering damage and impact of biocontrol agent *A. hygrophila* on seedling recruitment (Sun et al., 2020). Therefore, warming changes the interactions between invasive *A. philoxeroides* and native *A. sessilis* and shifts the plant community from invader-dominated to native-dominated, but only in the presence of *A. hygrophila* as a result of the disproportionate increases in herbivory on the invader (Sun et al., 2020). It was also reported that increases in rainfall will affect biocontrol of ragwort (*Jacobaea vulgaris*) by ragwort flea beetle (*Longitarsus jacobaeae*) by extending the areas where there is poor control in inland western districts in both the North and South Island and good control in Gisborne and Northland regions (Gerard et al., 2010, 2013).

Different studies reported that plants have increases in growth rate and invest more in carbon-based defences when grown at elevated $CO_2$. This has some implications for the management of

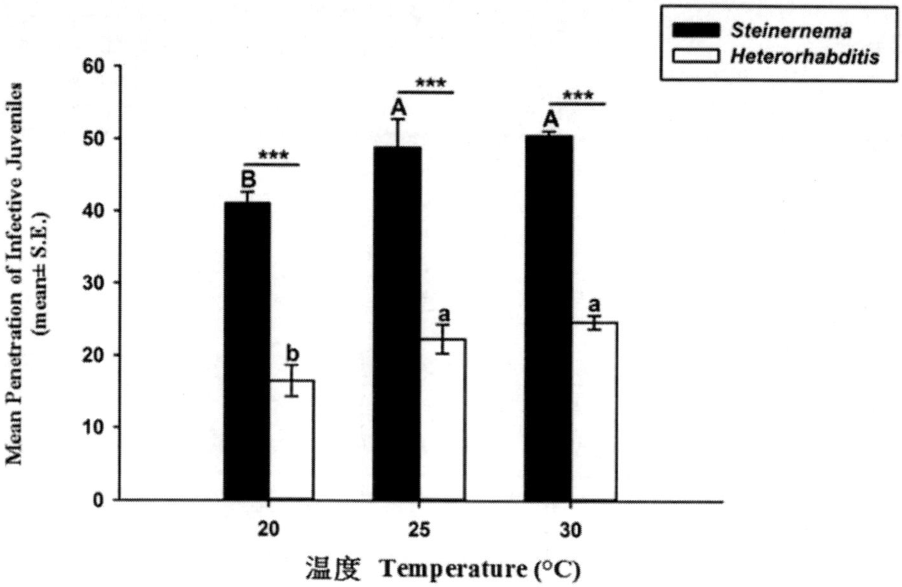


**FIGURE 9.2** Mean penetration of infective juveniles of *Steinernema carpocapsae* and *Heterorhabditis bacteriophora* at different temperatures into third instar larvae of *Octodonta nipae*. Means with the same letter are not significantly different.

**Source: Sanda et al. (2020).**

invasive alien species, especially using biological control, mainly using plant-feeding insects. In the study of Baso et al. (2021), four invasive aquatic weeds were grown at elevated $CO_2$ and subjected to herbivory by their target biological control agents. It was shown that the insect feeding was reduced in three of the weeds species at elevated $CO_2$ except for *Salvinia molesta*; this implies that the weeds species will become more difficult to manage using biological control in the future (Baso et al., 2021; Reeves, 2017). It was reported in a meta-analysis how an elevated level of $CO_2$ affects plant-herbivore interactions; it was found that elevated $CO_2$ decreased insect pest abundance on the treated plants (Stiling & Cornelissen, 2007). Nonetheless, the decline in herbivory can be attributed to an elevated C:N ratio, leading to a subsequent reduction in insect size, fecundity, and reproductive efforts (Robinson et al., 2012; Baso et al., 2021).

### 9.4.5 Effect of Multi-Climate-Factor on Biocontrol Agents

There are inadequate studies on combined effects of precipitation, $CO_2$ and temperature on biological control of pests of crops. According to Wang et al. (2014), plants growing under temperature extremes, decreased precipitation and elevated $CO_2$ will provide different nutritional resources for invertebrate hosts by indirectly influencing the fitness of predators and parasitoids feeding on these hosts. This fitness will decrease as the quality of the host plant decreases. For instance, *Leis axyridis* Pallas is the predator of *A. gossypii* and was reported to consume more prey at higher $CO_2$ (Chen et al., 2005). Chen et al. (2005) found a little impact of elevated $CO_2$ on its predator, *Harmonia axyridis*, but increased abundance of the *Aphidius picipes* (Nees), resulting in increased aphid parasitism in elevated $CO_2$. Therefore, the biocontrol efficiency of the parasitoid against aphid may be enhanced in elevated $CO_2$ and make generalist predators more effective in controlling pests (Sun et al., 2011). It should be noted that there is a linkage between temperature and atmospheric $CO_2$, which either acts to synergize or counteract the effects of biocontrol-plant interactions (van Doan et al., 2021). For example, elevated $CO_2$ increased success of two endophagous larval agents

(*Octotoma championi* and *O. scabripennis*) by counteracting leaf loss of *Lantana camara* caused by warming alone. Thus, the elevated $CO_2$ allowed for water savings by plants that prevented the leaf loss, which was responsible for reduced survival under warming treatments alone. Therefore, multiple factor experiments are possible under field conditions. According to van Doan et al. (2021), entomopathogenic had slightly reduced the survival rate when exposed to four-day future climatic conditions (RCP 8.5), entailing a 28% decrease in precipitation, a 3.4°C raise in temperature, and a 400 ppm increase in $CO_2$ levels but had no effect on the survival of wolf spider, a ladybug and a parasitoid wasp.

## 9.5 CONCLUSION

In this chapter, it was elucidated that climate change affects both pests and their biological control agents in many ways. In one way, the climate change aided the spread of agricultural pests at a faster rate around the world. In addition, the increase in relative humidity allows these pests to easily adapt to new environments and to be very active over longer periods of time. On the other hand, climate changes impact rate of infection, efficacy of biocontrol agents in the new environment, development time, reproduction, geographical distributions and emergence of biocontrol agent. Specifically, a rise in temperature causes significant variation in overlap between natural enemies and pests in space and time. Thus, this hugely affects agricultural production and by extension reduces yields and profits. Therefore, it is recommended that when planning for any biological control program, climatic variables need to be considered in relation to pests and natural enemies in the agroecosystem.

## REFERENCES

Adamczyk, J. J., Adams, L. C., & Hardee, D. D. (2001). Field efficacy and seasonal expression profiles for terminal leaves of single and double *Bacillus thuringiensis* toxin cotton genotypes. *Journal of Economic Entomology*, *94*, 1589–1593.

Angilletta, M. J., Jr., Steury, T. D., & Sears, M. W. (2004). Temperature, growth rate, and body size in ectotherm: Fitting pieces of a life-history puzzle. *Integrative & Comparative Biology*, *44*, 498–509.

Atkinson, D. (1994). Temperature and organism size—a biological law for ectotherm? *Advances in Ecological Research*, *25*, 1–58.

Aydn, H., & Susurluk, A. (2005). Competitive abilities of the entomopathogenic nematodes *Steinernema feltiae* and *Heterorhabditis bacteriophora* in the same host at different temperatures. *Turkish Journal of Biology*, *29*, 35–39.

Barton, B. T., & Schmitz, O. J. (2009). Experimental warming transforms multiple predator effects in a grassland food web. *Ecology Letters*, *12*(12), 1317–1325.

Baso, N. C., Coetzee, J. A., Ripley, B. S., & Hill, M. P. (2021). The effects of elevated atmospheric $CO_2$ concentration on the biological control of invasive aquatic weeds. *Aquatic Botany*, *170*, 103348.

Bilgrami, A. L., & Gaugler, R. (2007). Effects of various stress factors on heat tolerance by *Heterorhabditis bacteriophora* and *Steinernema carpocapsae*. *Nematology*, *9*(2), 161–167.

Boggs, C. L. (2016). The fingerprints of global climate change on insect populations. *Current Opinion in Insect Science*, *17*, 69–73.

Boivin, G., & Lagacé, M. (1999). Effet de la taille sur la fitness de *Trichogramma evanescens* (Hymenoptera: Trichogrammatidae). Annales de la Société. *Entomologique de France*, *35*, 371–378.

Both, C., van Asch, M., Bijlsma, R. G., van den Burg, A. B., & Visser, M. E. (2009). Climate change and unequal phenological changes across four trophic levels: Constraints or adaptations? *Journal of Animal Ecology*, *78*, 73–83.

Calatayud, P. A., Polanía, M. A., Seligmann, C. D., & Bellotti, A. C. (2002). Influence of water-stressed cassava on *Phenacoccus herreni* and three associated parasitoids. *Entomologia Experimentalis et Applicata*, *102*, 163–175.

Chen, S., Li, J., Han, X., & Moens, M. (2003). Effect of temperature on the pathogenicity of entomopathogenic nematodes (*Steinernema* and *Heterorhabditis* spp.) to *Delia radicum*. *BioControl*, *48*, 713–724.

Chen, F., Ge, F., & Parajulee, M. N. (2005). Impact of elevated $CO_2$ on tri-trophic interaction of *Gossypium hirsutum*, *Aphis gossypii*, and *Leis axyridis*. *Environmental Entomology*, *34*(1), 37–46.

Chidawanyika, F., Mudavanhu, P., & Nyamukondiwa, C. (2019). Global climate change as a driver of bottom-up and top-down factors in agricultural landscapes and the fate of host-parasitoid interactions. *Frontiers in Ecology and Evolution*, *7*, 80.

Crowder, D. W., & Harwood, J. D. (2014). Promoting biological control in a rapidly changing world. *Biological Control*, *75*, 1–7.

Devlet, A. (2021). Modern agriculture and challenges. *Frontiers in Life Sciences and Related Technologies*, *2*(1), 21–29.

Diez, J. M., D'Antonio, C. M., Dukes, J. S., Grosholz, E. D., Olden, J. D., Sorte, C. J. B., Blumenthal, D. M., Bradley, B. A., Early, R., Ibanez, I., Jones, S. J., Lawler, J. J., & Miller, L. P. (2012). Will extreme climatic events facilitate biological invasions? *Frontiers in Ecology and the Environment*, *10*, 249–257.

Dong, Y., Han, P., Niu, C., Zappalá, L., Amiens-Desneux, E., Bearez, P., Lavoir, A. V., Biondi, A., & Desneux, N. (2018). Nitrogen and water inputs to tomato plant do not trigger bottom-up effects on a leafminer parasitoid through host and non-host exposures. *Pest Management Science*, *74*, 516–522. https://doi.org/10.1002/ps4750

Eigenbrode, S. D., Davis, T. S., & Crowder, D. W. (2015). Climate change and biological control in agricultural systems: Principles and examples from North America. In *Climate change and insect pests* (pp. 119–135). Wallingford: CABI.

FAO. (2020). *Food and agriculture organization plant pests and diseases in the context of climate change and climate variability, food security and biodiversity risks*. www.fao.org/3/nb088/nb088.pdf. Accessed 12 January 2020.

Fuhrer, J. (2003). Agroecosystem responses to combinations of elevated $CO_2$, ozone, and global climate change. *Agriculture, Ecosystems & Environment*, *97*(1–3), 1–20.

Furlong, M. J., & Zalucki, M. P. (2017). Climate change and biological control: The consequences of increasing temperatures on host-parasitoid interactions. *Current Opinion in Insect Science*, *20*, 39–44.

Furlong, M. J., Day, M. D., & Zalucki, M. P. (2021). Biological control and climate change. In *Biological control: Global impacts, challenges and future directions of pest management* Edited by Peter G. Mason. Clayton, VIC Australia: CSIRO Publishing. 220–235.

Gao, F., Chen, F. J., & Ge, F. (2010). Elevated $CO_2$ lessens predation of *Chrysopa sinica* on *Aphis gossypii*. *Entomologia Experimentalis Et Applicata*, *135*, 135–140.

Gao, F., Zhu, S., Sun, Y., Du, L., Parajulee, M., Kang, L., & Ge, F. (2009). Interactive effects of elevated $CO_2$ and cotton cultivar on tri-trophic interaction of *Gossypium hirsutum*, *Aphis gossyppii*, and *Propylaea japonica*. *Environmental Entomology*, *37*, 29–37.

Georgis, R., Koppenhöfer, A. M., Lacey, L. A., Bélair, G., Duncan, L. W., Grewal, P. S., Samish, M., Tan, L., Torr, P., & van Tol, R. W. H. M. (2006). Successes and failures in the use of parasitic nematodes for pest control. *Biological Control*, 38(1), 103–123.

Gerard, P. J., Barringer, J. R. F., Charles, J. G., Fowler, S. V., Kean, J. M., Phillips, C. B., Tait, A. B., and Walker, G. P. (2013). Potential effects of climate change on biological control systems: Case studies from New Zealand. *BioControl*, *58*, 149–162.

Gerard, P. J., Kean, J. M., Phillips, C. B., Fowler, S. V., Withers, T. M., Walker, G. P., & Charles, J. G. (2010). Possible impacts of climate change on biocontrol systems in New Zealand. Report for MAF Pol Project, 0910-11689.

Gilardi, G., Gisi, U., Garibaldi, A., & Gullino, M. L. (2017). Effect of elevated atmospheric $CO_2$ and temperature on the chemical and biological control of powdery mildew of zucchini and the Phoma leaf spot of leaf beet. *European Journal of Plant Pathology*, *148*, 229–236.

Gillespie, J. P., Kanost, M. R., & Trenczek, T. (1997). Biological mediators of insect immunity. *Annual Review of Entomology*, *42*, 611–643.

Gillespie, D. R., Quiring, D. J. M., Foottit, R. G., Foster, S. P., & Acheampong, S. (2009). Implications of phenotypic variation of *Myzus persicae* (Hemiptera: Aphididae) for biological control on greenhouse pepper plants. *Journal of Applied Entomology*, *133*(7), 505–511. https://doi.org/10.1111/j.1439-0418.2008.01365.x

Gillooly, J. F., Brown, J. H., West, G. B., Savage, V. M., & Charnov, E. L. (2001). Effects of size and temperature on metabolic rate. *Science*, *293*, 2248–2251.

Godfray, H. C. J., Beddington, J. R., Crute, I. R., Haddad, L., Lawrence, D., Muir, J. F Pretty, J., Robinson, S., Thomas, S. M., and Toulmin, C. (2010). Food security: The challenge of feeding 9 billion people. *Science*, *327*(5967), 812–818.

Grace, D., Bett, B., Lindahl, J., & Robinson, T. (2015). *Climate and livestock disease: Assessing the vulnerability of agricultural systems to livestock pests under climate change scenarios*. CCAFS Working Paper no. 116. Copenhagen: CGIAR Research Program on Climate Change, Agriculture and Food Security (CCAFS).

Hamilton, J. G., Dermody, O., Aldea, M., Zangerl, A. R., Rogers, A., Berenbaum, M. R., & Delucia, E. H. (2005). Anthropogenic changes in tropospheric composition increase susceptibility of soybean to insect herbivory. *Environmental Entomology*, *34*(2), 479–485.

Harmon, J. P., Moran, N. A., & Ives, A. R. (2009). Species response to environmental change: Impacts of food web interactions and evolution. *Science*, *323*, 1347–1350.

Hentley, W. T., Vanbergen, A. J., Hails, R. S., Jones, T. H., & Johnson, S. N. (2014). Elevated atmospheric $CO_2$ impairs aphid escape responses to predators and conspecific alarm signals. *Journal of Chemical Ecology*, *40*, 1110–1114.

IPPC Secretariat, Gullino, M. L., Albajes, R., Al-Jboory, I., Angelotti, F., Chakraborty, S., Garrett, K. A., Hurley, B. P., Juroszek, P., Makkouk, K., & Pan, X. (2021). *Scientific review of the impact of climate change on plant pests*. FAO on behalf of the IPPC Secretariat.

Juroszek, P., Racca, P., Link, S., Farhumand, J., & Kleinhenz, B. (2020). Overview on the review articles published during the past 30 years relating to the potential climate change effects on plant pathogens and crop disease risks. *Plant Pathology*, *69*(2), 179–193.

Kozlov, M. V., Lanta, V., Zverev, V., & Zvereva, E. L. (2015). Global patterns in background losses of woody plant foliage to insects. *Global Ecology and Biogeography*, *24*(10), 1126–1135.

Lalramliana, Y. A. K. (2016). Effects of storage temperature on survival and infectivity of three indigenous entomopathogenic nematodes strains (Steinernematidae and Heterorhabditidae) from Meghalaya, India. *Journal of Parasitic Diseases*, *16*(1), 21–25.

Laws, A. N. (2017). Climate change effects on predator-prey interactions. *Current Opinion in Insect Science*, *23*, 28–34.

Lu, X. M., Siemann, E., He, M. Y., Wei, H., Shao, X., & Ding, J. Q. (2017). Climate warming increases biological control agent impact on a non-target species. *Ecology Letters*, *20*, 112.

Luedeling, E., Steinmann, K. P., Zhang, M., Brown, P. H., Grant, J., & Girvetz, E. H. (2011). Climate change effects on walnut pests in California. *Global Change Biology*, *17*(1), 228–238.

Lykogianni, M., Bempelou, E., Karamaouna, F., & Aliferis, K. A. (2021). Do pesticides promote or hinder sustainability in agriculture? The challenge of sustainable use of pesticides in modern agriculture. *Science of the Total Environment*, *795*, 148625.

Magan, N., & Medina, A. (2018). Climate change and resilience of biocontrol agents for mycotoxin control. In *한국균학회 학술대회* (p. 41).

Mah, A. Y., Chapman, D. A., Markowitz, E. M., & Lickel, B. (2020). Coping with climate change: Three insights for research, intervention, and communication to promote adaptive coping to climate change. *Journal of Anxiety Disorders*, *75*, 102282.

Mahar, A. N., Munir, M., Elawad, S., Gowen, S. R., & Hague, N. G. M. (2005). Pathogenicity of bacterium, *Xenorhabdus nematophila* isolated from entomopathogenic nematode (*Steinernema carpocapsae*) and its secretion against *Galleria mellonella* larvae. *Journal of Zhejiang University Science*, *6*(6), 457–463.

Mbande, A., Tedder, M., & Chidawanyika, F. (2019a). Differential life-history responses in *Neolema abbreviata*, a biological control agent for *Tradescantia fluminensis* under water and nitrogen gradients. *Arthropod Plant Interact*, *13*, 57–70. doi:10.1007/s11829-018-9630-0

Mbande, A., Tedder, M., & Chidawanyika, F. (2019b). Offspring diet supersedes the transgenerational effects of parental diet in a specialist herbivore *Neolema abbreviata* under manipulated foliar nitrogen variability. *Insect Science*. https://doi.org/10.1111/1744-7917.12644

Miller, L. P., Matassa, C. M., & Trussell, G. C. (2014). Climate change enhances the negative effects of predation risk on an intermediate consumer. *Global Change Biology*, *20*, 3834–3844.

Motes, W. C. (2010). *Modern agriculture and its benefits—trends, implications and outlook*. Washington, DC: Global Harvest Initiative.

Müller-Schärer, H., & Schaffner, U. (2008). Classical biological control: Exploiting enemy escape to manage plant invasions. *Biological Invasions*, *10*, 859–874.

Murrell, E. G., & Barton, B. T. (2017). Warming alters prey density and biological control in conventional and organic agricultural systems. *Integrative and Comparative Biology*, *57*(1), 1–13.

Naranjo, S. E., & Ellsworth, P. C. (2009). Fifty years of the integrated control concept: Moving the model and implementation forward in Arizona. *Pest Management Science: Formerly Pesticide Science*, *65*(12), 1267–1286.

Nespolo, R. F., Artacho, P., & Castaneda, L. E. (2007). Cyclic gas-exchange in the Chilean red cricket: Inter-individual variation and thermal dependence. *Journal of Experimental Biology*, *210*, 668–675.

Norris, R. J., Memmott, J., & Lovell, D. J. (2002). The effect of rainfall on the survivorship and establishment of a biocontrol agent. *Journal of Applied Ecology*, 226–234.

Paper, M. K., Righetti, T., Raubenheimer, S. L., Coetzee, J. A., Sosa, A. J., Ripley, B. S. R., & Hill, M. P. (2022). Effects of elevated $CO_2$ on feeding responses of biological control agents of *Pontederia crassipes*. *Entomologia Experimentalis et Applicata*. https://doi.org/10.1111/eea.13289
Pigliucci, M. (2005). Evolution of phenotypic plasticity: Where are we going now? *Tree*, *20*, 481–486.
Reeves, J. L. (2017). Climate change effects on biological control of invasive plants by insects. *CABI Reviews*, *12*, 1–8.
Reiners, S., & Petzoldt, C. H. C. (2000). *Integrated crop and pest management guidelines for commercial vegetable production* (No. 632 R2753i Ej. 1 019918, p. 105). Cornell University, New York, 14850, United State of America.
Robinson, E. A., Ryan, G. D., & Newman, J. A. (2012). A meta-analytical review of the effects of elevated $CO_2$ on plant-arthropod interactions highlights the importance of interacting environmental and biological variables. *New Phytologist*, *194*(2), 321–336.
Roitberg, B. D., Boivin, G., & Vet, L. E. M. (2001). Fitness, parasitoids, and biological control: An opinion. *Canadian Entomologist*, *133*, 429–438.
Sanda, N. B., Wudil, B. S., & Hou, Y. M. (2020). Effect of temperatures on the pathogenicity of *Entomophathogenic* nematodes (*Steinernema carpocapsae* and *Heterorhabditis bacteriophora*) for biocontrol of beetles. *Nigerian Journal of Entomology*, *36*, 88–95. https://doi.org/10.36108/NJE/0202/63.01.01
Sanda, N. B., & Dong H. (2018). Effect of temperature variability influence on longevity and fecundity of biocontrol agent *Trichogramma dendrolimi* (Hymenoptera: Trichogrammatidae). *Lapai International Journal of Agriculture*, *1*(1), 59–65.
Sanda, N. B., & Sunusi, M. (2014). Fundamentals of biological control of pests. *International Journal of Clinical & Biological Sciences*, *1*(6), 1–11.
Sanda, N. B., Dong, H., & Cong, B. (2014a). Principles of biological control in integrated pest management. *International Journal of Applied Research and Technology*, *3*(1), 104–116.
Sanda, N. B., Dong, H., & Mustapha, S. (2014b). Biological control of insect pests in West Africa: A review. *International Journal of Applied Research and Technology*, *3*(9), 39–45.
Sanda, N. B., Muhammad, A., Ali, H., & Hou, Y. (2018). Entomopathogenic nematode *Steinernema carpocapsae* surpasses the cellular immune responses of the hispid beetle, *Octodonta nipae* (Coleoptera: Chrysomelidae). *Microbial Pathogenesis*, *124*, 337–345.
Sanda, N. B., & Hou, Y. (2023). The symbiotic bacteria—*Xenorhabdus nematophila* all and *Photorhabdus luminescens* H06 strongly affected the phenoloxidase activation of nipa palm hispid, *Octodonta nipae* (Coleoptera: Chrysomelidae) larvae. *Pathogens*, *12*, 506. https://doi.org/10.3390/pathogens12040506
Schmitz, O., & Barton, B. (2014). Climate change effects on behavioral and physiological ecology of predator-prey interactions: Implications for conservation biological control. *Biological Control*, *75*, 87–96.
Schuldt, A., Bruelheide, H., Buscot, F., Assmann, T., Erfmeier, A., Klein, A.-M., Ma, K., Scholten, T., Staab, M., Wirth, C., Zhang J., and Wubet, T. (2017). Belowground top-down and aboveground bottom-up effects structure multitrophic community relationships in a biodiverse forest. *Scientific Reports*, *7*, 4222. https://doi.org/10.1038/241598-017-04619-3
Selvaraj, S., Ganeshamoorthi, P., & Pandiaraj, T. (2013). Potential impacts of recent climate change on biological control agents in agro-ecosystem: A review. *International Journal of Biodiversity and Conservation*, *5*(12), 845–852.
Sibly, R. M., & Atkinson, D. (1994). How rearing temperature affects optimal adult size in ectotherms. *Functional Ecology*, *8*, 486–493.
Sisodia, S., & Singh, B. N. (2002). Effect of temperature on longevity and productivity in *Drosophila ananassae*: Evidence for adaptive plasticity and trade-off between longevity and productivity. *Genetica*, *114*, 95–102.
Stiling, P., & Cornelissen, T. (2007). How does elevated carbon dioxide ($CO_2$) affect plant-herbivore interactions? A field experiment and meta-analysis of $CO_2$-mediated changes on plant chemistry and herbivore performance. *Global Change Biology*, *13*(9), 1823–1842.
Sun, Y. C., Yin, J., Chen, F. J., Wu, G., & Ge, F. (2011). How does atmospheric elevated $CO_2$ affect crop pests and their natural enemies? Case histories from China. *Insect Science*, *18*(4), 393–400.
Sun, Y., Ding, J., Siemann, E., & Keller, S. R. (2020). Biocontrol of invasive weeds under climate change: Progress, challenges and management implications. *Current Opinion in Insect Science*, *38*, 72–78.
Thomas, M. B., & Blanford, S. (2003). Thermal biology in insect-pathogen interactions. *Trends in Ecology and Evolution*, *18*, 344–350.
Thurman, J. H., Crowder, D. W., & Northfield, T. D. (2017). Biological control agents in the Anthropocene: Current risks and future options. *Current Opinion in Insect Science*, *23*, 59–64.

Tylianakis, J. M., & Binzer, A. (2014). Effects of global environmental changes on parasitoid—host food webs and biological control. *Biological Control*, *75*, 77–86.
van Doan, C., Pfander, M., Guyer, A. S., Zhang, X., Maurer, C., & Robert, C. A. (2021). Natural enemies of herbivores maintain their biological control potential under short-term exposure to future $CO_2$, temperature, and precipitation patterns. *Ecology and Evolution*, *11*(9), 4182–4192.
Van Driesche, R. G., Lyon, S., Sanderson, J. P., Bennett, K. C., Stanek, E. J., III, & Zhang, R. (2008). Greenhouse trials of *Aphidius colemani* (Hymenoptera: Braconidae) banker plants for control of aphids (Hemiptera: Aphididae) in greenhouse spring floral crops. *Florida Entomologist*, *91*(4), 583–591.
Wang, Q., Zhang, M., Wang, S., Ma, Q., & Sun, M. (2014). Changes in temperature extremes in the Yangtze River Basin, 1962–2011. *Journal of Geographical Sciences*, *24*, 59–75. doi:10.1007/s11442-014-1073-7
Welch, K. D., & Harwood, J. D. (2014). Temporal dynamics of natural enemy-pest interactions in a changing environment. *Biological Control*, *75*, 18–27.
Werling, B. P., Dickson, T. L., Isaacs, R., Gaines, H., Gratton, C., Gross, K. L., Liere, H., Malmstrom, C. M., Meehan, T. D., Ruan, L., & Robertson, B. A. (2023). Perennial grasslands enhance biodiversity and multiple ecosystem services in bioenergy landscapes. *Proceedings of the National Academy of Sciences*, *111*(4), 1652–1657.
Williams, R. S., Lincoln, D. E., & Norby, R. J. (2003). Development of gypsy moth larvae feeding on red maple saplings at elevated $CO_2$ and temperature. *Oecologia*, *137*, 114–122.

# 10 Climate Change and Biodiversity of Beneficial and Social Insects

*Asim Abbasi, Jannat Zaka, Laiba Sadaqat, Hifza Inam, Laiba Kousar, Sadiqa Bibi, Kinza Bibi, and Mariam Khalid Abbasi*

## 10.1 INTRODUCTION

Loss of biodiversity is a major worrying concern for scientists because it may lead to ecological community failure with concomitant ill effects on various ecological functions and services necessary for the survival of humankind (Steffen et al. 2015). Across the globe, the noteworthy drivers for measuring biodiversity loss are overexploitation, environmental pollution, habitat destruction/fragmentation, biological intrusions and anthropogenic climate change (Venter et al. 2016). Rapid industrialization has already warmed planet earth by almost 1.1°C and is projected to warm it further by 2°C–5°C by 2100 unless proper checks are not maintained on the emission of greenhouse gases (IPCC 2021). The adverse effects of climate change not only lead to species extinction but also alter their assemblages, abundance, distribution patterns, composition and intraspecific interactions (Steinbauer et al. 2018; Schleuning et al. 2020; Raven and Wagner 2021). Moreover, climate change may also act either additively or synergistically with other factors to aggravate impacts on biodiversity (Verheyen and Stoks 2019; Halsch et al. 2021).

A substantial amount of recent literature exhibits that a number of insect populations are rapidly declining across many parts of the globe, although trends may vary among different functional groups or taxa depending on their geographical locations (Goulson 2019; Crossley et al. 2020; Didham et al. 2020; Janzen and Hallwachs 2021). These rapid declines in insect populations are of grave concern as terms like "insect apocalypse" are emerging and now being increasingly used by scientists and media to demonstrate this lethal phenomenon (Jarvis 2018; Warren et al. 2021). Current variations in the demographics of different important functional groups like biological control agents, social insects, nutrient recyclers and pollinators is currently considered noteworthy enough to merit profound concern (García-Robledo et al. 2020). Insects hold a prominent position in biodiversity (Stork 2018; Wagner et al. 2021) and contribute in different ecological systems by providing diverse ecological services (Dangles and Casas 2019). Besides providing resources for higher trophic levels including humans (Ramos-Elorduy 2009; Schowalter 2013), insects also provide a score of other economically important ecosystem services such as nutrient recycling, biological control of insect pests and pollination (Schowalter et al. 2018).

## 10.2 BENEFICIAL INSECTS

Beneficial insects offer a range of diverse ecological services such as pollination, soil formation, natural control of different harmful insect pests and nutrient cycling. Among the beneficial insect fauna, plant pollinators and natural enemies such as predators and parasitoids are of prime importance due to their abilities to offer pollination services and control of harmful insect pests (Lead et al. 2009; Aagustí et al. 2019) (Figure 10.1). However, the components of Global Environmental Change (GEC) such as fluctuated patterns of rainfall, enhanced atmospheric carbon dioxide ($CO_2$)


DOI: 10.1201/9781003382089-10

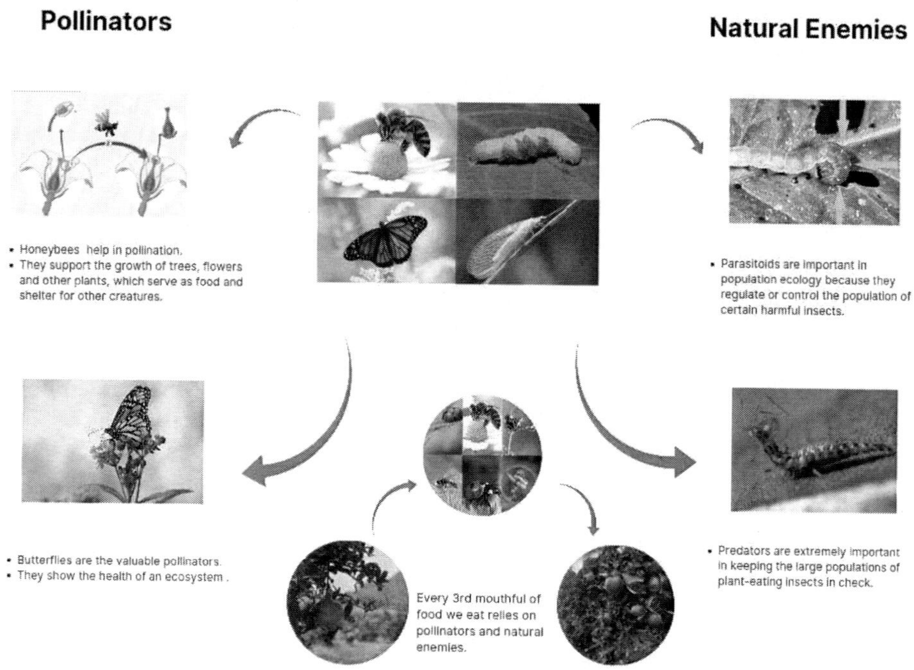


**FIGURE 10.1** Ecological services provided by beneficial insects.

level and rising temperature can significantly affect the morphology, phenology and distribution patterns of plants and their interaction with pollinators and natural enemies (Adebesin et al. 2017).

The detailed information regarding major beneficial insect groups and their vulnerability against climatic extremes is given in the following sections.

## 10.3 INSECT POLLINATORS

The survival of human life on planet earth depends on a number of ecological services provided by insects (Myers et al. 2000; Sidhu and Mehta 2008). Among these services, pollination is of great importance as it is required for proper fruit setting of crop plants (Kremen et al. 1993; Kannan and James 2009). Pollination is a natural ecological process of transmission of pollens possessing the male gametes, from anther to stigma of same and different flower, thereby enabling fertilization (Dhakal 2003; Murugan 2006). It is anticipated that insect pollination services have an annual approximate worth of $166 billion (Gallai et al. 2009). However, according to Intergovernmental Science Policy Platform on Biodiversity and Ecosystem Services (IPBES) the annual worth is closer to 235–577 billion USD. The majority of the flowering plants and 75% of all the food crops used directly by humans rely on insect pollination for their fertilization and proper fruit and seed setting (Ricketts et al. 2008). Other scientific studies also highlighted the role of entomophily in global food production as insect pollinators augment the quality and yield of different crops including vegetables, fruits, oilseeds, spices and fodders (Partap and Partap 1997; Losey and Vaughan 2006; Thapa 2006; Klein et al. 2007; Pudasaini 2014). Moreover, the stability of an ecosystem and global food security largely depend on the quality and quantity of insect pollinators (FAO 2008).

## 10.4 EFFECT OF CLIMATE CHANGE ON INSECT POLLINATORS

Currently humans are facing an era of pollination crisis primarily due to depletion of natural resources, rapidly increasing human population and anthropogenic climate change. Pollination is one of the 15 major ecological services provided by insects and is currently under severe threat due to gradually increasing negative impacts of climate change (Costanza et al. 1987; Millennium Ecosystem Assessment 2005; Biesmeijer et al. 2006; Memmott et al. 2007; Sachs 2008). Climate change not only alleviates the pollinating capability of bees, butterflies and moths but also alters their population abundance, geographical ranges, phenology and synchronization with plants (Steffan-Dewenter et al. 2005; FAO 2008; Schweizer et al. 2008; Hegland et al. 2009).

Among different pollinators, honeybees, the best-known pollinators, offer highly efficient pollination services for a range of different seasoned field crops due to their extreme floral fidelity (Calderone 2012). However, the physiology and behaviour of honeybees are greatly influenced by climate change. Rising temperature not only affects quality and quantity of floral resources but also mitigates development ability and harvesting potential of honeybees. Similarly, upsurging temperature also disturbs the foraging period of insects which results in reduced floral visits by pollinators (Corbet et al. 1993; Pudasaini 2014). The adaptation of the Saharan bee, *Apis mellifera* sahariensis to extreme heat shock conditions relies on sufficient floral water, the major part of which is utilized in regulating brood temperature (34°C–35°C) and raising their neonates. In a warmer climate, bees will be unable to get the required water and consequently they will die. As predicted by the global climatic change, some regions of the world will become drier in the near future, which will lead to the loss of not only bee food and their honey but also the pollination services they provide for stable floral diversity (Corbet et al. 1993; Conte and Navajas 2008).

Moreover, the synchronization of phenological stages of flowering plants with biological events of insect pollinators is a prerequisite for successful pollination (Rosenzweig et al. 2008). However, these biological synchronizations are greatly affected by the current events of climate change which ultimately mitigated the extent of insect pollination (Ingram et al. 1996; Kudo et al. 2004; Van Asch and Visser 2007; Post and Inouye 2008; Ricketts et al. 2008). Many bees evade the chance of pollinating plants due to the early flower blooming due to the increasing temperature. Thus, when the majority of the bees emerge the competition among them for the valuable nectar resources becomes more intense due to the limited number of flowers available in the field (Memmott et al. 2007; Thomson 2010). Furthermore, rising temperature may also lead to a possible early decline of floral resources during a growing season. This may significantly affect the efficiency of the bumblebee queen that usually requires an adequate and constant supply of pollen and nectar throughout their growing season for maintenance of a healthy colony (Aldridge et al. 2011).

The trophic mismatch or phenological decoupling happens when plant and pollinators dwelling in the same habitat use diverse environmental signals to adjust their biological events, as these signals are not equally affected by different climatic events (Gonzalez et al. 2009). Plant responses are mostly day-length dependent, whereas bee emergence is often encouraged by the consecutive number of days above a threshold temperature (Gonzalez et al. 2009).

Elevated temperature can also have impact on plant height, flower size, flower production, floral phenology as well as the composition and production of pollen and nectar which in turn have an influence on the fitness of various plant pollinators (Wahid et al. 2007; Liu et al. 2012). Moreover, elements of global climate change can alter the production of floral biogenic volatile organic compounds (BVOCs), which are important in attracting pollinators and maintaining plant-pollinator mutualistic interactions (Adebesin et al. 2017).

Moreover, decreased precipitation and increased $CO_2$ are two other important factors of climate change that influence the insect pollinators via flowering plants. Plants grown in high $CO_2$ environments usually have transformed floral traits such as pollen protein concentration (Ziska et al. 2016) and nectar composition (Hoover et al. 2012), which might have an impact on the fitness of the pollinators (Hoover et al. 2012). Similarly, synthetic nectar produced from flowers grown under elevated $CO_2$ levels results in shorter lifespan of *Bombus terrestris* workers (Hoover et al. 2012). Likewise, Ziska et al. (2016) reported that less protein in goldenrod pollen may have an adverse effect on bee performance. Similarly, drought is another major phenomenon of climate change affecting all levels of biodiversity, and its severity is anticipated to increase in the near future across all parts of the globe (Pederson et al. 2010; Dai 2013; Cook et al. 2015). Plant growth and some vital plant traits which are important for pollination as well as interactions with other species can be significantly influenced by the availability of water. For instance, drought stress decreased the nectar volume and flower size in *Epilobium angustifolium* commonly known as fireweed (Carroll et al. 2001), and in a wild species of tobacco called *Nicotiana quadrivalvis* (Halpern et al. 2010). Considering changes in quality of plant traits caused by drought, the level of herbivory and its effects are also expected to change (Huberty and Denno 2004). Indeed, in comparison with other factors of climate change, drought is likely to have the most pronounced effects on herbivory (Scherber et al. 2013). Additionally, pollination and herbivory do not happen separately, but these are simultaneous processes. Plant herbivory is a ubiquitous phenomenon and is expected to increase as the temperatures rise (Curran et al. 2010; Kessler et al. 2011). Herbivore feeding has recently been shown to affect populations of pollinators through flower volatile emissions (Theis et al. 2009; Pareja et al. 2012; Bruinsma et al. 2014). If the potential of pollinators to discover the right flowers in the environment is affected in response to climate change, agriculture and plant communities are expected to experience grave reproductive penalties (Farre-Armengol et al. 2013).

## 10.5 NATURAL ENEMIES

The regulation of pest populations through biological control is often accomplished by natural enemies, which can induce direct mortality in pests and disrupt their feeding activities, ultimately mitigating crop damage (Belmain et al. 2013; MacFadyen et al. 2015). Several insect orders contain predaceous predators, which are categorized as free-living, mobile, conspicuous, and are typically larger than their insect prey. These predators have the ability to consume multiple prey items throughout their life cycle (Getanjaly et al. 2015). The suppression of pest damage to crops and regulation of pest populations is achieved through the action of predators. Various predators, including praying mantids, spiders, earwigs, true bugs, ladybird beetles, ground beetles, lacewings and hoverfly larvae, are commonly known to prey on insect pests (Cork et al. 2009; Lee and Landis 2001).

Parasitoids are predominantly found in two insect orders, namely Hymenoptera and Diptera, and are known to have a more specialized host range than predators. These insects have free-living adult stages and actively seek out hosts, parasitizing different life stages of their prey depending on the species of parasitoid. Host location is a complex task, but parasitoids have evolved to use a combination of chemical, visual, tactile, thermal and auditory cues to detect the presence or potential presence of hosts. Parasitoids may deposit one or more eggs on or inside their host, and the immature parasitoid feeds on the host to complete development, ultimately killing the host and emerging as a free-living adult (Getanjaly et al. 2015).

Natural enemies in agricultural landscapes can prevent economically damaging pest levels. Diverse populations of natural enemies increase the effectiveness of pest control through different phenology. Integrated pest management programs can manipulate natural enemies through Classical Biological Control and Augmentative Biological Control methods (Getanjaly et al. 2015). In addition to providing ecological benefits, many predators and parasitoids also serve as valuable pollinators by feeding on nectar or pollen in their adult stage (Jeffs and Lewis 2013). Predatory flies, hover flies, predatory wasps, lacewings, lady beetles and predatory bugs are among the types of parasitoids and predators that play a double role by serving as pollinators in addition to providing natural control of insect pests (Elmqvist et al. 2010; Nicholls and Altieri 2013; Altieri et al. 2015).

## 10.6 EFFECT OF CLIMATE CHANGE ON NATURAL ENEMIES

The effectiveness of biological control through use of natural enemies relies heavily on several factors, including the location of their habitat, the ability to locate their hosts, their potential to effectively evade or manipulate the host's immune system, and their ability to continuously adapt to the host's changing environment. Temperature has a significant impact on all of these traits, making it crucial to understand the effects of temperature on natural enemies for the development of effective pest management programs (Harrington et al. 2001; Sentis et al. 2013).

### 10.6.1 Parasitoids

The effects of subjecting parasitoids to extreme temperatures have been extensively documented. These effects can take the form of either lethal or sublethal outcomes, both of which play a significant role in shaping the life-history traits of parasitoids and their effectiveness in agricultural landscapes. Exposure to extremely high or low temperatures, for instance, can lead to parasitoid mortality (Chown and Nicolson 2004; Hance et al. 2007). The reasons behind this phenomenon may include irreparable damage to the cells or alterations in physical structures resulting from the formation of intracellular or extracellular ice, particularly in the case of extreme low temperatures. Additionally, freezing has the potential to disrupt metabolic processes and cause stress related to osmotic or oxygen levels (Turnock and Fields 2005). The impact of temperature extremes on parasitoids can result in lethal effects, even at temperatures above freezing, which can disrupt the interactions between hosts and parasitoids, ultimately affecting the ecological services provided by the latter. Furthermore, parasitoids may also experience sublethal effects due to extreme temperatures, which can lead to failed biological control efficacy. For instance, restricted degree-day accumulation and longer generation times are associated with low temperatures (Lysyk 2004; Bale and Walters 2001).

In order to effectively regulate pest populations, parasitoids must possess certain attributes, such as high fecundity and proficient host searching and finding abilities, as well as long lifespans. However, exposure to extreme temperatures can hinder these attributes, resulting in negative impacts on biological control. Exposure to low temperatures has been found to decrease both adult longevity and fecundity, resulting in reduction of the ecological services provided by parasitoids (Levie et al. 2005; Pandey and Johnson 2005; Foerster and Doetzer 2006). In addition, exposure to low temperatures reduces the mobility of both male and female parasitoids, which in turn decreases their effectiveness in locating potential mates and hosts (Denlinger and Lee 1998). Parasitoids rely on endosymbiont bacteria such as *Wolbachia* and *Buchnera* species for their normal functioning. However, exposure to extreme temperatures can negatively affect or even kill these bacteria, which can have consequences on parasitoid fitness and activity (Thomas and Blanford 2003). Moreover, temperature stress has been found to increase the occurrence of morphological deformities in insects. Low temperatures directly impact differentiating tissues, disrupt hormonal balance and may result in physical deformities. Studies indicate that the severity of birth deformities is positively correlated with the duration of temperature stress (Tezze and Botto 2004).

Micha et al. (2000) reported that when plants are attacked by herbivores, they produce synomones that attract natural enemies of the herbivores. However, climate change–induced temperature increases have been shown to have a negative impact on synomone blends, leading to failed parasitoid recruitment. In addition, many parasitoids have a narrow optimal temperature range for perceiving synomones. For instance, *Cotesia plutellae* optimally responds to synomones at temperatures between 30°C and 35°C (Reddy et al. 2002). As a result, temperature fluctuations and extremes can hamper the ability of parasitoids to detect and respond to cues related to host location and habitat, ultimately affecting their effectiveness in pest control. Additionally, insects incur high costs of survival under extreme heat stress, as evidenced by the reduced efficiency of mitochondria in converting carbohydrates to energy in *Manduca sexta*. As a consequence, the decline in mitochondrial capacity may lead to a decrease in juvenile stages, such as larval growth rates. This could potentially impact parasitoid phenology, abundance and host parasitization efficiency (Martinez et al. 2017).

Bale et al. (2002) have suggested that climate-induced shifts in geographic range may occur in parasitoids, depending on their physiological adaptations and dispersal behaviour. They argued that non-diapausing insects with rapid development are more likely to expand their range under climate warming, as compared to diapausing and slow-developing parasitoids that require low temperatures for diapause induction. In addition, the availability of resources, photoperiods, predation

**TABLE 10.1**
**Impact of Climate Change on Beneficial Insect Fauna**

| Climatic Factor | Insects | Effect | References |
|---|---|---|---|
| | | **Pollinators** | |
| High temperature | Honeybee | Loss of synchronization between plant-pollinator interactions | Gupta et al. 2019 |
| Heat stress | Dragonfly specie (*Erythemis collocata*) | Reduction in wing size which affects flight performance and dispersal capacity | Filazzola et al. 2021 |
| Drought | Honeybee and butterflies | Drastically reduced per-plant-pollinator visitation rate due to decrease in plant height, flower number and nectar volume | Crowley et al. 2021 |
| Elevated $CO_2$ | Honeybee | Shift in plant-pollinator network resulting in reduced forage resources for certain pollinators<br>Loss of synchronization between plant-pollinator interaction | Crowley et al. 2021 |
| | | **Natural enemies** | |
| Elevated $CO_2$ | *Diaeretiella rapae* (a specialized parasitoid of an aphid host) | Fifty percent reduction in population size | Nayak et al. 2020 |
| High temperature | Green lacewings | Reduction in survival and fecundity | Promrak and Rattanakul 2017 |
| High temperature | *Manduca sexta* | Reduction in mitochondrial efficiency which has negative effects on phenology, abundance and efficacy of parasitoid | Martinez et al. 2017 |

by natural enemies and intra- and interspecific competition can also influence these range shifts (Gutierrez et al. 2010; Walther et al. 2002). Populations and food webs may adjust to changes in biogeographic distribution patterns through various mechanisms, including demographic patterns, physiological and phenotypic plasticity, and natural selection (Webster et al. 2016). These adjustments can have significant consequences on parasitism and other ecological interactions (Feldman et al. 2017). Agroecosystems could be affected by climate-induced biogeography in two hypothetical ways. Firstly, if parasitoid species migrate to more favourable habitats, it could lead to a decrease in parasitoid diversity and an increase in pest pressure, especially if pests do not follow similar migration patterns as their parasitoids. This could also result in limited host availability for specialist parasitoids, leading to reduced population growth and eventual extinction. Secondly, changes in distribution patterns could benefit agroecosystems by introducing new parasitoid species that increase the parasitization of pests. Furthermore, migration of pests from unsuitable climates could reduce pressure and increase yield. However, this scenario is unlikely since dormant species may become more prevalent in the absence of competition. Thus, the impacts of climate-induced biogeography on parasitoids and pests could have multifaceted consequences, affecting crop productivity and biodiversity conservation in different ways (Table 10.1).

### 10.6.2 Predators

The effects of climate change on species can range from changes in individual traits to population-level effects that alter predator-prey interactions. These interactions can be affected by shifts in plant traits, which may be mediated by climate change. Specifically, changes in nutrient content and concentrations of secondary compounds can modify plant quality and influence predator-prey dynamics. As a result, climate change is likely to modify these traits and impact plant quality, potentially altering the outcomes of predator-prey interactions (Zvereva and Kozlov 2006; Ode et al. 2014). Climate change–induced modifications in plant traits have the potential to impact herbivore foraging behaviours and nutrient content, which can in turn influence predator-prey interactions. Moreover, changes in plant phenology, distribution and abundance resulting from climate change are also likely to play a significant role in mediating these interactions between predators and prey. Together, these effects of climate change on plant communities and associated herbivore and predator populations can have far-reaching consequences for ecosystem dynamics (Ainsworth and Long 2005; Cleland et al. 2007). Climate change–related abiotic factors, including temperature, have the potential to alter species traits, including development time, oviposition rates and behavioural characteristics (Puentes et al. 2015; Bodlah et al. 2016; Xi et al. 2016). These changes in species traits can, in turn, influence the outcome of predator-prey interactions. By mediating the behaviour and physiology of predator and prey species, the effects of climate change on species traits may have cascading effects on ecosystem dynamics (Brown et al. 2004; Laparie and Renault 2016).

Changes in local community composition resulting from climate change have the potential to significantly impact predator-prey interactions over both spatial and temporal scales. As species shift their ranges in response to changing environmental conditions, new communities may arise, disrupting current interactions and establishing new ones. Furthermore, changes in phenology triggered by climate change can also affect species interactions, especially if interacting species respond differently to these changes, leading to a mismatch in their occurrences over time. These alterations in community composition and species interactions may have far-reaching implications for ecosystem dynamics and functioning (Rafferty et al. 2013). Likewise, changes in atmospheric $CO_2$ and temperature can impact the production of plant volatile organic compounds (VOCs). These compounds play a crucial role as chemical signals that facilitate herbivores in identifying suitable host plants, and they enable predators and parasitoids to locate prey and hosts, respectively. The production of VOCs may either increase or decrease due to the influence of $CO_2$ on plants (Penuelas and Staudt 2010) (Table 10.1).

## 10.7 SOCIAL INSECTS

Social insects, i.e. termites, ants and wasps, are key components of ecological communities and are important ecosystem service providers. These insects live together in large groups, communities, share food and feed and protect their young ones in colonies (Elizalde et al. 2020).

The pool of social insects is highly enriched with different invading species of termites and ants but recently an invasive wasp, i.e. Asian hornet, has also spread to different new localities at a higher pace (Lester and Beggs 2019). The majority of the social insects belong to the order Hymenoptera or Blattodea (Sumner et al. 2018). Social insects belong to a highly diverse taxonomic group that comprises a significant proportion of individuals and biomass in terrestrial habitats relative to vertebrates and other arthropods. As a result, they exert a substantial influence on the organization and operation of biological communities. Social insects are proficient environmental engineers, thus modifying the physical surroundings and influencing the accessibility of resources to other living beings. Social insects are involved in a score of important community activities varying from soil-turners, harvester of dead woods, seed dispersers and nest builders (Alonso and Agosti 2000; Kambhampati and Eggleton 2000; Kaspari 2000) (Figure 10.2).

### 10.7.1 Termites

Termites are the small eusocial insects, mainly distributed in the subtropical and tropical regions of the globe (Bignell and Eggleton 2000; Brune 2014; Bonachela et al. 2015) and constitute about 10% of all animal biomass (van Huis 2017). Termites are a highly influential group of arthropod decomposers which exert an important ecological effect on farming and non-farming biological

**FIGURE 10.2** Ecological services provided by social insects.

systems. They play a crucial role in the global carbon cycle, as well as in the decomposition process and breakdown of nutrient-rich cellulose (Bignell et al. 2011; Traore et al. 2015). Moreover, termite activities increase accessibility of nutrients, enhance soil fertility and improve infiltration of water, ultimately leading to higher crop yields (Evans et al. 2011; Lagendijk et al. 2016; Govorushko 2019). Termites have also been used by humans in artwork and literature and also as medicine and food (van Huis 2017). Furthermore, termites have also been acknowledged as weathering agents due to their capacity to alter minerals chemically (Jouquet et al. 2002; Sako et al. 2009).

### 10.7.2 Ants

Ants are a globally pervasive insect group that has successfully adapted to nearly all terrestrial niches and roles (Parker and Kronauer 2021). Their eusocial lifestyle, which involves living in colonies and dividing labour between sterile workers and reproductive castes, is a fundamental aspect of ant biology. Across the world, ants are regarded as crucial contributors to various ecological processes, including but not limited to seed dispersal, pest management and soil bioturbation. Additionally, ants are recognized as key players in shaping invertebrate communities, both through predation and competition (Del Toro et al. 2012). Ants have a significant impact on nutrient availability in ecosystems as they consume plant material, detritus and seeds while also engaging in bioturbation. Additionally, numerous ant species are involved in seed harvesting, which can heavily influence plant diversity and composition through seed predation (MacMahon et al. 2000; De Almeida et al. 2020).

### 10.7.3 Wasps

The order Hymenoptera comprises a paraphyletic assemblage of wasps, which encompasses roughly 103,000 extant species (Branstetter et al. 2017). Among these described species, approximately

70% are parasitic wasps which are actively involved in suppression of different agricultural insect pests (Quicke 2015; Wang et al. 2019). In addition, it is noteworthy that the aculeate wasps, which comprise a total of 33,000 species, exhibit a social behaviour and are categorized into the Vespid subfamilies, namely Vespinae, Polistinae and Stenogastrinae (Lester 2018; Sumner et al. 2018). Aculeate wasps are known to provide significant ecological benefits by functioning as insect predators. The hunting behaviour exhibited by predatory and parasitic aculeate wasps is believed to have significant implications for the regulation of insect populations in various habitats globally. This clearly depicts that they play a major role in stabilizing the functioning of an ecosystem by regulating the population of biodiversity (Grissell 2010). The wasp species are integral to the pollination of many plants, with active involvement in nearly 1,000 plant-pollinator interactions. About 164 plant species rely heavily on wasps for their survival and pollination. Among these, interactions between thyniid wasps and orchids are of prime importance, without which the orchids might not survive due to poor pollination. Wasps also exhibit a high degree of generalization in their pollination networks and can serve as efficient secondary pollinators. This is particularly significant as it ensures the continuity of pollination even in the absence of primary pollinators (Hallett et al. 2017). Wasps also indirectly support the ecological efficiency of biological systems through their involvement in the processes of carrion decomposition and seed dispersal. Furthermore, wasps also serve as scavengers and play important roles in the recycling of different nutrients and decomposition of dead organisms (Chen et al. 2017; Burge and Beck 2019).

## 10.8 EFFECT OF CLIMATE CHANGE ON SOCIAL INSECTS

Social insects are considered quasi-sessile as their long-distance aerial movements are only possible through alates. However, once established their nesting sites will be either fixed or limited to a small locality which makes their nests potentially more vulnerable to even micro-changes in climate (Menzel and Feldmeyer 2021). The belowground nesting sites of certain social insects like ants and termites remain susceptible to extreme climatic events such as floods and heavy rains (Figure 10.3).

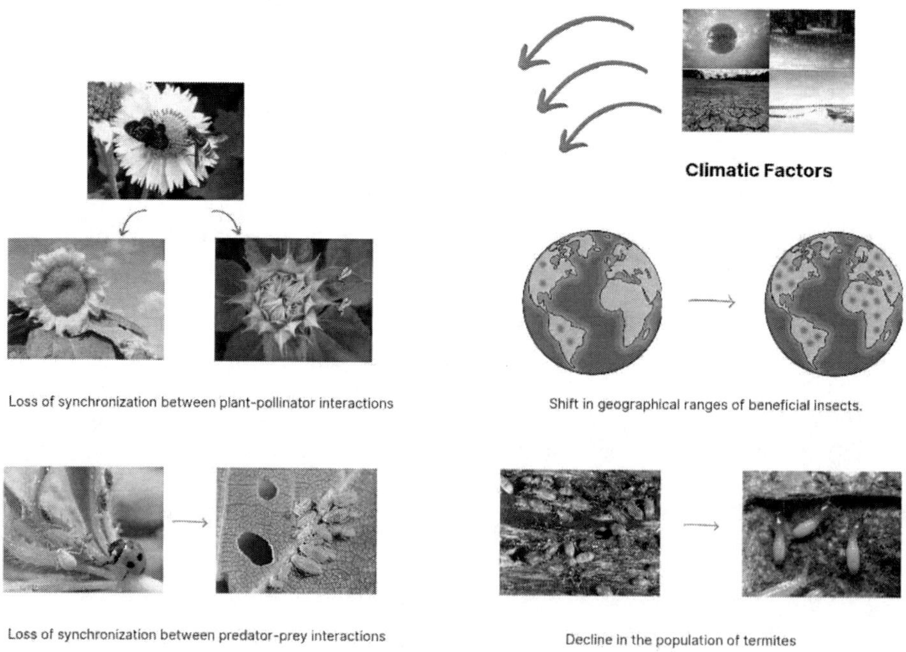


**FIGURE 10.3** Effect of climate change on beneficial and social insects.

**TABLE 10.2**
**Impact of Climate Change on Social Insects**

| Climatic Factor | Insects | Effect | References |
|---|---|---|---|
| High temperature | Ants | Destabilization of ant communities, limit foraging periods | Pelini et al. 2012 |
| | | Disrupted biotic interaction with other grassland species | Wills and Landis 2018 |
| | | Negative effects on ant's survival, fecundity, generation time and dispersal | |
| High temperature | Social wasps | Reduction in wing size which had negative impact on foraging efficiency, escape ability and competition outputs among others | Polidori et al. 2020 |
| Rainfall | Termites | Termite diversity decreased in rainforests | Clement et al. 2021 |
| Increased desiccation | Termites | Water loss in termites which results in change in morphology and size reduction | Woon et al. 2019 |

Drought is another major consequence of climate change and its severity is increasing with every passing day. The exacerbated effects of drought with rising temperature enhance metabolic rates of insects and ultimately cause desiccation (Gibbs et al. 2003). However, this water loss can be adjusted by the cuticular hydrocarbon layer of insects which acts as a desiccation barrier (Menzel et al. 2018; Menzel and Feldmeyer 2021). Nonetheless, social insects with non-flying workers such as termites and ants are more vulnerable to drought periods as they cannot fly to distant water sources. Termite distributions and foraging activity in harvester ants largely depend on water availability (Friedman et al. 2019; Woon et al. 2019). Moreover, young queens are also vulnerable to drought as they cannot penetrate the hard and dry soil for their nesting sites (Menzel and Feldmeyer 2021) (Table 10.2).

Similarly, warmer temperatures tend to influence ant populations directly through effects on survival, fecundity, generation time and dispersal (Pelini et al. 2012). Ants are highly sensitive to temperature variations and are considered cryophobic or thermophilic. Moreover, at extreme high and low temperatures, ant performance and metabolism drop considerably (Hurlbert et al. 2008; Berman et al. 2010; Jenkins et al. 2011) (Table 10.2). Furthermore, it was also noted that seasonal variations in temperature force ants to relocate their established nesting sites to more suitable thermal environments (Heller and Gordon 2006; McGlynn et al. 2010). However, most of the social insects develop mechanisms to regulate nest climate. Nests are usually constructed in such ways that support optimal aeration and climate inside (Jones and Oldroyd 2006; King et al. 2015; Tschinkel 2015). Another way to regulate nest climate is through fanning or carrying water into the nest by workers. However, in this case suitable nearby water resources were needed which are often difficult to locate in drought periods (Weidenmuller 2004; Klingner et al. 2005; Kovac et al. 2009; Friedman et al. 2019).

## 10.9 CONSERVATION OF INSECTS IN RESPONSE TO CLIMATE CHANGE

Conserving biodiversity of beneficial and social insects is very important for humans due to their diverse ecological services which they provide for us. In this regard integrated interventions must be followed to conserve an adequate number of these beneficial insects in the ecosystem. Some of the important aspects regarding biodiversity conservation of beneficial and social insects are provided in the following sections.

### 10.9.1 Developing Appreciation for Insects

Improving communication about the significance of these crucial insect groups to humanity is the first step towards better appreciation and conservation. Valuing them is essential, as it provides a basis for our actions and helps establish standards and directions. One way to achieve this is by emphasizing the importance of personal and collective well-being and using conservation strategies that prioritize relational values. By focusing on these values, we can increase the effectiveness of our efforts to protect these insect groups (Chan et al. 2016). The significance of these insect groups lies in the interdependence and shared destiny that exist within their ecosystems. Valuing insects is an ethical imperative and goes beyond mere economic considerations (Samways 2017; Basset and Lamarre 2019). However, we must not ignore the fact that certain insect species are medically significant or pose a threat as invasive alien organisms (Bradshaw et al. 2016). Conversely, some insects hold immense practical value as natural enemies of forestry and agricultural pests (Hajek et al. 2016).

Despite the many admirable qualities of insects, it is important to acknowledge that many people perceive them as unattractive, insignificant, aggressive, indestructible and carriers of diseases (Nash 2004). Identifying nonhuman characteristics that make insects more appealing to humans is a vital step in fostering a closer relationship between insects and humans and promoting the joint destiny of these two groups (Lorimer 2007). Nevertheless, we must also be aware that our tendency towards anthropomorphism can introduce biases that might affect our approach to wildlife conservation as a whole (Manfredo et al. 2020). Using insect icons and flagship species can be immensely beneficial in our efforts to enhance insect conservation on a global scale (Barua et al. 2012).

### 10.9.2 Ecological Assessment in Insect Conservation

The aim of the global conservation initiatives is to conserve and manage 17% of the global terrestrial area for supporting insect biodiversity. The conserved areas must be ecologically representative and well managed bearing all the necessary requirements to support diverse insect biodiversity. However, it will be difficult to manage all the vulnerable insect species with this approach. The functional connectivity between protected areas enables range expansion and helps in better conservation of threatened species. These areas must also include specialized sites for irreplaceable insect fauna. Moreover, protected areas must be extended on a global scale outside the designated borders using widespread ecological networks which will be highly effective in conserving insect biodiversity at a global level (Samways et al. 2020).

### 10.9.3 Insect Indicators

The presence or absence of certain insects within a community illustrates the effect of climatic changes on insect habitats, state of insect diversity, stability of communities or ecosystem in which they dwell and biotic and abiotic factors of that particular area. Their presence or particular number helps us to gauge the stress level of that system, or retrieval from stressors following interventions. These indications are usually based on some frameworks having specified goals or some reference values which demonstrate the state of an ecosystem (Samways et al. 2020).

## 10.10 CONCLUSION

Currently, many insect groups or taxa are facing severe declines in both tropical and temperate ecosystems primarily due to the rapidly mounting wave of global climate change. It is important to recognize that human-mediated changes to the environment such as pollution, habitat fragmentation/loss and introduction of invasive species are the main drivers of current insect declines around the globe. Apart from their direct effects, climate change will also intensify the negative effects of other factors such as human land use and hinder the insect's ability to adapt or avoid multiple anthropogenic stressors. In such cases, migration to new locality will not be fruitful for insects as these localities have already been changed into unsuitable habitats by human land use. Climate change may have a different effect on each species in food webs ultimately leading to loss of valuable resources due to phenological mismatches. The decline in the insect population will not affect their immediate ecological services but will affect those species which are relying on these insects for their survival. Hence, the role of climate change on insect decline needs incessant attention due to their vital role in broader ecosystem services. By restoring ecological stability in farming ecosystems and by conserving beneficial insect fauna, substantially mitigating societal benefits can be improved and human well-being can be achieved. Keeping in view the unabated effects of climatic extremes, it is important to make certain insect habitat locations into "climate-proof" places in which insects can ride out extreme climatic events with ease. Moreover, corridors should be maintained that allow insects to find more suitable places to avoid adverse effects of climate change. Insects should be conserved for the prosperity of our future generations, and we should adopt immediate measures at local, regional and global levels to conserve their diversity.

## REFERENCES

Aagustí, M. B., Sutter, L., Bocci, G., Moonen, A. C., Cresswell, J., Entling, M., Giffard, B., Jacot, K., Jeanneret, J., Hollan, J., Pfister, S., Pintér, O., Veromann, E., Winkler, K., and Albrecht, M. 2019. The potential of different semi-natural habitats to sustain pollinators and natural enemies in European agricultural landscapes. *Agriculture Ecosystems and Environment* 279. https://doi.org/10.1016/j.agee.2019.04.009.

Adebesin, F., Widhalm, J. R., Boachon, B., Lefèvre, F., Pierman, B., Lynch, J. H., and Porter, J. A. 2017. Emission of volatile organic compounds from petunia flowers is facilitated by an ABC transporter. *Science* 356: 1386–1388.

Ainsworth, E. A., and Long, S. P. 2005. What have we learned from 15 years of free-air CO2 enrichment (FACE)? A meta-analytic review of the responses of photosynthesis, canopy. *New Phytology* 165: 351–371.

Aldridge, G., Inouye, D. W., Forrest, J. R. K., Barr, W. A., and Miller-Rushing, A. J. 2011. Emergence of a mid-season period of low floral resources in a montane meadow ecosystem associated with climate change. *Journal of Ecology* 99: 905–913.

Alonso, L. E., and Agosti, D. 2000. Biodiversity studies, monitoring, and ants: an overview. In: Ants Đ Standard Methods for Measuring and Monitoring Biodiversity, Eds. D. Agosti, J. D. Majer, L. E. Alonso, and T. R. Schultz, Washington, DC and London: Smithsonian Institution Press. pp. 1–8.

Altieri, M. A., Nicholls, C. I., Gillespie, M., Waterhouse, B., Wratten, S., Gbèhounou, G., and Gemmill-Herren, B. 2015. Crops, Weeds and Pollinators Understanding Ecological Interaction for Better Management. Rome: FAO. p. 95.

Bale, J. S. B., Masters, G. J., Hodkinson, I. D., Awmack, C., Bezemer, T. M., and Brown, V. K. 2002. Herbivory in global climate change research: direct effects of rising temperature on insect herbivores. *Global Change in Biology* 8: 1–16.

Bale, J. S., and Walters, K. F. A. 2001. Overwintering biology as a guide to the establishment potential of non-native arthropods in the UK. In: Environment and Animal Development: Gens, Life Histories and Plasticity, Eds. D. Atkinson and M Thorndyke, Oxford: Bios. pp. 343–354.

Barua, M., Gurdak, D. J., Ahmed, R. A., and Tamuly, J. 2012. Selecting flagships for invertebrate conservation. *Biodiversity Conservation* 21: 1457–1476.

Basset, Y., and Lamarre, G. P. A. 2019. Toward a world that values insects. *Science* 364: 1230–1231.

Belmain, S., Haggar, J., Holt, J., and Stevenson, P. 2013. Managing Legume Pests in Sub-Saharan Africa: Challenges and Prospects for Improving Food Security and Nutrition through Agro-Ecological Intensification. Chatham Maritime: Natural Resources Institute, University of Greenwich. p. 34.

Berman, D. I., Alfimov, A. V., Zhigulskaya, Z. A., and Leirikh, A. N. 2010. Overwintering and Cold Hardiness of Ants in the Northeast of Asia. Sofia-Moscow: Pensoft. p. 294.

Biesmeijer, J. C., Roberts, S. P. M., Reemer, M., Ohlemuller, R., Edwards, M., and Peeters, T. 2006. Parallel declines in pollinators and insect-pollinated plants in Britain and the Netherlands. *Science* 313: 351–354.

Bignell, D. E., Roisin, Y., and Lo, N. 2011. Biology of Termites: A Modern Synthesis. Dordrecht: Springer. p. 576.

Bignell, D. E., and Eggleton, P. 2000. Termites in ecosystems. In: Termites: Evolution, Sociality, Symbiosis, Ecology, Eds. T. Abe, D. E. Bignell, and H. Higashi, Dordrecht: Kluwer Academic Publisher. pp. 363–387.

Bodlah, M. A., Zhu, A. X., and Liu, X. D. 2016. Host choice, settling and folding leaf behaviors of the larval rice leaf folder under heat stress. *Bull Entomology Research* 106: 809–817.

Bonachela, J. A., Pringle, R. M., Sheffer, E., Coverdale, T. C., Guyton, J. A., and Caylor, K. K. 2015. Termite mounds can increase the robustness of dryland ecosystems to climatic change. *Science* 347: 651–655.
Bradshaw, C. J. A., Leroy, B., Bellard, C., Albert, C., Roiz, D., Barbet, M. M., Fournier, A., Salles, J. M., Simard, F., and Courchamp, F. 2016. Massive yet grossly underestimated global costs of invasive insects. *Nature Commun*ication 7: 12986.
Branstetter, M. G., Danforth, B. N., Pitts, J. P., Faircloth, B. C., Ward, P. S., Buffington, M. L., Gates, M. W., Kula, R. R., and Brady, S. G. 2017. Phylogenomic insights into the evolution of stinging wasps and the origins of ants and bees. *Current Biology* 27: 1019–1025.
Brown, J. H., Gillooly, J. F., Allen, A. P., Savage, V. M., and West, G. B. 2004. Toward a metabolic theory of ecology. *Ecology* 85: 1771–1789.
Bruinsma, M., Lucas-Barbosa, D., ten Broeke, C. J. M., van Dam, N. M., van Beek, T. A., Dicke, M., and van Loon, J. J. A. 2014. Folivory affects composition of nectar, floral odor and modifies pollinator behavior. *Journal of Chemical Ecology* 40: 39–49.
Brune, A. 2014. Symbiotic digestion of lignocellulose in termite guts. *Nature Reviews Microbiology* 12: 168–180.
Burge, D. O., and Beck, J. J., 2019. Dispersal of spicebush (*Calycanthus occidentalis*) by yellow jackets (genus Vespula; Hymenoptera: Vespidae). *Madrono* 66: 41–46.
Calderone, N. W. 2012. Insect pollinated crops, insect pollinators and US agriculture: trend analysis of aggregate data for the period 1992–2009. *PLOS ONE* 7: e37235.
Carroll, A. B., Pallardy, S.G., and Galen, C. 2001. Drought stress, plant water status, and floral trait expression in fireweed (*Epilobium angustifolium*) (Onagraceae). *American Journal of Botany* 88: 438–446.
Chan, K. M. A., Balvanera, P., Benessaiah, K., Chapman, M., Díaz, S., and Gómez-Baggethun 2016. Opinion: why protect nature? Rethinking values and the environment. *PNAS* 113: 1462–1465.
Chen, G., Wang, Z. W., Qin, Y., and Sun, W. B. 2017. Seed dispersal by hornets: an unusual insect-plant mutualism. *Journal of Integrative Plant Biology* 59: 792–796.
Chown, S. L., and Nicolson, S. 2004. Insect Physiological Ecology: Mechanisms and Patterns. Oxford: Oxford University Press.
Cleland, E. E., Chuine, I., Menzel, A., Mooney, H. A., and Schwartz, M. D. 2007. Shifting plant phenology in response to global change. *Trends Ecological Evolution* 22: 357–365.
Clement, R. A., Flores-Moreno, H., Cernusak, L. A., Cheesman, A. W., Yatsko, A. R., Allison, S. D., Egglepon P., and Zanne, A. E. 2021. Assessing the Australian termite diversity anomaly, how habitat and rainfall affect termite assemblages. *Frontiers in Ecology and Evolution* 9: 657444.
Conte, Y. L., and Navajas, M., 2008. Climate change: impact on honey bee populations and diseases. *Revue scientifique et technique (International Office of Epizootics)* 27: 499–510.
Cook, B. I., Ault, T. R., and Smerdon, J. E. 2015. Unprecedented 21st century drought risk in the American Southwest and Central Plains. *Science Advances* 1: e1400082.
Corbet, S. A., Fussell, M., Ake, R., Fraser, A., Gunson, C., Savage, A., and Smith, K. 1993. Temperature and the pollinating activity of social bees. *Ecological Entomology* 18: 17–30.
Cork, A., Dobson, H., Grzywacz, D., Hodges, R., Orr, A., and Stevenson, P. 2009. Review of Pre- and Post-Harvest Pest Management for Pulses with Special Reference to East and Southern Africa. London: Natural Resources Institute, University of Greenwich.
Costanza, R., d'Arge, R., de Groot, R., Farber, S., Grasso, M., Hannon, B., Limburg, K., Naeem, S., O'Neill, R. V., Paruelo, J., Raskin, R. G., Sutton, P., and van den Belt, M. 1987. The value of the world's ecosystem services and natural capital. *Nature* 387: 253–260.
Crossley, M. S., Meier, A. R., Baldwin, E. M., Berry, L. L., Crenshaw, L. C., Hartman, G. L., Lagos-Kutz, D., Nichols, D. H., Patel, K., and Varriano, S. 2020. No net insect abundance and diversity declines across us long term ecological research sites. *Nature Ecology and Evolution* 4: 1368–76.
Crowley, L. M., Sadler, J. P., Pritchard, J., and Hayward, S. A. L. 2021. Elevated $CO_2$ impacts on plant-pollinator interactions: a systematic review and free air carbon enrichment field study. *Insects* 12: 512.
Curran, E. D., Labandeira, C. C., and Wilf, P. 2010. Fossil insect folivory tracks paleotemperature for six million years. *Ecological Monographs* 80: 547–567.
Dai, A. 2013. Increasing drought under global warming in observations and models. *Nature Climate Change* 3: 52–58.
Dangles, O., and Casas, J. 2019. Ecosystem services provided by insects for achieving sustainable development goals. *Ecosystem Services* 35: 109–15.
De Almeida, T., Blight, O., Mesléard, F., Bulot, A., Provost, E., and Dutoit, T. 2020. Harvester ants as ecological engineers for Mediterranean grassland restoration: impacts on soil and vegetation. *Biological Conservation* 245: 108547.

Del Toro, I., Ribbons, R. R., and Pelini, S. L. 2012. The little things that run the world revisited: a review of ant-mediated ecosystem services and disservices (Hymenoptera: Formicidae). *Myrmecological News* 17: 133–146.

Denlinger, D. L., and Lee, R. E. 1998. Physiology of cold sensitivity. In: Temperature Sensitivity in Insects and Application in Integrated Pest Management, Eds. G. J. Hallman and D. L. Denlinger, Boulder, CO: Westview. pp. 55–96.

Dhakal, G. C. 2003. A comparative study of *Apis cerana* F. and *Apis mellifera* L. On pollination of *Brassica campestris var. Toria* and *Fagopyrum esculentum* M. at Rampur, Chitwan. M. Sc. In Entomology Thesis TU. Chitwan: IAAS Rampur. p. 76.

Didham, R. K., Barbero, F., Collins, C. M., Forister, M. L., Hassall, C., Leather, S. R., Packer, L., Saunders, M. E., and Stewart, A. J. 2020. Spotlight on insects: trends, threats and conservation challenges. *Insect Conservation and Diversity* 13: 99–102.

Elizalde, L., Marina, A., Xavier, A., Paul, E., Inara, R. L., María, N. L., Agustín S., Victoria, W., and Gabriela I. P. 2020. The ecosystem services provided by social insects: traits, management tools and knowledge gaps. *Biological Reviews* 95: 1418–1441.

Elmqvist, T., Malty, E., Barker, T., Mortimer, M., Perrings, C., Aronson, J., and Norberg, J. 2010. Biodiversity, ecosystems and ecosystem services. In: The Economics of Ecosystems and Biodiversity: The Ecological and Economic Foundations, Ed. P. Kumar, London: Earthscan. pp. 41–111.

Evans, T. A., Dawes, T. Z., Ward, P. R., and Lo, N. 2011. Ants and termites increase crop yield in a dry climate. *Nature Communications* 2: 262.

FAO, 2008. Food and Agriculture Organization of the United Nations. Rapid Assessment of Pollinators' Status. Rome: FAO.

Farre-Armengol, G., Filella, I., Llusia, J., and Penuelas, J. 2013. Floral volatile organic compounds: between attraction and deterrence of visitors under global change. *Perspectives in Plant Ecology, Evolution and Systematics* 15: 56–67.

Feldman, R. E., Peers, M. J. L., Pickles, R. S. A., Thornton, D., and Murray, D. L. 2017. Climate driven range divergence among host species affects range-wide patterns of parasitism. *Global Ecology Conservation* 9: 1–10.

Filazzola, A., Matter, S. F., and MacIvor, J. S. 2021. The direct and indirect effects of extreme climate events on insects. *Science of the Total Environment* 769: 145161.

Foerster, L. A., and Doetzer, A. K. 2006. Cold storage of the egg parasitoids *Trissolcus basalis* (Wollaston) and *Telenomus podisi* Ashmead (Hymenoptera: Scelionidae). *Journal of Biological Control* 36: 232–237.

Friedman, D. A., Greene, M. J., and Gordon, D. M. 2019. The physiology of forager hydration and variation among harvester ant (*Pogonomyrmex barbatus*) colonies in collective foraging behavior. *Science Republicans* 9: 1–9.

Gallai, N., Salles, J. M., Settele, J., and Vaissiere, B. E. 2009. Economic valuation of the vulnerability of world agriculture confronted with pollinator decline. *Ecological Economics* 68: 810–821.

García-Robledo, C., Baer, C. S., Lippert, K., and Sarathy V. 2020. Evolutionary history, not ecogeographic rules, explains size variation of tropical insects along elevational gradients. *Functional Ecology* 34: 2513–23.

Getanjaly, Rai, V. L., Sharma, P., and Kushwaha, R. 2015. Beneficial insects and their value to agriculture. *Research Journal of Agriculture and Forestry Sciences* 3: 25–30.

Gibbs, A. G., Fukuzato, F., and Matzkin, L. M. 2003. Evolution of water conservation mechanisms in Drosophila. *Journal Experimental Biology* 206: 1183–1192.

Gonzalez, A. M. M., Dalsgaard, B., Ollerton, J., Timmermann, A., Olesen, J. M., Andersen, L., and Tossas, A. G. 2009. Effects of climate on pollination networks in the west indies. *Journal of Tropical Ecology* 25: 493–506.

Goulson, D. 2019. The insect apocalypse, and why it matters. *Current Biology* 29: 967–971.

Govorushko, S. 2019. Economic and ecological importance of termites: a global review. *Entomological Science* 22: 21–35.

Grissell, E. 2010. Bees, Wasps, and Ants: The Indispensable Role of Hymenoptera in Gardens. Portland, OR: Timber Press.

Gupta, H., Tiwari, C., and Diwakar, S. 2019. Butterfly diversity and effect of temperature and humidity gradients on butterfly assemblages in a sub-tropical urban landscape. *Tropical Ecology* 60: 150–158.

Gutierrez, A. P., Ponti, L., and Gilioli, G. 2010. Climate change effects on plant-pest-natural enemy interactions. In: Handbook of Climate Change and Agroecosystems: Impacts, Adaptation, and Mitigation, Eds. D. Hillel and C. Rosenzweig, London: Imperial College Press. pp. 209–237.

Hajek, A. E., Hurley, B. P., Kenis, M., Garnas, J. R., Bush, S. J., and Wingfield, M. J. 2016. Exotic biological control agents: a solution or contribution to arthropod invasions? *Biological Invasions* 18: 953–969.

Hallett, A. C., Mitchell, R. J., Chamberlain, E. R., and Karron, J. D. 2017. Pollination success following loss of a frequent pollinator: the role of compensatory visitation by other effective pollinators. *AoB Plants* 9: 20.

Halpern, S. L., Adler, L. S., and Wink, M. 2010. Leaf herbivory and drought stress affect floral attractive and defensive traits in Nicotiana quadrivalvis. *Oecologia* 163: 961–971.

Halsch, C. A., Shapiro, A. M., Fordyce, J. A., Nice, C. C., Thorne, J. H., Waetjen, D. P., and Forister, M. L. 2021. Insects and recent climate change. *Proceedings of the National Academy of Sciences* 118: e2002543117.

Hance, T., van Baaren, J., Vernon, P., and Boivin, G. 2007. Impact of extreme temperatures of parasitoids in a climate change perspective. *Annual Review of Entomology* 52: 107–126.

Harrington, R., Fleming, R. A., and Woiwod, I. 2001. Climate change impacts on insect management and conservation in temperate regions. Can they be predicted? *Agriculture for Meteorology* 3: 233–240.

Hegland, S. J., Nielsen, A., Lazaro, A., Bjerknes, A., and Totland, O. 2009. How does climate warming affect plant-pollinator interactions? *Ecology Letters* 12: 184–195.

Heller, N. E., and Gordon, D. M. 2006. linking temporal and spatiel scales in the study of an Argentine ant invasion. *Biological Invasions* 8: 501–507.

Hoover, S. E. R., Ladley, J. J., Shchepetkina, A. A., Tisch, M., Gieseg, S. P., and Tylianakis, J. M. 2012. Warming, $CO_2$, and nitrogen deposition interactively affect a plant-pollinator mutualism. *Ecology Letters* 15: 227–234.

Hubert, A. F., and Denno, R. F. 2004. Plant water stress and its consequences for herbivorous insects: a new synthesis. *Ecology* 85: 1383–1398.

Hurlbert, A. H., Ballantyne, F., and Powell, S. 2008. Shaking a leg and hot to trot: the effects of body size and temperature on running speed in ants. *Ecological Entomology* 33: 144–154.

Ingram, M., Nabhan, G., and Buchmann, S. 1996. Our forgotten pollinators: protecting the birds and bees. *Global Pesticide Campaigner* 6: 1–12.

IPCC. 2021. Climate change 2021: the physical science basis. In: Contribution of Working Group-I to the Sixth Assessment Report of the Intergovernmental Panel on Climate Change, Eds. V. Masson-Delmotte, P. Zhai, A. Pirani, S. L. Connors, C. Péan, S. Berger, N. Caud, Y. Chen, L. Goldfarb, M. I. Gomis, M. Huang, K. Leitzell, E. Lonnoy, J. B. R. Matthews, T. K. Maycock, T. Waterfield, O. Yelekçi, R. Yu, and B. Zhou, Cambridge and New York, NY: Cambridge University Press.

Janzen, D. H., and Hallwachs, W. 2021. To us insectometers, it is clear that insect decline in our Costa Rican tropics is real, so let's be kind to the survivors. *Proceedings of the National Academy of Sciences* 118: e2002546117.

Jarvis, B. 2018. The insect apocalypse is here. *The New York Times*, 27.

Jeffs, C. T., and Lewis, O. T. 2013. Effects of climate warming on host-parasitoid interactions. *Ecological Entomology* 38: 209–218.

Jenkins, C. N., Sanders, N. J., Andersen, A. N., Arnan, X., Brühl, C. A., Cerda, X., Ellison, A. M., Fisher, B. L., Fitzpatrick, M. C., Gotelli, N. J., Gove, A. D., Guénard, B., Lattke, J. E., Lessard, J.-P., McGlynn, T. P., Menke, S. B., Parr, C. L., Philpott, S. M., Vasconcelos, H. L., and Dunn, R. R. 2011. Global diversity in light of climate change: the case of ants. *Diversity and Distributions* 17: 652–662.

Jones, J. C., and Oldroyd, B. P. 2006. Nest thermoregulation in social insects. *Advances in Insect Physics* 33: 153–191.

Jouquet, P., Lepage, M., and Velde, B. 2002. Termite soil preferences and particle selection: strategies related to ecological requirements. *Insectes Sociaux* 49: 1–7.

Kambhampati, S., and Eggleton, P. 2000. Taxonomy and phylogeny of termites. In: Termites: Evolution, Sociality, Symbioses, Ecology, Dordrecht: Springer. pp. 1–23.

Kannan, R., and James, D. A. 2009. Effects of climate change on global diversity: a review of key literature. *Tropical Ecology* 50: 31–39.

Kaspari, M. 2000. A primer on ant ecology. In: Ants Ð Standard Methods for Measuring and Monitoring Biodiversity, Eds. D. Agosti, J. D. Majer, L. E. Alonso, and T. R. Schultz, Washington, DC and London: Smithsonian Institution Press. pp. 9–24.

Kessler, A., Halitschke, R., and Poveda, K. 2011. Herbivory-mediated pollinator limitation: negative impacts of induced volatiles on plant-pollinator interactions. *Ecology* 92: 1769–1780.

King, H., Ocko, S., and Mahadevan, L. 2015. Termite mounds harness diurnal temperature oscillations for ventilation. *Proceedings of the National Academy of Sciences of the United States of America* 112: 11589–11593.

Klein, A. M., Vaissere, B. E., Cane, J. H., Steffan-Dewenter, I., Cunnigham, S. A., Kremen, C., and Tscharntke, T. 2007. Importance of pollinators in changing landscapes for world crops. *Proceedings of the Royal Society B: Biological Sciences* 274: 303–313.

Klingner, R., Richter, K., Schmolz, E., and Keller, B. 2005. The role of moisture in the nest thermoregulation of social wasps. *Naturwissenschaften* 92: 427–430.

Kovac, H., Stabentheiner, A., and Schmaranzer, S. 2009. Thermoregulation of water foraging wasps (*Vespula vulgaris* and *Polistes dominulus*). *Journal of Insect Physiology* 55: 959–966.

Kremen, C., Colwell, R. K., Erwin, T. L., Murphy, D. D., Noss, R. F., and Sanjayan, M. A. 1993. Terrestrial arthropod assemblages: their use in conservation planning. *Conservation Biology* 7: 796–808.

Kudo, G., Ishikawa, Y., Kasagi, T., and Kosuge, S. 2004. Does seed production of spring ephemerals decrease when spring comes early? *Ecological Research* 19: 255–259.

Lagendijk, D., Davies, A., Eggleton, P., and Slotow, R. 2016. No evidence for an elephant-termite feedback loop in Sand Forest, South Africa. *Biological Conservation* 203: 125–133.

Laparie, M., and Renault, D. 2016. Physiological responses to temperature in *Merizodus soledadinus* (Col., Carabidae), a subpolar carabid beetle invading sub-Antarctic islands. *Polar Biology* 39: 35–45.

Lead, C., Kumar, P., Brondizio, E., Elmqvist, T., Gatzweiler, F., Gowdy, J., and Reyers, B. 2009. The Economics of Ecosystems and Biodiversity: The Ecological and Economic Foundation. London and Washington, DC: Earthscan.

Lee, J. C. T., and Landis, D. A. 2001. Natural Enemies in Field Crops: A Guide to Biological Control. East Lansing, MI: Michigan State University Extension.

Lester, P. J. 2018. The Vulgar Wasp: The Story of a Ruthless Invader and Ingenious Predator. Wellington: Victoria University Press.

Lester, P. J., and Beggs, J. R. 2019. Invasion success and management strategies for social *Vespula wasps*. *Annual Review of Entomology* 64: 51–71.

Levie, A., Vernon, P., and Hance, T. 2005. Consequences of acclimation on survival and reproductive capacities of cold-stored mummies of *Aphidius rhopalosiphi* (Hymenoptera: Aphidiinae). *Journal of Economic Entomology* 98: 704–708.

Liu, Y., Mu, J., Niklas, K. J., Li, G., and Sun, S. 2012. Global warming reduces plant reproductive output for temperate multi-inflorescence species on the Tibetan plateau. *New Phytologist* 195: 427–436.

Lorimer, J. 2007. Nonhuman charisma. *Environment and Planning D: Society and Space* 25: 911–932.

Losey, D., and Vaughan, M. 2006. The economic value of ecological services provided by insects. *Bioscience* 15: 311–324.

Lysyk, T. J. 2004. Effects of cold storage on the development and survival of three species of parasitoids (Hymenoptera: Pteromalidae) of house fly, *Musca domestica* L. *Environmental Entomology* 33: 823–831.

MacFadyen, S., Davies, A. P., and Zalucki, M. P. 2015. Assessing the impact of arthropod natural enemies on crop pests at the field scale. *Insect Science* 22: 20–34.

MacMahon, J. A., Mull, J. F., and Crist, T. O. 2000. Harvester ants (*Pogonomyrmex* spp.): Their community and ecosystem influences. *Annual Review of Ecology and Systematics* 31: 265–291.

Manfredo, M. J., Urquiza-Haas, E. G., Don Carlos, A. W., Bruskotter, J. T., and Dietsch, A. M. 2020. How anthropomorphism is changing the societal context of modern wildlife conservation. *Biological Conservation*. https://doi.org/10.1016/j.biocon.2019.108297.

Martinez, E., Menze, M. A., and Agosta, M. J. 2017. Reduced mitochondrial efficiency explains mismatched growth and metabolic rate at supraoptimal temperatures. *Physiology Biochemistry Zoology* 90: 294–298.

McGlynn, T. P., Dunn, T., Wayman, E., and Romero, A. 2010. A thermo-phile in the shade: light-directed nest relocation in the Costa Rican ant *Ectatomma ruidum*. *Journal of Tropical Ecology* 26: 559–562.

Memmott, J., Craze, P. G., Waser, N. M., and Price, M. V. 2007. Global warming and the disruption of plant-pollinator interactions. *Ecology Letters* 10: 710–717.

Menzel, F., and Feldmeyer, B. 2021. How does climate change affect social insects? *Current Opinion in Insect Science* 46: 10–15.

Menzel, F., Zumbusch, M., and Feldmeyer, B. 2018. How ants acclimate: impact of climatic conditions on the cuticular hydrocarbon profile. *Functional Ecology* 32: 657–666.

Micha, S. G., Kistenmacher, S., Molck, G., and Wyss, U. 2000. Tritrophic interactions between cereals, aphids and parasitoids: discrimination of different plant host complexes by *Aphidius rhopalosiphi* (Hymenoptera: Aphidiidae). *European Journal of Entomology* 97: 539–543.

Millennium Ecosystem Assessment. 2005. Ecosystems and Human Wellbeing: Synthesis. Washington, DC: Island Press.

Murugan, K. 2006. Bio-diversity of insects. *Current Science* 91: 1602–1603.

Myers, N., Mittermeier, R. A., Mittermeier, C. G., Da-Fonseca, G. A. B., and Kent, J. 2000. Biodiversity hotspots for conservation priorities. *Nature* 403: 853–858.

Nash, S. 2004. Desperately seeking charisma: improving the status of invertebrates. *BioScience* 54: 487–494.

Nayak, S. B., Rao, K. S., and Ramalakshmi, V. 2020. Impact of climate change on insect pests and their natural enemies. *International Journal of Ecology and Environmental Sciences* 2: 579–584.

Nicholls, C. I., and Altieri, M. A. 2013. Plant biodiversity enhances bees and other insect pollinators in agroecosystems. A review. *Agronomy for Sustainable Development* 33: 257–274.

Ode, P. J., Johnson, S. N., and Moore, B. D. 2014. Atmospheric change and induced plant secondary metabolites—are we reshaping the building 425 blocks of multi-trophic interactions? *Current Opinion in Insect Science* 5: 57–65.

Pandey, R. R., and Johnson, M. W. 2005. Effects of cold storage on *Anagyrus ananatis* Gahan (Hymenoptera: Encyrtidae). *Biological Control* 35: 9–16.

Pareja, M., Qvarfordt, E., Webster, B., Mayon, P., Pickett, J., Birkett, M., and Glinwood, R. 2012. Herbivory by a phloem-feeding insect inhibits floral volatile production. *PLOS ONE* 7: e31971.

Parker, J., and Kronauer, D. J. 2021. How ants shape biodiversity. *Current Biology* 31: 1208–1214.

Partap, U., and Partap, T. 1997. Managed crop pollination. The missing dimension of mountain crop productivity. Discussion paper series No. MFS 97/1, ICIMOD, Kathmandu, Nepal. p. 26.

Pederson, G. T., Graumlich, L. J., Fagre, D. B., Kipfer, T., and Muhlfeld, C. C. 2010. A century of climate and ecosystem change in Western Montana: what do temperature trends portend? *Climatic Change* 98: 133–154.

Pelini, S. L., Diamond, S. E., MacLean, H., Ellison, A. M., Gotelli, N. J., Sanders, N. J., and Dunn, R. R. 2012. Common garden experiments reveal uncommon responses across temperatures, locations, and species of ants. *Ecology and Evolution* 2: 3009–3015.

Penuelas, J., and Staudt, M. 2010. BVOCs and global change. *Trends in Plant Sci*ence 15: 133–144.

Polidori, C., Gutierrez-Canovas, C., Sanchez, E., Tormos, J., Castro, L., and Sanchez-Fernandez, D. 2020. Climate change-ribbon body size shrinking in a social wasp. *Ecological Entomology* 45: 130–141.

Post, E. S., and Inouye, D. W. 2008. Phenology: response, driver, and integrator. *Ecology* 89: 319–320.

Promrak, J., and Rattanakul, C. 2017. Effect of increased global temperatures on biological control of green lacewings on the spread of mealybugs in a cassava field: a simulation study. *Advances in Difference Equations* 161. https://doi.org/10.1186/s13662-017-1218-y.

Pudasaini, R. 2014. Survey, monitoring and effect of pollination on rapeseed (*Brassica campestris* Var. Toria) production in Chitwan, Nepal. M. Sc. In Entomology Thesis TU. IAAS Rampur, Chitwan, Nepal. p. 77.

Puentes, A., Torp, M., Weih, M., and Bjorkman, C. 2015. Direct effects of elevated temperature on a tritrophic system: salix, leaf beetles and predatory bugs. *Arthropod-Plant Interaction* 9: 567–575.

Quicke, D. L. J. 2015. The Braconid and Ichneumonid Parasitoid Wasps: Biology, Systematics, Evolution and Ecology. Hoboken, NJ: Wiley-Blackwell Publishing Ltd.

Rafferty, N. E., CaraDonna, P. J., Burkle, L. A., Iler, A. M., and Bronstein, J. L. 2013. Phenological overlap of interacting species in a changing climate: an assessment of available approaches. *Ecological Evolution* 3: 3183–3193.

Ramos-Elorduy, J. 2009. Anthropo-entomophagy: cultures, evolution and sustainability. *Entomological Research* 39: 271–288.

Raven, P. H., and Wagner, D. L. 2021. Agricultural intensification and climate change are rapidly decreasing insect biodiversity. *Proceedings of the National Academy of Sciences* 118: e2002548117.

Reddy, G. V. C., Holopainen, J. K., and Guerrero, A. 2002. Olfactory responses of *Plutella xylostella* natural enemies to host pheromone, larval frass, and green leaf cabbage volatiles. *Journal of Chemical Ecology* 28: 131–143.

Ricketts, T. H., Regetz, J., Steffan-Dewenter, I., Cunningham, S. A., Kremen, C., Bogdanski, A., Gemmil-Herren, B., Greenleaf, S. S., Klein, A. M., Mayfield, M. M., Morandin, L. A., Ochieng, A., and Viana, B. F. 2008. Landscape effects on crop pollination services: are there general patterns? *Ecology Letters* 11: 499–515.

Rosenzweig, C., Karoly, D., Vicarelli, M., Neofotis, P., Wu, Q., and Casassa, G. 2008. Attributing physical and biological impacts to anthropogenic climate change. *Nature* 453: 353–357.

Sachs, J. 2008. Common Wealth: Economics for a Crowded Planet. London: Allen Lane.

Sako, A., Mills, A. J., and Roychoudhury, A. N. 2009. Rare earth and trace element geochemistry of termite mounds in central and northeastern Namibia: mechanisms for micro-nutrient accumulation, *Geoderma* 153: 1–2.

Samways, M. J. 2017. Reconciling ethical and scientific issues for insect conservation. In: Insect Biodiversity: Science and Society, second ed. Volume 1. Eds. R. G. Foottit and P. H. Adler, Oxford: Wiley-Blackwell. pp. 747–766.

Samways, M. J., Philip, S., Barton, B., Filipe, C., Charl, D., Thomas, F., Caroline, S., Fukushima, R. G., Jan, C., Habelf, C. A., Hallmann, M. J., Hill, A. H., Lauri, K., Mackenzie, L., Kwak, D. M., Stefano, M., Jorge, A., Noriega, A. B., Orfinger, F. P., James, S. P., Fabio, O., Roque, J. S., John, P., Simaikay, N. E. Stork, F. S., Carlin, V., and Cardoso, P. 2020. Solution for humanity on how to conserve insects. *Biological Conservation*. https://doi.org/10.1016/j.biocon.2020.108427.

Scherber, C., Gladbach, D. J., and Stevnbak, K. 2013. Multi-factor climate change effects on insect herbivore performance. *Ecology and Evolution* 3: 1449–1460.

Schleuning, M., Neuschulz, E. L., Albrecht, J., Bender, I. M., Bowler, D. E., Dehling, D. M., Fritz, S. A., Hof, C., Mueller, T., and Nowak, L. 2020. Trait-based assessments of climate-change impacts on interacting species. *Trends in Ecology and Evolution* 35: 319–328.

Schowalter, T. D. 2013. Insects and Sustainability of Ecosystem Services. Boca Raton, FL: CRC Press.

Schowalter, T. D., Noriega, J. A., and Tscharntke, T. 2018. Insect effects on ecosystem services—introduction. *Basic and Applied Ecology* 26: 1–7.

Schweizer, O., Settle, J., Kudrna, O., Klotz, S., and Kuhn, I. 2008. Climate change can cause spatial mismatch of trophically interacting species. *Ecology* 89: 3472–3479.

Sentis, A., Hemptinne, J. L., and Brodeur, J. 2013. Effects of simulated heat 411 waves on an experimental plant-herbivore-predator food 412 chain. *Global Change Biology* 19: 833–842.

Sidhu, A. K., and Mehta, H. S. 2008. Role of butterflies in the natural ecosystem with special reference to high altitude (Pangi Valley, Himachal Pradesh). In Proceedings of International Conference on Climate Change, Biodiversity and Food Security in the South Asian Region, 3–4 November. Punjab State Council for Science and Technology, Chandigarh and United Nations Educational, Scientific and Cultural Organization, New Delhi. p. 36.

Steffan-Dewenter, I., Potts, S. G., and Packer, L. 2005. Pollinator diversity and crop pollination services are at risk. *Trends in Ecology and Evolution* 20: 651–652.

Steffen, W., Richardson, K., Rockström, J., Cornell, S. E., Fetzer, I., Bennett, E. M., Biggs, R., Carpenter, S. R., De Vries, W., and De Wit, C. E. 2015. Planetary boundaries: guiding human development on a changing planet. *Science* 347: 1259855.

Steinbauer, M. J., Grytnes, J. A., Jurasinski, G., Kulonen, A., Lenoir, J., Pauli, H., Rixen, C., Winkler, M., Bardy-Durchhalter, M., and Barni, E. 2018. Accelerated increase in plant species richness on mountain summits is linked to warming. *Nature* 556: 231–234.

Stork, N. E. 2018. How many species of insects and other terrestrial arthropods are there on earth? *Annual Review of Entomology* 63: 31–45.

Sumner, S., Law, G., and Cini, A. 2018. Why we love bees and hate wasps. *Ecological Entomology* 43: 836–845.

Tezze, A. A., and Botto, E. N. 2004. Effect of cold storage on the quality of *Trichogramma nerudai* (Hymenoptera: Trichogrammatidae). *Biological Control* 30: 11–16.

Thapa, R. B. 2006. Impact Assessment of Beekeeping Programme: A Case Study of Selected VDCs of Kaski District. Lalitpur: Nepal ICIMOD.

Theis, N., Kesler, K., and Adler, L. S. 2009. Leaf herbivory increases floral fragrance in male but not female *Cucurbita pepo* subsp. texana (Cucurbitaceae) flowers. *American Journal of Botany* 96: 897–903.

Thomas, M. B., and Blanford, S. 2003. Thermal biology in insect parasite interactions. *Trends Ecology Evolution* 18: 344–350.

Thomson, J. D. 2010. Flowering phenology, fruiting success and progressive deterioration of pollination in an early-flowering geophyte. *Philosophical Transactions of the Royal Society B* 365: 3187–3199.

Traore, S., Tigabu, M., Jouquet, P., Ou´edraogo, S. J., Guinko, S., and Lepage, M. 2015. Long-term effects of *Macrotermes* termites, herbivores and annual early fire on woody undergrowth community in Sudanian woodland, Burkina Faso. *Flora-Morphology, Distribution, Functional Ecology of Plants* 211: 40–50.

Tschinkel, W. R. 2015. The architecture of subterranean ant nests: beauty and mystery underfoot. *Journal of Bioeconomics* 17: 271–291.

Turnock, W. J., and Fields, P. G. 2005. Winter climate and cold hardiness in terrestrial insects. *European Journal of Entomology* 102: 561–576.

Van Asch, M., and Visser, M. E. 2007. Phenology of forest caterpillars and their host trees: the importance of synchrony. *Annual Review of Entomology* 52: 37–55.

van Huis, A. 2017. Cultural significance of termites in sub-Saharan Africa. *Journal of Ethnobiology and Ethnomedicine* 13: 1–12.

Venter, O., Sanderson, E. W., Magrach, A., Allan, J. R., Beher, J., Jones, K. R., Possingham, H. P., Laurance, W. F., Wood, P., and Fekete, B. M. 2016. Sixteen years of change in the global terrestrial human footprint and implications for biodiversity conservation. *Nature Communications* 7: 1–11.

Verheyen, J., and Stoks, R. 2019. Current and future daily temperature fluctuations make a pesticide more toxic: contrasting effects on life history and physiology. *Environmental Pollution* 248: 209–218.
Wagner, D. L., Grames, E. M., Forister, M. L., Berenbaum, M. R., and Stopak, D. 2021. Insect decline in the Anthropocene: death by a thousand cuts. *Proceedings of the National Academy of Sciences* 118: e2023989118.
Wahid, A., Gelani, S., Ashraf, M., and Foolad, M. R. 2007. Heat tolerance in plants: an overview. *Environmental and Experimental Botany* 61: 199–223.
Walther, G. R., Post, E., Convey, P., Menzel, A., Parmesan, C., and Beebee, T. J. C. 2002. Ecological responses to recent climate change. *Nature* 416: 389–395.
Wang, Z. Z., Liu, Y. Q., Shi, M., Huang, J. H., and Chen, X. X. 2019. Parasitoid wasps as effective biological control agents. *Journal of Integrative Agriculture* 18: 705–715.
Warren, M. S., Maes, D., van Swaay, C. A., Goffart, P., Van Dyck, H., Bourn, N. A., Wynhoff, I., Hoare, D., and Ellis, S. 2021. The decline of butterflies in Europe: problems, significance, and possible solutions. *Proceedings of the National Academy of Sciences* 118: e2002551117.
Webster, M. S., Colton, M. A., Darling, E. S., Armstrong, J., Pinsky, M. L., and Knowlton, N. 2016. Who should pick the winners of climate change? *Trends in Ecology Evolution* 32: 167–173.
Weidenmuller, A. 2004. The control of nest climate in bumblebee (*Bombus terrestris*) colonies: interindividual variability and self-reinforcement in fanning response. *Behavioral Ecology* 15: 120–128.
Wills, B. D., and Landis, D. A. 2018. The role of ants in north temperate grasslands. *Oecologia* 186: 323–338.
Woon, J. S., Boyle, M. J. W., Ewers, R. M., Chung, A., and Eggleton, P. 2019. Termite environmental tolerances are more linked to desiccation than temperature in modified tropical forests. *Insects Sociaux* 66: 57–64.
Xi, X. Q., Wu, X. W., Nylin, S., and Sun, S. C. 2016. Body size response to warming: time of the season matters in a tephritid fly. *Oikos* 125: 386–394.
Ziska, L. H., Pettis, J. S., Edwards, J., Hancock, J. E., Tomecek, M. B., Clark, A., Dukes, J. S., Loladze, I., and Polley, H. W. 2016. Rising atmospheric $CO_2$ is reducing the protein concentration of a floral pollen source essential for North American bees. *Proceedings of the Royal Society B* 283: 20160414.
Zvereva, E. L., and Kozlov, M. V. 2006. Consequences of simultaneous elevation of carbon dioxide and temperature for plant-herbivore interactions: a meta-analysis. *Global Change Biology* 12: 27–41.

# 11 Effect of Climate Change on Insect Pollinator

*Inzamam Ul Haq, Shahbaz Ali, Azad Ali, and Habib Ali*

## 11.1 INTRODUCTION

About 87.5% of angiosperms in nature require animal pollination to reproduce (Ollerton et al., 2011). There are about 349,000 species of pollinating animals known worldwide, with insects accounting for the highest proportion (Ollerton, 2017). About three-quarters of the world's crops are estimated to rely on animal pollination (Potts et al., 2010). According to statistics, in 2005, the contribution of pollinating insects to the global economy was approximately $215 billion, accounting for 9.5% of the global agricultural output value (Gallai et al., 2009). In China, the number of angiosperm species accounts for about 10% of the total number of angiosperm species worldwide, and the diversity of pollinating insects is also extremely rich. For example, the number of known bumblebee species in China accounts for 50% of the total species worldwide (Huang and An, 2018). China has a long history of farming and beekeeping. Liu Pengfei et al. (2011) selected 36 main crops in China to evaluate the economic value of bee pollination in China. The study showed that the average annual value of bee pollination between 2006 and 2008 was as high as 304.22 billion yuan, equivalent to 12.30% of China's total agricultural output value. It can be seen that the establishment of the interaction between plants and pollinating insects is of great significance for the sustainable development of traditional agriculture and biodiversity conservation (Ren et al., 2014, 2018).

The study of plant-pollinating insect interactions began in the first half of the 19th century, starting with the observation and description of species during the Darwin period and developing to today's comprehensive research with evolution as the core to explore the occurrence history, maintenance mechanism, and ecological protection of species diversity as the ultimate goal. With global warming, the pollination crisis caused by the sharp decline in the diversity and population size of pollinating insects has attracted worldwide attention to ecological and food security (Klein et al., 2007; Potts et al., 2010; Ren et al., 2014; Garibaldi et al., 2016). Therefore, the study of plant pollination insect interactions has received widespread and sustained attention (Ren et al., 2018). Establishing a relationship between plants and pollinating insects requires both parties to match in terms of time, space, and functional traits to establish effective pollination communication signals (pollination attraction) and flower-visiting relationships. The main impact of climate warming on plants and pollinating insects is that it affects the geographical distribution of plants and pollinating insects in space, advances or delays the flowering and fruiting stages of plants in time, and changes the active period of pollinating insects (Heland et al., 2009), which in turn changes the spatiotemporal and trait matching of two-way interactions, affecting the structure and function of their pollination networks (Menéndez et al., 2007) (Figure 11.1). There have been discussions and reflections on some theoretical frameworks for the disintegration and mismatch of plant-pollinator interactions caused by climate change (Memmott et al., 2007; Hegland et al., 2009). This chapter summarizes the characteristics and progress of research in this field, especially the impact of climate warming on plants and pollinating insects, emphasizes the importance of plant-pollinating insect interaction networks and their spatiotemporal evolution, as well as the study of interaction relationships and functional trait recombination, and puts forward suggestions and prospects for future research.

DOI: 10.1201/9781003382089-11

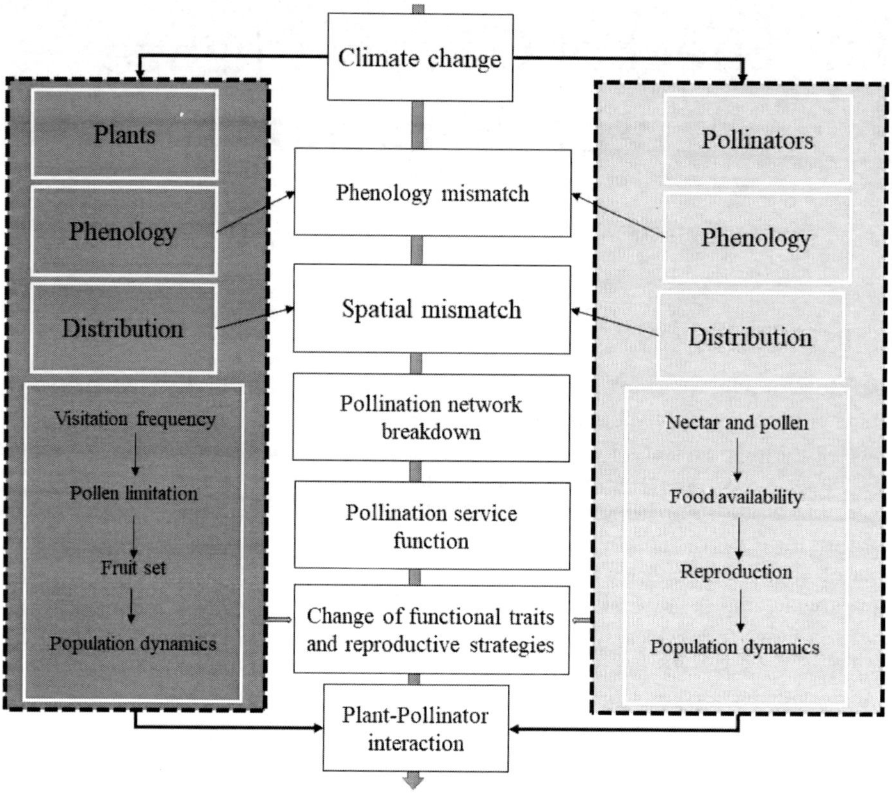


**FIGURE 11.1** Effect of climate change on plants and pollinator insects and their interaction.

## 11.2 EFFECTS OF CLIMATE CHANGE ON THE DISTRIBUTION AND PHENOLOGY OF PLANTS AND POLLINATING INSECTS

### 11.2.1 Geographical Distribution of Plants

The geographical distribution of plants is affected by climate change and the living environment of plants changes. They can change their distribution areas by diffusing to more suitable habitats or adapt to environmental changes by changing their reproductive characteristics. Generally, temperature rise leads to changes in most species' distribution range, and most plants' distribution area shows a trend of migration to high-latitude (Cavanaugh et al., 2014) and high-altitude (Kelly and Goulden, 2008; Lenoir et al., 2008) regions. Some studies have also pointed out that the distribution areas of many animals and plants have undergone significant changes in recent decades, but the observed changes do not fully follow the predicted trend of moving towards high latitude and altitude. The distribution range of locally distributed species, especially those growing in polar regions and mountain tops, has experienced a significant contraction (Parmesan, 2006).

Currently, the analysis of changes in existing plant distribution areas includes two aspects: (a) comparing historical data sets with current data. For example, by comparing the distribution of 171 forest plants from sea level to 2,600 m in Western Europe from 1905 to 1985 and 1986 to 2005, Lenoir et al. (2008) found that climate warming caused a significant increase in the optimal altitude of species, affecting not only the central distribution area of plants but also their marginal areas. Kelly and Goulden (2008) conducted a comparative study of vegetation changes at an altitude of 2,314 m in the Santa Rosa Mountains of Southern California from 1977 to 2006–2007, indicating that with climate warming, the average elevation of dominant plants in the community increased by 65 m, which may be the result of regional climate change. (b) Another aspect is using models to predict the impact of climate change on species distribution based on historical data. For example, He et al. (2019a) predicted the distribution areas of ten species of the genuses *Cyanathus* and *Primula* unique to the Himalayan Hengduan Mountains and found that under the climate warming scenario, the distribution altitude of all studied species would move upward and the latitude would move northward to occupy habitats similar to their previous habitats. In addition, by comparing specimen data with altitude records, it was found that the average altitude of the seven species of *Meconopsis* endemic to the region increased by 302.3 m before 1970 (1922–1969) and after 1970 (1970–2016), and it was predicted that the suitable climate space for all species would continue to move upward (including northward and westward) by 2070, and its distribution range might shrink. Xu Xiaoting et al. (2008) used Diva geographic information system (GIS) software and a species distribution model to estimate the potential distribution area of the rare and endangered plant *Taxus chinensis*. They found that under the climate change scenario, the scope of the basic habitat of *T. chinensis* was expanded, but its optimal habitat was fragmented. Zhang Lei et al. (2011) used three atmospheric circulation models and six species distribution models to predict the response of *Pinus massoniana* to climate change. They found that under climate change, the distribution area of *P. massoniana* will gradually migrate northward, and the area of potential distribution areas will increase progressively in the future. Under different models and climate scenarios, the migration distance and area change amplitude of potential distribution areas of *P. massoniana* are different. Currently, most existing studies in China mainly use species distribution models to predict changes in plant geographical distribution based on climate data, but these studies do not consider the interaction between plants and pollinating insects as a comprehensive factor in the model simulation. Most previous model prediction studies have assumed that all populations of a species respond uniformly to environmental changes, but some studies have pointed out that species may have local adaptation in different climatic environments (Hällfors et al., 2016). Recently, Chen et al. (2020) used the MaxEnt model to predict the changes in the distribution area of the endemic species *Paeonia mairei* in the Hengduan Mountains and the Qinling Mountains under different climate scenarios in southwest China and found that different populations have significantly different responses to climate change. If all populations are predicted as a whole, a narrow niche will be generated, and some suitable distribution areas may not be predicted. Therefore, species' possible local environmental adaptability is also a problem worth considering in model prediction research.

### 11.2.2 Plant Phenology

There are differences in the response of different plant phenology to climate warming. As temperature increases, the flowering period of some plants may be advanced (Pyke et al., 2016; Yang Liping et al., 2020) or delayed (Wang Dongni et al., 2020), while the flowering period of some plants does not significantly change (Scaven and Rafferty, 2013). Bertin (2008) summarized published studies on the response of plant phenology to climate change in Europe, North America, and Asia and found that there are geographical differences in the pattern and extent of plant phenology change. The flowering season in spring is usually early, while autumn is usually delayed, and the degree of change is different. Wang Ming and Sang Weiguo (2020), based on meteorological data from Dongling Mountain in Beijing from 2003 to 2014 and phenological observation data of three trees and three shrubs in the warm temperate zone, showed that the growing season of all species was shortened. The overall phenology showed a trend of delaying in spring and advancing in autumn. Li Dengke and Wang Zhao (2020) studied vegetation phenology in the Qinling Mountains based on climate data from 2001 to 2016. They found that phenology generally exhibits a zonal differentiation pattern from south to north during the initial, growth, and final stages and that phenology is correlated with altitude.

Elevated temperature not only changes the initial flowering period and the entire flowering period of a plant but also may change the reproductive pattern of the plant at the community level and cause the flowering and fruiting stages of the plant to differentiate toward both ends of the growing season, leading to significant segmentation in the flowering and fruiting transition period (Sherry et al, 2007). In order to predict the impact of global warming on the success of plant reproduction, a study has reported that in a cold temperate deciduous forest in northern Japan, a comparative analysis of the flowering periods of short-lived plants in extremely warm spring (2002) and early spring in normal years found that all plants observed had early flowering in extremely warm spring, and the seed-setting rate of *Corydalis ambigua*, which relies on pollination by bumblebees (*Bombus* spp.), decreased sharply in all its populations. The number of *Gagia lutea* that rely on a small bee (*Halictus* spp.) for pollination has also significantly decreased. However, early flowering does not affect the seed-setting rate of *Adonis ramosa* and *Anemone flaccida* pollinated by flies. These studies indicate that the impact of global warming on the reproductive success of early spring ephemeral plants varies depending on the type of pollinating insect, with a more significant impact on the reproductive success of plants that rely on bee pollination (Kudo et al., 2004).

Recent studies have shown that the spatiotemporal changes in plant phenology may be more complex than previously recognized, and plants can adjust their flowering time based on the spatiotemporal changes in the microclimate (Rafferty et al., 2020). In addition, the response of plant phenology to climate warming has significant consistency among different species (Thackeray et al., 2016). Theobald et al. (2017) found that the flowering/flowering duration of 48 subalpine plants in the Rainier Mountains of the United States was related to snow melt time, temperature, and soil moisture through their research on the relationship between the flowering period and climate between 2010 and 2015. Although there were significant differences in phenological sensitivity among species, it is noteworthy that all species they studied had earlier flowering times, and this consistent response is known as "phenological recombination." It is helpful further to understand the common changes in phenology of different species using the interannual changes in climate.

### 11.2.3 Geographical Distribution of Pollinating Insects

Like plants, the changes in the distribution of flying pollinating insects have also received more attention, especially for groups sensitive to climate change, such as butterflies and bumblebees. Some studies have shown that climate warming leads to changes in insect distribution areas (Konvicka et al., 2003; Pyke et al., 2016). In addition to changes in species' life cycle, climate warming has also led to the expansion of the range of pollinating insects moving toward polar and higher-altitude regions while maintaining stability near the equatorial edge (Kerr et al., 2015). According to research reports, the altitude distribution range of 16 butterfly species in central Spain increased by an average of 212 m in 30 years from 1967 to 2004. These changes mean that the habitat area of these butterfly species has decreased by an average of one-third, and it is expected that the loss of habitat area in the next century will reach 50% to 80%. Many species have suffered habitat loss due to climate change, which may threaten their long-term survival opportunities (Wilson et al., 2005). Warren et al. (2001) evaluated the distribution area size and species richness of 46 butterfly species in northern England and found that population richness is closely related to changes in their distribution. The dual pressures of habitat change and climate change may lead to a decline in the number of endemic species in the region. Aguirre-Gutiérrez et al. (2017) simulated the distribution areas of 398 species of pollinating insects (bees, butterflies, and aphids) in the Netherlands using species distribution models and data sets over 60 years (1951–2014). They found that the distribution of pollinating insects changed over time and pointed out that species distribution models calibrated based on current climatic conditions may be difficult to fully predict the impact of environmental factors on future changes in species distribution areas.

It is noteworthy that the genus Bombus is one of the important pollinating insects in the northern hemisphere ecosystem (Bingham and Orthner, 1998; Williams et al., 2017). Bumblebees are typical groups that adapt to high mountains and low temperatures and are extremely sensitive to rising temperatures. Kerr et al. (2015) predicted the changes in the distribution area of bumblebees in North America and Europe. They found that bumblebees cannot migrate to the polar regions with increasing temperature, but the distribution area in the south continues to decrease. They predicted that bumblebees would migrate to high altitudes. In addition, since 1974, the average observed altitude of bumblebee species in southern Europe has increased by about 300 m, and the migration of different species along the altitude gradient varies greatly. Habitat changes in high-altitude areas may lead to a significant decrease in the number of mountain bumblebee species.

### 11.2.4 Pollinating Insect Phenology

Changes in seasonal activity patterns (phenology) in response to warming produced some of the earliest and most well-documented evidence of climate-change consequences on individuals (Table 11.1). Butterflies have long piqued the interest of both novice and experienced

**TABLE 11.1**
**Effects of Climate Change on Insect Pollinators**

| Type of Climate-Change Impact | Response Variable | Demonstrated Relationship between Rising Temperature and Response Variable | Pollinator Taxa Studied | Reference |
|---|---|---|---|---|
| **Direct** | Phenology (emergence or first appearance) | Usually negative (earlier) | Bees, butterflies | Forister and Shapiro (2003); Gordo and Sanz (2005, 2006); Sparks et al. (2010); Bartomeus et al. (2011); Ovaskainen et al. (2013) |
| | Phenology (flight-season duration) | Positive | Butterflies | Roy and Sparks (2000) |
| | Voltinism | Positive (more generations) | Butterflies | Roy and Sparks (2000); Altermatt (2010) |
| | Activity level (feeding or foraging rate) | Positive and negative | Bees, butterflies | Herrera (1995); Cameron et al. (1996); Higgins et al. (2014) |
| | Body size | Negative | Bees, hawkmoths | Davidowitz et al. (2004); Radmacher and Strohm (2010) |
| | Development rate | Positive and negative | Bees, butterflies | Stephen (1965); Kingsolver et al. (2011); O'Neill et al. (2011); Radmacher and Strohm (2011) |
| | Energetic expenditures or mass loss | Usually positive | Bees, butterflies | Sgolastra et al. (2010); Stuhldreher et al. (2014) |
| | Mortality rate | Positive and negative | Bees, butterflies, fig wasps | Whitfield and Richards (1992); Bosch and Kemp (2003, 2004); Radmacher and Strohm (2011); Jevanandam et al. (2013); Stuhldreher et al. (2014); Bennett et al. (2015a) |
| **Indirect** | Flower production | Positive and negative | NA | Mu et al. (2015) |
| | Nectar production (per flower) | Positive and negative | NA | Mu et al. (2015); Takkis et al. (2015) |
| | Pollen production (per flower) | Negative | NA | Prasad et al. (2002); Pressman et al. (2002); Koti et al. (2005) |
| | Temporal overlap with flowers | Negative | Fies (Muscidae, Chironomidae) | Høye et al. (2013) |
| | Spatial overlap with host plants | Negative (simulated) | Butterflies | Schweiger et al. (2008, 2012) |
| | Activity or prevalence of natural enemies | Positive and negative | Bees, butterflies | Stone and Willmer (1989); van Nouhuys and Lei (2004); James (2005) |

entomologists alike, and as a result, they have presented some of the best and most reliable data for evaluating the impacts of climate change on insect phenology. These data show that, since the 1970s, several butterfly species in the United States and Europe have appeared earlier in the year, up to 16 days prior per decade (range: 16 days prior to 4 days later per decade), because of increasing temperatures (Forister and Shapiro, 2003; Gordo and Sanz, 2005, 2006; see Parmesan, 2007, for review and synthesis of earlier studies). Long-term data from Japan show no progress in phenology, but this appears to be an artifact of the butterfly under study's negative population growth rather than a true failure to shift seasonal activity in response to climate change (Ellwood et al., 2012).

The phenological development of other pollinating species in temperate climates has also been reported to be progressing. Honeybees (*Apis mellifera*) in Spain and Poland (but not Austria; Scheifinger et al., 2005) have been making their first appearances in the spring increasingly earlier (by up to 13.5 days/decade) since the 1970s or 1980s. This trend has been observed in both countries since the 1970s or 1980s (Gordo and Sanz, 2006; Sparks et al., 2010). Phenology had progressed by nearly ten days since 1880, at a maximum rate of 1.4 day/decade (for *Colletes inequalis*); the majority of the advance (nearly seven days) had occurred since 1970. Bartomeus et al. (2011) studied ten spring-active bee species native to North America and found that phenology had progressed by nearly ten days since 1880. During the 1970s, the first sightings of bumblebees (*Bombus* spp.) at a Russian locale have also been taking place sooner, at a pace of 3.1 days each decade (Ovaskainen et al., 2013). All of these studies found a strong correlation between springtime temperatures and bee phenologies, which suggests that warmer springs are causing earlier activity, earlier development, or earlier termination of diapause in these bees. These bees included species that overwinter either as non-diapausing adults (*Apis*), as diapausing adults (*Osmia*, *Bombus* spp.), or as larvae (*Andrena*, *Colletes* spp.). Iler et al. (2013), on the other hand, observed that there was no discernible shift in the phenology of syrphid flies between the years 1992 and 2011 at the research location located in the Rocky Mountains of Colorado, United States. Yet, this lack of trend reflected the fact that during the course of the analyzed time period, there was no discernible shift in the directional change of climatic parameters. Phenology is not insensitive to climate. In point of fact, syrphid phenology was shown to have a significant correlation with the date of snowmelt, which has been gradually moving up in this area over the course of many years (Iler et al., 2013).

In addition to starting out their flying seasons earlier in the spring, several species of pollinators have also extended the time that they spend actively foraging and breeding. As a result of the increase in temperature, some butterfly species in the United Kingdom have extended their flight seasons (Roy and Sparks, 2000). The most substantial increases (in some cases by more than ten days per decade) have been seen in species that have been able to fit an additional generation into the prolonging growing season. This pattern has been observed even in some univoltine species, which are species that only have one generation per year. However, the most substantial increases have been seen in species that have been able to do so (Roy and Sparks, 2000). In the case of European butterflies and moths, the occurrence of second (or subsequent) generations has significantly risen in the time after 1980 in comparison with the decades that came before it (Altermatt, 2010). However, wintertime bumblebee (*Bombus terrestris*) activity has recently been reported in UK cities, apparently in response to warm temperatures and flowering of cultivated plants in urban gardens. Bumblebees (*Bombus* spp.) in northern Europe and North America typically have an annual colony life cycle; this means that they have a single queen-to-queen generation per year (Stelzer et al., 2010). There is a possibility that the ordinarily univoltine British subspecies of *B. terrestris* has sought to produce a second generation with these winter-active bees. It is not known at this time if the second generation will be able to generate new reproductives (males and queens) effectively (Stelzer et al., 2010).

There are not many long-term data available on the phenologies of tropical pollinators, and in environments where there are no distinct growing seasons, it may be difficult to even simply characterize phenology, much alone record long-term changes (see Valtonen et al., 2013b). In addition, since it is less probable for phenology to be a function of temperature at lower latitudes (Wolda, 1988), it is doubtful that a warmer climate would generate a consistent (and hence clearly discernible) influence on the phenology of pollinators.

## 11.3 EFFECTS OF CLIMATE CHANGE ON THE RELATIONSHIP BETWEEN PLANTS AND POLLINATING INSECTS

### 11.3.1 Plants—The Spatial Mismatch of Pollinating Insects

Climate change can disrupt interactions at species' trophic level because different species do not necessarily respond to such changes in similar ways. From the altitude gradient, observing the spatial changes in species distribution caused by temperature changes is easier. Studies have shown that historical climate change has spatially affected plant pollination and insect interactions (Sosa-Pivatto et al., 2017). The phenomenon of species migration from low-altitude to high-altitude regions caused by temperature rise is relatively common (Johnson et al., 2010). Due to the phenology of species in high-latitude and high-altitude regions being later than those in low-latitude and low-altitude regions, pollinating insects can migrate to higher latitudes or altitudes when a nutritional mismatch occurs. The difference in migration speed between plants and pollinating insects may cause mismatches in their geographical distribution.

The distribution of their pollinating insects often limits the distribution of plants that rely on pollinating insects for successful reproduction. Recently, Tsiftsis and Djordjević (2020) used a species distribution model to study orchid plants (*Ophrys argolica*, *O. delphinensis*) and their pollinating insects (*Anthrophora tractita*). It was found that if the distribution of pollinating insects was considered, the potentially suitable area of the two orchids would be reduced by about 52% and 73%, respectively. Under future climate change, some populations of orchids may lack pollinating insects. Gorostiague et al. (2018) simulated the distribution regions of 11 species of cactus in the central Andes. They predicted the possibility of a mismatch between plants and pollinating insects under changing distribution regions. It was found that there was no spatial mismatch between the two. Currently, most of these studies are based on theoretical research and model predictions. The spatial mismatch between plants and pollinating insects caused by climate warming is rarely observed (Gérard et al., 2020). It should be emphasized that many large insects have a flight range of up to 1 km. At the same time, existing evidence indicates that most of the observed spatial migrations, whether of plants or pollinating insects, are hundreds of meters (Wolf and Moritz, 2008).

### 11.3.2 Plants–Pollinating Insect Phenological Mismatch

Climate change directly affects the interaction between plants and pollinating insects, potentially disrupting the overlap between the plant reproductive period and the flying season of pollinating insects, thereby altering the opportunities for interaction between animals and plants. Studies have shown that decreased fecundity or longevity of pollinating insects may reduce population density or growth rate, thereby increasing the risk of extinction (Memmott et al., 2007). When pollinating insects' activities are not synchronized with the flowering time of plants, the overlapping time of pollination interactions will be shortened, weakening or even disintegrating the plant–pollinating insect relationship.

In the past century, global warming has advanced the initial flowering period of forest plants in the western prairie of Illinois by about four days, and the seasonal activity period of some pollinating insects has advanced by about four days per degree Celsius in temperate regions on average. Without compensation mechanisms, phenological changes caused by climate change may disrupt the time overlap between pollinating insects' feeding and their flower food resources. This reduction in the overlap time between the two sides of the interaction reduces the width of the foraging niche of pollinating insects, especially those with narrow and specialized niches, which are prone to interference from phenological mismatches (Memmott et al., 2007). Kudo and Ida (2013) investigated the early spring short-lived plant Yanhusuo in Hokkaido, Japan. The plant bloomed 10 to 14 days (from mid-April to mid-May) after snowmelt, and the pollinating insect was a bumblebee. Their phenological monitoring and statistical analysis of the plant seed-setting rate showed that Yanhusuo and bumblebees had different sensitivities to early spring, resulting in a phenological mismatch, which reduced the plant seed-setting rate. Therefore, the phenological mismatch is the main limiting factor for reproducing early spring short-lived plants.

In addition, Doi et al. (2008) found that in the past few decades, the flowering time of plants has advanced, while the emergence time of butterflies has been delayed. About 30–40 days before flowering, plants are strongly affected by temperature, while butterflies are less affected by temperature, and their optimal adaptation periods to temperature are different. The temperature during the sensitive period of plants has risen sharply since 1953, while the temperature during the sensitive period of butterflies has not changed much. This difference in climate sensitivity affects the phenological match between plum plants and pollinating insects. Rafferty and Ives (2011) studied the impact of changes in the initial flowering period of 14 perennial plants in southern Wisconsin, United States, on pollinating insect pollination. They found that in the past 70 years, plants that are not restricted by pollinating insects have exhibited early flowering, while plants that are restricted by pollinating insects have not changed. This study is different from existing reports of plant pollination insect phenology mismatches. The activities of flowering and pollinating insects in some plants are controlled by their biological clocks. Research on the interaction between flowering and pollinating insect activity cycles in plants on a fine diurnal time scale has not received sufficient attention (Fründ et al., 2011; Knop et al., 2018).

### 11.3.3 Construction and Reorganization of Pollination Networks

The distribution and phenological mismatches between plants and pollinating insects caused by climate change can disrupt their original interactions, thereby affecting the stability of the pollination network structure. Burkle et al. (2013) used historical data sets to quantify the impact of global changes over the past 120 years on the interplay between plants and pollinators in the temperate forest understory of Illinois, United States. The study found that 50% of bee pollinators experienced changes in their interactions, network structure, and function, which can be attributed to phenological mismatches between plants and bees, nonrandom bee species extinction, and changes in the spatial coexistence of existing species. At the same time, the quantity and quality of pollination services have decreased over time, and the adaptability of pollination network structures to future changes is gradually decreasing. Suppose species have different abilities to track climate change through migration and diffusion. In that case, new species combinations may form, and the disintegration and reorganization interaction network may change species richness, distribution, and extinction probability under climate change, increasing the risk of species extinction (Gilman et al., 2010).

In addition to some specialized pollination relationships, many plants form generalized connections with pollinating insects; that is, a plant can be accessed by different pollinating insects at different times and locations and can achieve the same pollination efficiency. Some studies have shown that as altitude increases, the decrease in species will lead to a decrease in the abundance of pollinating insects. In high-altitude regions, pollinating insects have a more generalized selection of plants, with a greater width of their nutrient niche (Lara Romero et al., 2019). Research on the structure of pollination networks has shown that many species can sustain their survival by rapidly responding to environmental changes. Although the species of their reciprocal partners may change (Burkle and Alarcón, 2011), some specialized pollinating insects may disappear, but some generalized pollinating insects can replace them in pollination and achieve recombination. In addition, wind pollination will to some extent replace or supplement the demand for alpine plants for pollinators which helps to cushion the impact of climate warming on plant-pollinator interactions.

## 11.4 EFFECTS OF CLIMATE CHANGE ON FUNCTIONAL TRAITS OF PLANTS AND POLLINATING INSECTS, AND PLANT REPRODUCTION

### 11.4.1 Changes in Plant Functional Traits

Under natural conditions, phenotypic changes occur in different plant populations. Whether these changes in traits are caused by climate change, are caused by their plasticity, or are the result of rapid evolution has attracted scholars' attention (Franks et al., 2014). As climate changes, plant populations may not be able to adapt to new environments, choosing to migrate and track suitable living conditions to maintain their optimal state of adaptation. However, this choice is limited and may result in populations being out of sync with other environmental factors, such as photoperiod, or out of sync with key pollinating insects visiting flowers (McNamara et al., 2011). Under extreme climatic conditions, early spring warming can directly cause damage to the floral structure of a large number of wild plants, thereby affecting the seed-setting rate of plants (Ren Zongxin et al., 2014).

Many plants may respond to climate change through phenotypic plasticity or adaptive evolution of functional traits (Hoffmann and Sgro, 2011; Anderson and Gezon, 2015). Dai et al. (2017) investigated natural populations of *Pedicularis siphonantha* at different altitude gradients in northwest Yunnan and conducted experiments in homogeneous gardens at low altitudes. The results showed that plants transplanted to low altitudes had earlier flowering periods, shorter flower lifespans, and reduced pollen yield. In contrast, insect visits to flowers increased, resulting in increased seed yield. This plasticity is helpful for plants to adjust their resource allocation strategies before and after pollination and may be conducive to adapting to environmental changes. Franks et al. (2014) discussed factors that may affect phenotypic plasticity and evolution of plants, including patterns of environmental change, life-history characteristics of species (including generations and reproductive systems), and the extent and direction of gene flow. Their summary and analysis of 38 research cases indicate that all studies have shown phenotypic plasticity or adaptive response in plants, and 26 have shown that both strategies exist simultaneously. However, due to climate change or other selection pressures, many issues still need to be further explored on how the transformation of the breeding system of plants that rely on pollinating insects responds to climate change.

### 11.4.2 Changes in Functional Traits of Pollinating Insects

Recent studies have shown that the decline in the diversity and population size of some long-beaked pollinating insects threatens matching traits in pollination systems. Miller-Struttmann et al. (2015) measured the beak length changes of two alpine bumblebees using bumblebee specimens collected in the central Rocky Mountains of the United States from 1966 to 1980 and 2012 to 2014. It was found that the beak length of the two species of bumblebees has become shorter over the past 40 years, while the flower tube length associated with pollination has not significantly changed. It is speculated that the reasons for the change in the beak length of bumblebees may be the decrease in body size, competition with subalpine invasive bumblebee species, and reduction in food resources. Due to the decrease in nectar and pollen resources of flowering plants, it is beneficial for generalized bumblebees to forage, and the shortened beak is more suitable for visiting plants with short corolla tubes. With fewer nectar resources, the dominant position of the long-beaked bumblebee is weakened, which may promote the rapid evolution of the short-beaked bumblebee species.

### 11.4.3 Impact on the Pollination Network

The matching of functional traits among species in a pollination network affects their interactions. Suppose a functional trait of a plant or pollinating insect is missing. In that case, it may affect its functional diversity, causing a contraction in the niche of plant-pollinating insect interactions at the trophic level, which means that a decrease in the functional diversity of species at one trophic level will lead to increased competition among crops at another trophic level (Gagic et al., 2015; Albrecht et al., 2018). When assessing the impact of climate change on biodiversity, although models for predicting the response of individual species to climate change have been widely used, the interactions between organisms are often ignored, and the extent to which climate change affects the interactions between species is not very clear. Schleuning et al. (2016) combined species distribution models with ecological network analysis to test the potential impact of climate change on more than 700 species of animal and plant pollination networks in central Europe. A study examining the climatic niche breadth of plants, their pollinating insects (bees, butterflies, aphids), and seed-dispersing birds revealed that pollinating insects engaged with plant species of limited diversity tend to possess narrower niches, rendering them more susceptible to the impacts of climate change. On the contrary, the degree of specialization of plants is independent of their niche width and vulnerability. This important difference has an important impact on predicting their extinction risk under climate change because the cascade effect from plants to pollinating insects is likely to trigger the extinction of pollinating insects in their interactions and may exacerbate the impact of climate change on biodiversity. Therefore, considering the interactions between species is very important for accurately predicting the impact of climate change on pollinating insects. The distribution of different pollinating functional groups in environmental gradients and their responses to climate change are different (Schleuning et al., 2016). However, research in this area is still relatively scarce.

### 11.4.4 Impact on Plant Reproduction

Under climate change, changes in plant-pollinator interactions affect the reproductive success of plants (Richman et al., 2020). Eckert et al. (2010) analyzed 22 studies involving 27 plants and found that due to habitat changes caused by human interference and climate change, the proportion of seeds produced by plant outcrossing significantly decreased. In addition, continuous changes in snowmelt time and growing season temperature have significant impacts on plant reproduction and population dynamics in the Arctic and alpine regions (Kudo, 1991; Kudo et al., 2004; Kudo and Ida, 2013; Totland, 1997; Totland and Alatalo, 2002). In addition to temperature and precipitation, changes in the photoperiod also affect plant reproduction (Walker et al, 2019). In addition, alien plant invasion affects the reproductive success of native plants (Hernández-Castellano et al., 2020). Warming experiments found that the invasive plant *Calluna vulgaris* responded more quickly to phenology, especially for invasive species with higher flower density, which increased their pollination competition against native plants (Giejsztowt et al., 2020). On the one hand, climate change may increase the risk of extinction for species that cannot migrate to new habitats or have low environmental adaptability (Grazer and Martin, 2012); on the other hand, it may stimulate changes in the genetic material of the species to adapt to the environment (Parmesan, 2006; Visser, 2008). Currently, the role of plant microevolution in addressing climate change still needs in-depth research.

## 11.5 EFFECT OF FITNESS: POLLINATION LIMITATIONS

The health level of pollination function is often measured by whether the diversity of pollinating insects can maintain plant reproduction or crop yield. Outcrossing plants rely on pollinating insects to transmit pollen (gametes) for sexual reproduction, and insufficient quantity or quality of pollen can reduce the success rate of plant reproduction, which is referred to as "pollination restriction" (Ashman et al., 2004). Pollen reception is a key indicator of plant reproductive success. Through artificial pollination experiments, it is possible to estimate the extent of pollination restrictions. Suppose the seed-setting rate of plants that undergo artificial cross-pollination is higher than that of plants that undergo natural pollination. In that case, it indicates that their reproductive success is affected by pollination restrictions, not by abiotic factors (Bennett et al., 2018). Previous studies have shown that most plants have pollination restrictions. Benjamin and Winfree (2014) studied the degree of pollination restriction of two blueberry varieties in southern New Jersey, United States, and found significant pollen restrictions on cultivated blueberries. The availability of pollinating insects in some alpine plants decreases with increasing altitude (García-Camacho and Totland, 2009; Zhao and Wang, 2015). Currently, climate change or human interference is exacerbating the level of pollination restrictions (IPBES, 2016).

Pollination restrictions affect the level of natural selection of plant flower traits by affecting reproductive fitness (Harder and Aizen, 2010; Johnston and Bartkowska, 2017). Suppose the population size of a species is limited by seed production. In that case, pollination restrictions may also reduce the richness of the species to varying degrees, thereby changing the species composition and structure of the community (Ashman et al., 2004). Therefore, it is necessary to deeply study the dynamics and potential consequences of pollination restrictions caused by climate change. It should be noted that the GloPL database has collected basic data for assessing the intensity and variability of pollination restrictions on 1,265 wild plants worldwide (Bennett et al., 2018), providing an important basis for further research.

## 11.6 FUTURE PROSPECTS

Organisms in most regions of the world are affected by climate change, and these trends may continue in the coming decades (He et al, 2019a, 2019b). In the past, a large number of model predictions have been made for the distribution changes of plants and pollinating insects under climate change (Elith et al., 2006; Bose et al., 2016; He et al., 2019a, 2019b). In recent years, research has gradually considered species' interaction in species distribution models (Pellissier et al., 2013). With the development of species distribution prediction methods, potential climate change impacts are being considered in model predictions in an increasingly realistic manner (Bellard et al., 2012). In order to make model predictions more accurate, high-quality basic data related to biology need to be considered in model predictions (Grazer and Martin, 2012). Currently, the main data sources for domestic model prediction include database data, flora data, including the Flora of China and local flora, and the China Species Information System and remote sensing image extraction (Liu Xiaotong et al., 2019).

In addition to studying the temporal and spatial responses of plants and pollinating insects to climate change, plant phenotypic plasticity and rapid adaptive evolution also require in-depth exploration. Understanding plant plasticity and its ability to regulate its own temperature is crucial for formulating conservation strategies (Van der Kooi et al., 2019). Some insects can adjust and mitigate the impact of climate change through phenotypic adaptation and rapid evolutionary response. Still, there is currently a lack of research on the mechanisms of insect genetic changes that may be triggered by climate change (Bale et al., 2002). The response of plant phenotypes and breeding systems to climate change is relatively difficult to detect, and it is difficult to detect population fluctuations that occur before plant distribution areas migrate. Empirical data on the likely response rates and types of plants, as well as the prediction of plant and trait specificity when climatic conditions continue to change, are still relatively scarce. Therefore, it is necessary to carry out more field observation and control experiments to detect population changes over time with a large sample size. At the same time, it is necessary to establish a relevant standard database to comprehensively study and predict the response of plant populations to climate change (Franks et al., 2014).

The issues and suggestions that need attention in future work are as follows: (a) research covering multiple scales of biodiversity; (b) long-term observation of plant-pollinating insect interaction networks; (c) evaluation of reproductive fitness of important indicator species; (d) changes and recombination of functional traits in plants and pollinating insects, in particular, the diverse observation and accumulation of phenological data across altitude gradients; and (e) assessment and protection of important groups of plants and pollinating insects. Plants that rely on pollinating insects or have specialized pollination systems are the first to be affected by fluctuations in the number of pollinating insect populations, and it is necessary to strengthen the evaluation and protection of these important groups.

## REFERENCES


Aguirre-Gutiérrez J, Kissling WD, Biesmeijer JC, WallisDeVries MF, Reemer M, Carvalheiro LG (2017) Historical changes in the importance of climate and land use as determinants of Dutch pollinator distributions. Journal of Biogeography, 44, 696–707.

Albrecht J, Classen A, Vollstädt MGR, Mayr A, Mollel NP, Costa DS, Dulle HI, Fischer M, Hemp A, Howell KM, Kleyer M, Nauss T, Peters MK, Tschapka M, Steffan-Dewenter I, Böhning-Gaese K, Schleuning M (2018) Plant and animal functional diversity drive mutualistic network assembly across an elevational gradient. Nature Communications, 9, 3177.

Altermatt F (2010) Climatic warming increases voltinism in European butterflies and moths. Proceedings of the Royal Society B: Biological Sciences, 277(1685), 1281–1287.

Anderson JT, Gezon ZJ (2015) Plasticity in functional traits in the context of climate change: A case study of the subalpine forb *Boechera stricta* (Brassicaceae). Global Change Biology, 21, 1689–1703.

Ashman TL, Knight TM, Steets JA, Amarasekare P, Burd M, Campbell DR, Dudash MR, Johnston MO, Mazer SJ, Mitchell RJ, Morgan MT, Wilson WG (2004) Pollen limitation of plant reproduction: Ecological and evolutionary causes and consequences. Ecology, 85, 2408–2421.

Bale JS, Masters GJ, Hodkinson ID, Awmack C, Bezemer TM, Brown VK, Butterfield J, Buse A, Coulson JC, Farrar J, Good JEG, Harrington R, Hartley S, Jones TH, Lindroth RL, Press MC, Symrnioudis I, Watt AD, Whittaker JB (2002) Herbivory in global climate change research: Direct effects of rising temperature on insect herbivores. Global Change Biology, 8, 1–16.

Bartomeus I, Ascher JS, Wagner D, Danforth BN, Colla S, Kornbluth S, Winfree R (2011) Climate-associated phenological advances in bee pollinators and bee-pollinated plants. Proceedings of the National Academy of Sciences of the United States of America, 108, 20654–20659.

Bellard C, Bertelsmeier C, Leadley P, Thuiller W, Courchamp F (2012) Impacts of climate change on the future of biodiversity. Ecology Letters, 15, 365–377.

Benjamin FE, Winfree R (2014) Lack of pollinators limits fruit production in commercial blueberry (*Vaccinium corymbosum*). Environmental Entomology, 43, 1574–1583.

Bennett JM, Steets JA, Burns JH, Durka W, Vamosi JC, Arceo-Gómez G, Burd M, Burkle LA, Ellis AG, Freitas L, Li J, Rodge JG, Wolowski M, Xia J, Ashman TL, Knight TM (2018) GloPL, a global data base on pollen limitation of plant reproduction. Scientific Data, 5, 180249.

Bennett MM, Cook KM, Rinehart JP, Yocum GD, Kemp WP, Greenlee KJ (2015a) Exposure to suboptimal temperatures during metamorphosis reveals a critical developmental window in the solitary bee, *Megachile rotundata*. Physiological and Biochemical Zoology: Ecological and Evolutionary Approaches, 88, 508–520.

Bertin RI (2008) Plant phenology and distribution in relation to recent climate change. Journal of the Torrey Botanical Society, 135, 126–146.

Bingham RA, Orthner AR (1998) Efficient pollination of alpine plants. Nature, 391, 238–239.

Bosch J, Kemp WP (2003) Effect of wintering duration and temperature on survival and emergence time in males of the orchard pollinator *Osmia lignaria* (Hymenoptera: Megachilidae). Environmental Entomology, 32, 711–716.

Bosch J, Kemp WP (2004) Effect of pre-wintering and wintering temperature regimes on weight loss, survival, and emergence time in the mason bee *Osmia cornuta* (Hymenoptera: Megachilidae). Apidologie, 35, 469–479.

Bose R, Munoz F, Ramesh BR, Pelissier R (2016) Past potential habitats shed light on the biogeography of endemic tree species of the Western Ghats biodiversity hotspot, south India. Journal of Biogeography, 43, 899–910.

Burkle LA, Alarcón R (2011) The future of plant-pollinator diversity: Understanding interaction networks across time, space and global change. American Journal of Botany, 98, 528–538.

Burkle LA, Marlin JC, Knight TM (2013) Plant-pollinator interactions over 120 years: Loss of species, co-occurrence, and function. Science, 339, 1611–1615.

Cameron SA, Whitfield JB, Hulslander SL, Cresko WA, Isenberg SB, King RW (1996) Nesting biology and foraging patterns of the solitary bee *Melissodes rustica* (Hymenoptera: Apidae) in northwest Arkansas. Journal of the Kansas Entomological Society, 69, S260–S273.

Cavanaugh KC, Kellner JR, Forde AJ, Gruner DS, Parker JD, Rodriguez W, Feller IC (2014) Poleward expansion of mangroves is a threshold response to decreased frequency of extreme cold events. Proceedings of the National Academy of Sciences, USA, 111, 723–727.

Chen QH, Yin YJ, Zhao R, Yang Y, da Silva JAT, Yu XN (2020) Incorporating local adaptation into species distribution modeling of *Paeonia mairei*, an endemic plant to China. Frontiers in Plant Science, 10, 1717.

Dai WK, Amboka GM, Kadiori EL, Wang QF, Yang CF (2017) Phenotypic plasticity of floral traits and pollination adaption in an alpine plant *Pedicularis siphonantha* D. Don when transplanted from higher to lower elevation in Eastern Himalaya. Journal of Mountain Science, 14, 1995–2002.

Davidowitz G, D'Amico LJ, Nijhout HF (2004) The effects of environmental variation on a mechanism that controls insect body size. Evolutionary Ecology Research, 6, 49–62.

Doi H, Gordo O, Katano I (2008) Heterogeneous intra-annual climatic changes drive different phenological responses at two trophic levels. Climate Research, 36, 181–190.

Eckert CG, Kalisz S, Geber MA, Sargent R, Elle E, Cheptou PO, Goodwillie C, Johnston MO, Kelly JK, Moeller DA, Porcher E, Ree RH, Vallejo-Marín M, Winn AA (2010) Plant mating systems in a changing world. Trends in Ecology and Evolution, 25, 35–43.

Elith J, Graham CH, Anderson RP, Dudik M, Ferrier S, Guisan A, Hijmans RJ, Huettmann F, Leathwick JR, Lehmann A, Li J, Lohmann LG, Loiselle BA, Manion G, Moritz C, Nakamura M, Nakazawa Y, Overton JM, Peterson AT, Phillips SJ, Richardson K, Scachetti-Pereira R, Schapire RE, Soberon J, Williams S, Wisz MS, Zimmermann NE (2006) Novel methods improve prediction of species' distributions from occurrence data. Ecography, 29, 129–151.

Ellwood ER, Diez JM, Ibáñez I, Primack RB, Kobori H, Higuchi H, Silander JA (2012) Disentangling the paradox of insect phenology: Are temporal trends reflecting the response to warming? Oecologia, 168, 1161–1171.

Forister ML, Shapiro AM (2003) Climatic trends and advancing spring flight of butterflies in lowland California. Global Change Biology, 9, 1130–1135.

Franks SJ, Weber JJ, Aitken SN (2014) Evolutionary and plastic responses to climate change in terrestrial plant populations. Evolutionary Applications, 7, 123–139.

Fründ J, Dormann CF, Tscharntke T (2011) Linné's floral clock is slow without pollinators-flower closure and plant-pollinator interaction webs. Ecology Letters, 14, 896–904.

Gagic V, Bartomeus I, Jonsson T, Taylor A, Winqvist C, Fischer C, Slade EM, Steffan-Dewenter I, Emmerson M, Potts SG, Tscharntke T, Weisser W, Bommarco R (2015) Functional identity and diversity of animals predict ecosystem functioning better than species-based indices. Proceedings of the Royal Society B: Biological Sciences, 282, 20142620.

Gallai N, Salles JM, Settele J, Vaissière BE (2009) Economic valuation of the vulnerability of world agriculture confronted with pollinator decline. Ecological Economics, 68, 810–821.

García-Camacho R, Totland Ø (2009) Pollen limitation in the alpine: A meta-analysis. Arctic, Antarctic, and Alpine Research, 41, 103–111.

Garibaldi LA, Carvalheiro LG, Vaissiere BE, Gemmill-Herren, B, Hipolito J, Freitas BM, Ngo HT, Azzu N, Saez A, Astrom J, An JD, Blochtein B, Buchori D, Garcia FJC, da Silva FO, Devkota K, Ribeiro MD, Freitas L, Gaglianone MC, Goss M, Irshad M, Kasina M, Pacheco AJS, Kiill LHP, Kwapong P, Parra GN, Pires C, Pires V, Rawal RS, Rizali A, Saraiva AM, Veldtman R, Viana BF, Witter S, Zhang H (2016) Mutually beneficial pollinator diversity and crop yield outcomes in small and large farms. Science, 351, 388–391.

Gérard M, Vanderplanck M, Wood T, Michez D (2020) Global warming and plant-pollinator mismatches. Emerging Topics in Life Sciences, 4, 77–86.

Giejsztowt J, Classen AT, Deslippe JR (2020) Climate change and invasion may synergistically affect native plant reproduction. Ecology, 101, e02913.

Gilman SE, Urban MC, Tewksbury J, Gilchrist GW, Holt RD (2010) A framework for community interactions under climate change. Trends in Ecology & Evolution, 25, 325–331.

Gordo O, Sanz JJ (2005) Phenology and climate change: A long-term study in a Mediterranean locality. Oecologia, 146, 484–495.

Gordo O, Sanz JJ (2006). Temporal trends in phenology of the honey bee *Apis mellifera* (L.) and the small white *Pieris rapae* (L.) in the Iberian Peninsula (1952–2004). Ecological Entomology, 31, 261–268.

Gorostiague P, Sajama J, Ortega-Baes P (2018) Will climate change cause spatial mismatch between plants and their pollinators? A test using Andean cactus species. Biological Conservation, 226, 247–255.

Grazer VM, Martin OY (2012) Investigating climate change and reproduction: Experimental tools from evolutionary biology. Biology, 1, 411–438.

Hällfors MH, Liao J, Dzurisin J, Grundel R, Hyvärinen M, Towle K, Wu GC, Hellmann JJ (2016) Addressing potential local adaptation in species distribution models: Implications for conservation under climate change. Ecological Applications, 26, 1154–1169.

Harder LD, Aizen MA (2010) Floral adaptation and diversification under pollen limitation. Philosophical Transactions of the Royal Society B: Biological Sciences, 365, 529–543.

He X, Burgess KS, Gao LM, Li DZ (2019a) Distributional responses to climate change for alpine species of *Cyananthus* and *Primula* endemic to the Himalaya-Hengduan Mountains. Plant Diversity, 41, 26–32.

He X, Burgess KS, Yang XF, Ahrends A, Gao LM, Li DZ (2019b) Upward elevation and northwest range shifts for alpine *Meconopsis* species in the Himalaya-Hengduan Mountains region. Ecology and Evolution, 9, 4055–4064.

Hegland SJ, Nielsen A, Lázaro A, Bjerknes AL, Totland Ø (2009) How does climate warming affect plant-pollinator interactions? Ecology Letters, 12, 184–195.

Hernández-Castellano C, Rodrigo A, Gómez JM, Stefanescu C, Calleja JA, Reverté S, Bosch J (2020) A new native plant in the neighborhood: Effects on plant-pollinator networks, pollination, and plant reproductive success. Ecology, 101, e03046.

Herrera CM (1995) Floral biology, microclimate, and pollination by ectothermic bees in an early-blooming herb. Ecology, 76, 218–228.

Higgins JK, MacLean HJ, Buckley LB, Kingsolver JG (2014) Geographic differences and microevolutionary changes in thermal sensitivity of butterfly larvae in response to climate. Functional Ecology, 28, 982–989.

Hoffmann AA, Sgro CM (2011) Climate change and evolutionary adaptation. Nature, 470, 479–485.

Høye TT, Post E, Schmidt NM, Trøjelsgaard K, Forchhammer MC (2013) Shorter flowering seasons and declining abundance of flower visitors in a warmer Arctic. Nature Climate Change, 3, 759–763.

Huang JX, An JD (2018) Species diversity, pollination application and strategy for conservation of the bumblebees of China. Biodiversity Science, 26, 486–497.

Iler AM, Inouye DW, Høye TT, Miller-Rushing AJ, Burkle LA, Johnston EB (2013) Maintenance of temporal synchrony between syrphid flies and floral resources despite differential phenological responses to climate. Global Change Biology, 19, 2348–2359.

IPBES (2016) Summary for policymakers of the assessment report of the Intergovernmental Science—Policy Platform on Biodiversity and Ecosystem Services on pollinators, pollination and food production. In: Intergovernmental Science—Policy Platform on Biodiversity and Ecosystem Services Deliverables of the 2014–2018 Work Programme (eds Potts SG, Imperatriz-Fonseca VL, Ngo HT, Biesmeijer JC, Breeze TD, Dicks LV, Garibaldi LA, Hill R, Settele J, Vanbergen AJ, Aizen MA, Cunningham SA, Eardley C, Freitas BM, Gallai N, Kevan PG, Kovács-Hostyánszk A, Kwapong PK, Li J, Li X, Martins DJ, Nates-Parra G, Pettis JS, Rader R, Viana BF), pp. 1–28. IPBES, Bonn.

James RR (2005) Temperature and chalkbrood development in the alfalfa leafcutting bee, *Megachile rotundata*. Apidologie, 36, 15–23.

Jevanandam N, Goh AGR, Corlett RT (2013) Climate warming and the potential extinction of fig wasps, the obligate pollinators of figs. Biology Letters, 9, 20130041.

Johnson DM, Büntgen U, Frank DC, Kausrud K, Haynes KJ, Liebhold AM, Esper J, Stenseth NC (2010) Climatic warming disrupts recurrent alpine insect outbreaks. Proceedings of the National Academy of Sciences, USA, 107, 20576–20581.

Johnston MO, Bartkowska MP (2017) Individual pollen limitation, phylogeny and selection. New Phytologist, 214, 909–912.

Jump AS, Penuelas J (2005) Running to stand still: Adaptation and the response of plants to rapid climate change. Ecology Letters, 8, 1010–1020.

Kelly AE, Goulden ML (2008) Rapid shifts in plant distribution with recent climate change. Proceedings of the National Academy of Sciences, USA, 105, 11823–11826.

Kerr JT, Pindar A, Galpern P, Packer L, Potts SG, Roberts SM, Rasmont P, Schweiger O, Colla SR, Richardson LL, Wagner DL, Gall LF, Sikes DS, Pantoja A (2015) Climate change impacts on bumblebees converge across continents. Science, 349, 177–180.

Kingsolver JG, Woods HA, Buckley LB, Potter KA, MacLean HJ, Higgins JK (2011) Complex life cycles and the responses of insects to climate change. Integrative and Comparative Biology, 51, 719–732.

Klein AM, Vaissiere BE, Cane JH, Steffan-Dewenter I, Cunningham SA, Kremen C, Tscharntke T (2007) Importance of pollinators in changing landscapes for world crops. Proceedings of the Royal Society B: Biological Sciences, 274, 303–313.

Knop E, Gerpe C, Ryser R, Hofmann F, Menz MHM, Trösch S, Ursenbacher S, Zoller L, Fontaine C (2018) Rush hours in flower visitors over a day-night cycle. Insect Conservation and Diversity, 11, 267–275.
Konvicka M, Maradova M, Benes J, Fric Z, Kepka P (2003) Uphill shifts in distribution of butterflies in the Czech Republic: Effects of changing climate detected on a regional scale. Global Ecology and Biogeography, 12, 403–410.
Koti S, Reddy KR, Reddy VR, Kakani VG, Zhao DL (2005) Interactive effects of carbon dioxide, temperature, and ultraviolet-B radiation on soybean (*Glycine max* L.) flower and pollen morphology, pollen production, germination, and tube lengths. Journal of Experimental Botany, 56, 725–736.
Kudo G (1991) Effects of snow-free period on the phenology of alpine plants inhabiting snow patches. Arctic and Alpine Research, 23, 436–443.
Kudo G, Ida TY (2013) Early onset of spring increases the phenological mismatch between plants and pollinators. Ecology, 94, 2311–2320.
Kudo G, Nishikawa Y, Kasagi T, Kosuge S (2004) Does seed production of spring ephemerals decrease when spring comes early? Ecological Research, 19, 255–259.
Lara-Romero C, Seguí J, Pérez-Delgado A, Nogales M, Traveset A (2019) Beta diversity and specialization in plant-pollinator networks along an elevational gradient. Journal of Biogeography, 46, 1598–1610.
Lenoir J, Gegout JC, Marquet PA, de Ruffray P, Brisse H (2008) A significant upward shift in plant species optimum elevation during the 20th century. Science, 320, 1768–1771.
Li DK, Wang Z (2020) Spatiotemporal variation of vegetation phenology and its response to climate in Qinling Mountains based on MCD12Q2. Ecology and Environmental Sciences, 29, 11–22.
Liu PF, Wu J, Li HY, Lin SW (2011) Economic values of bee pollination to China's agriculture. Scientia Agricultura Sinica, 44, 5117–5123.
Liu XT, Yuan Q, Ni J (2019) Research advances in modelling plant species distribution in China. Chinese Journal of Plant Ecology, 43, 273–283.
McNamara JM, Barta Z, Klaassen M, Bauer S (2011) Cues and the optimal timing of activities under environmental changes. Ecology Letters, 14, 1183–1190.
Memmott J, Craze PG, Waser NM, Price MV (2007) Global warming and the disruption of plant-pollinator interactions. Ecology Letters, 10, 710–717.
Menéndez R, González-Megías A, Collingham Y, Fox R, Roy DB, Ohlemüller R, Thomas CD (2007) Direct and indirect effects of climate and habitat factors on butterfly diversity. Ecology, 88, 605–611.
Miller-Struttmann NE, Geib JC, Franklin JD, Kevan PG, Holdo RM, Ebert-May D, Lynn AM, Kettenbach JA, Hedrick E, Galen C (2015) Functional mismatch in a bumble bee pollination mutualism under climate change. Science, 349, 1541–1544.
Mu J, Peng Y, Xi X, Wu X, Li G, Niklas KJ, Sun S (2015) Artificial asymmetric warming reduces nectar yield in a Tibetan alpine species of Asteraceae. Annals of Botany, 116, 899–906.
O'Neill KM, O'Neill RP, Kemp WP, Delphia CM (2011) Effect of temperature on post-wintering development and total lipid content of alfalfa leafcutting bees. Environmental Entomology, 40, 917–930.
Ollerton J (2017) Pollinator diversity: Distribution, ecological function, and conservation. Annual Review of Ecology, Evolution, and Systematics, 48, 353–376.
Ollerton J, Winfree R, Tarrant S (2011) How many flowering plants are pollinated by animals? Oikos, 120, 321–326.
Ovaskainen O, Skorokhodova S, Yakovleva M, Sukhov A, Kutenkov A, Kutenkova N, Shcherbakov A, Meyke E, del Mar Delgado M (2013) Community-level phenological response to climate change. Proceedings of the National Academy of Sciences of the United States of America, 110, 13434–13439.
Parmesan C (2006) Ecological and evolutionary responses to recent climate change. Annual Review of Ecology, Evolution, and Systematics, 37, 637–669.
Parmesan C (2007) Influences of species, latitudes and methodologies on estimates of phenological response to global warming. Global Change Biology, 13, 1860–1872.
Pellissier L, Rohr RP, Ndiribe C, Pradervand JN, Salamin N, Guisan A, Wisz M (2013) Combining food web and species distribution models for improved community projections. Ecology and Evolution, 3, 4572–4583.
Potts SG, Biesmeijer JC, Kremen C, Neumann P, Schweiger O, Kunin WE (2010) Global pollinator declines: Trends, impacts and drivers. Trends in Ecology & Evolution, 25, 345–353.
Prasad PVV, Boote KJ, Allen LH Jr, Thomas JMG (2002) Effects of elevated temperature and carbon dioxide on seed-set and yield of kidney bean (*Phaseolus vulgaris* L.). Global Change Biology, 8, 710–721.
Pressman E, Peet MM, Pharr DM (2002) The effect of heat stress on tomato pollen characteristics is associated with changes in carbohydrate concentration in the developing anthers. Annals of Botany, 90, 631–636.

Pyke GH, Thomson JD, Inouye DW, Miller TJ (2016) Effects of climate change on phenologies and distributions of bumble bees and the plants they visit. Ecosphere, 7, e01267.
Radmacher S, Strohm E (2010) Factors affecting offspring body size in the solitary bee Osmia bicornis (Hymenoptera, Megachilidae). Apidologie, 41, 169–177.
Radmacher S, Strohm E (2011) Effects of constant and fluctuating temperatures on the development of the solitary bee *Osmia bicornis* (Hymenoptera: Megachilidae). Apidologie, 42, 711–720.
Rafferty NE, Diez JM, Bertelsen CD (2020) Changing climate drives divergent and nonlinear shifts in flowering phenology across elevations. Current Biology, 30, 432–441.
Rafferty NE, Ives AR (2011) Effects of experimental shifts in flowering phenology on plant-pollinator interactions. Ecology Letters, 14, 69–74.
Ren ZX, Wang H, Peter B, Li DZ (2014) Insect pollination and self-incompatibility in edible and/or medicinal crops in southwestern China, a global hotspot of biodiversity. American Journal of Botany, 101, 1700–1710.
Ren ZX, Zhao YH, Liang H, Tao ZB, Tang H, Zhang HP, Wang H (2018) Pollination ecology in China from 1977 to 2017. Plant Diversity, 40, 172–180.
Richman SK, Levine JM, Stefan L, Johnson CA (2020) Asynchronous range shifts drive alpine plant-pollinator interactions and reduce plant fitness. Global Change Biology, 26, 3052–3064.
Roy DB, Sparks TH (2000) Phenology of British butterflies and climate change. Global Change Biology, 6, 407–416.
Scaven VL, Rafferty NE (2013) Physiological effects of climate warming on flowering plants and insect pollinators and potential consequences for their interactions. Current Zoology, 59, 418–426.
Scheifinger H, Koch E, Winkler H (2005) Results of a first look into the Austrian animal phenological records. Meteorologiche Zeitschrift, 14, 203–209.
Schleuning M, Fründ J, Schweiger O, Welk E, Albrecht J, Albrecht M, Beil M, Benadi G, Blüthgen N, Bruelheide H, Böhning-Gaese K, Dehling DM, Dormann CF, Exeler N, Farwig N, Harpke A, Hickler T, Kratochwil A, Kuhlmann M, Kühn I, Michez D, Mudri-Stojnić S, Plein M, Rasmont P, Schwabe A, Settele J, Vujić A, Weiner CN, Wiemers M, Hof C (2016) Ecological networks are more sensitive to plant than to animal extinction under climate change. Nature Communications, 7, 13965.
Schweiger O, Heikkinen RK, Harpke A, Hickler T, Klotz S, Kudrna O, Kühn I, Pöyry J, Settele J (2012) Increasing range mismatching of interacting species under global change is related to their ecological characteristics. Global Ecology and Biogeography, 21, 88–99.
Schweiger O, Settele J, Kudrna O, Klotz S, Kühn I (2008) Climate change can cause spatial mismatch of trophically interacting species. Ecology, 89, 3472–3479.
Sgolastra F, Bosch J, Molowny-Horas R, Maini S, Kemp WP (2010) Effect of temperature regime on diapause intensity in an adult-wintering Hymenopteran with obligate diapause. Journal of Insect Physiology, 56, 185–194.
Sherry RA, Zhou XH, Gu SL, Arnone JA, Schimel DS, Verburg PS, Wallace LL, Luo YQ (2007) Divergence of reproductive phenology under climate warming. Proceedings of the National Academy of Sciences, USA, 104, 198–202.
Sosa-Pivatto M, Cosacov A, Baranzelli MC, Iglesias MR, Espíndola A, Sérsic AN (2017) Do 120,000 years of plant-pollinator interactions predict floral phenotype divergence in *Calceolaria polyrhiza*? A reconstruction using species distribution models. Arthropod-Plant Interactions, 11, 351–361.
Sparks TH, Langowska A, Głazaczow A, Wilkaniec Z, Bien´kowska M, Tryjanowski P (2010) Advances in the timing of spring cleaning by the honeybee *Apis mellifera* in Poland. Ecological Entomology, 35, 788–791.
Stelzer RJ, Chittka L, Carlton M, Ings TC (2010) Winter active bumblebees (*Bombus terrestris*) achieve high foraging rates in urban Britain. PLOS ONE, 5, e9959.
Stephen WP (1965) Temperature effects on the development and multiple generations in the alkali bee, *Nomia melanderi* Cockerell. Entomologia Experimentalis et Applicata, 8, 228–240.
Stone GN, Willmer PG (1989) Warm-up rates and body temperatures in bees: The importance of body size, thermal regime and phylogeny. Journal of Experimental Biology, 147, 303–328.
Stuhldreher G, Hermann G, Fartmann T (2014) Cold-adapted species in a warming world—an explorative study on the impact of high winter temperatures on a continental butterfly. Entomologia Experimentalis et Applicata, 151, 270–279.
Takkis K, Tscheulin T, Tsalkatis P, Petanidou T (2015) Climate change reduces nectar secretion in two common Mediterranean plants. AoB Plants, 7, plv111.
Thackeray SJ, Henrys PA, Hemming D, Bell JR, Botham MS, Burthe S, Helaouet P, Johns DG, Jones ID, Leech DI, Mackay EB, Massimino D, Atkinson S, Bacon PJ, Brereton TM, Carvalho L, Clutton-Brock TH, Duck C, Edwards M, Elliott JM, Hall SJG, Harrington R, Pearce-Higgins JW, Høye TT, Kruuk LEB, Pemberton JM, Sparks TH, Thompson PM, White I, Winfield IJ, Wanless S (2016) Phenological sensitivity to climate across taxa and trophic levels. Nature, 535, 241–245.

Theobald EJ, Breckheimer I, HilleRisLambers J (2017) Climate drives phenological reassembly of a mountain wildflower meadow community. Ecology, 98, 2799–2812.

Totland Ø (1997) Effects of flowering time and temperature on growth and reproduction in *Leontodon autumnalis* var. taraxaci, a late flowering alpine plant. Arctic and Alpine Research, 29, 285–290.

Totland Ø, Alatalo JM (2002) Effects of temperature and date of snowmelt on growth, reproduction, and flowering phenology in the arctic/alpine herb, *Ranunculus glacialis*. Oecologia, 133, 168–175.

Tsiftsis S, Djordjević V (2020) Modelling sexually deceptive orchid species distributions under future climates: The importance of plant-pollinator interactions. Scientific Reports, 10, 10623.

Valtonen A, Molleman F, Chapman CA, Carey JR, Ayres MP, Roininen H (2013b) Tropical phenology: Bi-annual rhythms and interannual variation in an Afrotropical butterfly assemblage. Ecosphere, 4, 36.

Van der Kooi CJ, Kevan PG, Koski MH (2019) The thermal ecology of flowers. Annals of Botany, 124, 343–353.

Vázquez DP, Blüthgen N, Cagnolo L, Chacoff NP (2009) Uniting pattern and process in plant-animal mutualistic networks: A review. Annals of Botany, 103, 1445–1457.

van Nouhuys S, Lei GC (2004) Parasitoid-host metapopulation dynamics: The causes and consequences of phenological asynchrony. Journal of Animal Ecology, 73, 526–535.

Visser ME (2008) Keeping up with a warming world; assessing the rate of adaptation to climate change. Proceedings of the Royal Society B: Biological Sciences, 275, 649–659.

Walker WH, Meléndez-Fernández OH, Nelson RJ, Reiteret RJ (2019) Global climate change and invariable photoperiods: A mismatch that jeopardizes animal fitness. Ecology and Evolution, 9, 10044–10054.

Wang DN, Tang XL, Lian Y, Cui JL, Du CY, Ji YH (2020) Response of phenology of three typical deciduous trees to climate warming—Take Changbai Mountains as an example. Journal of Northeast Forestry University, 48(9), 51–55.

Wang M, Sang WG (2020) The change of phenology of tree and shrub in warm temperate zone and their relationships with climate change. Ecological Science, 39, 164–175.

Warren MS, Hill JK, Thomas JA, Asher J, Fox R, Huntley B, Royk DB, Telferk MG, Jeffcoate S, Harding P, Jeffcoate G, Willis SG, Greatorex-Daviesk JN, Moss D, Thomas CD (2001) Rapid responses of British butterflies to opposing forces of climate and habitat change. Nature, 414, 65–69.

Whitfield GH, Richards KW (1992) Temperature-dependent development and survival of immature stages of the alfalfa leafcutter bee, *Megachile rotundata* (Hymenoptera: Megachilidae). Apidologie, 23, 11–23.

Williams PH, Huang JX, An JD (2017) Bear wasps of the Middle Kingdom: A decade of discovering China's bumblebees. Antenna, 41, 21–24.

Wilson RJ, Gutierrez D, Gutierrez J, Martinez D, Agudo R, Monserrat VJ (2005) Changes to the elevational limits and extent of species ranges associated with climate change. Ecology Letters, 8, 1138–1146.

Wolda H (1988) Insect seasonality: Why? Annual Review of Ecology and Systematics, 19, 1–18.

Wolf S, Moritz RFA (2008) Foraging distance in *Bombus terrestris* L. (Hymenoptera: Apidae). Apidologie, 39, 419–427.

Xiao YA, Zhang SS, Yan XH, Dong M (2015) New advances in effects of global warming on plant-pollinator networks. Acta Ecologica Sinica, 35, 3871–3880.

Xu XT, Yang Y, Wang LS (2008) Geographic distribution and potential distribution estimation of *Pseudotaxus chienll*. Chinese Journal of Plant Ecology, 32, 1134–1145.

Yang LP, Zhang CH, Dai HY, Lü DB, Han J (2020) Effects of climate change on phenophase of woody plants at the northern foot of Yinshan Mountain. Journal of Meteorology and Environment, 36, 102–107.

Zhang L, Liu SR, Sun PS, Wang TL (2011) Comparative evaluation of multiple models of the effects of climate change on the potential distribution of *Pinus massoniana*. Chinese Journal of Plant Ecology, 35, 1091–1105.

Zhao ZG, Wang YK (2015) Selection by pollinators on floral traits in generalized *Trollius ranuncuoides* (Ranunculaceae) along altitudinal gradients. PLOS ONE, 10, e0118299.

# 12 Climate Change and Household Insect Pests

*Ammara Riaz, Muhammad Asad, Kashif Ali, Sana Ullah, Asma Ashraf, Saima Qadeer, Mawra Rasheed, Usama Abdullah, and Hina Zawar*

## 12.1 INTRODUCTION

Climate change is a critical issue that poses a major threat to the environment, public health, and global economic stability. The effects of climate change are being felt around the world, including rising sea levels, changes in rainfall patterns, and more intense storms. In addition, climate change also has a major impact on mental health, as it causes stress, depression, and anxiety. Climate change is one of the most pressing issues of the 21st century because its effects are not uniform, and less developed societies are usually more vulnerable to its impacts (Manzanedo & Manning, 2020; Cianconi et al., 2020).

Both statesmen and scientists agree that climate change needs to be addressed. According to a report by the Intergovernmental Panel on Climate Change (IPCC), decisions made today can greatly affect the extent of climate change in the future. The urgent need to act has been emphasized. Combating climate change requires institutional action at national and international levels, including implementing policies to reduce greenhouse gas emissions, promoting new sources of clean energy, and investing in sustainable infrastructure development (Manzanedo & Manning, 2020; IPCC, 2021).

Climate change is a complex and urgent global challenge, and predicting future changes in temperature is difficult due to the variety of factors that influence these changes. A major cause of this difficulty is the earth's ecosystem, in which various physical, chemical, and biological processes are intimately interrelated. All these factors interact, making temperature changes difficult to predict (Jansson & Hofmockel, 2020). In addition to direct effects, several factors such as greenhouse gas emissions, solar radiation, volcanic activity and land-use change can influence temperature changes (Piao et al., 2019). All these factors make it difficult to accurately predict future temperature fluctuations.

Predicting future climate behaviors is difficult due to various environmental factors. Climate systems can vary from region to region and time to time, making it difficult to make accurate predictions about the future. In addition, some weather systems, such as El Niño and La Niña, have significant effects on the environment, but it is not always possible to predict them (Rocklöv & Dubrow, 2020; Tramblay et al., 2020).

Insects are included in a group that is said to be highly intolerant of climate change. Many insect populations are declining due to climate change (Halsch et al., 2021). To decide the future of these insects, it is important to understand the reactions of these insects with new environmental conditions (Skendžić et al., 2021). Insect responses vary according to various environmental factors (Table 12.1) such as temperature, humidity, and food availability, and these responses may depend on the species and their life stages (Kellermann & van Heerwaarden, 2019). Therefore, assessing the immediate responses of insect populations is important for predicting their future (Birrell et al., 2020).

Insect populations, numbers, and distribution are affected by climate change (Lehmann et al., 2020). To understand the effects of heat on insect damage, it is important to understand the


DOI: 10.1201/9781003382089-12

**TABLE 12.1**
**Specific Climate Changes Affecting Insect Pests**

| Series No. | Climate Changes | References |
|---|---|---|
| 1 | Changes in temperature | Skendžić et al., 2021; Harvey et al., 2020: Tougeron et al., 2020 |
| 2 | Changes in precipitation | Venäläinen et al., 2020; Malhi et al., 2021 |
| 3 | Changes in extreme weather events | Harvey et al., 2020; Tougeron et al., 2020 |
| 4 | Changes in plant phenology | Trebicki, 2020; Chidawanyika et al., 2019 |
| 5 | Changes in host-parasitoid interactions | Chidawanyika et al., 2019 |
| 6 | Changes in pest distribution | Santana et al., 2019 |
| 7 | Changes in insect hybridization | Larson et al., 2019 |
| 8 | Changes in forest management practices | Harvey et al., 2020 |

environmental factors of insect populations (Lehmann et al., 2020). For example, diapause, a period of dormancy, is an important adaptive feature in insects that helps them survive unfavorable conditions such as winter and drought (Tougeron, 2019). Changes in diapause can affect insect population dynamics and alter their habitat (Tougeron, 2019).

Insects are highly responsive to heat stress, so predicting their response is particularly difficult. All insect species and populations have distinct physiological mechanisms and environmental influences (González-Tokman et al., 2020). Thermal plasticity in insects plays a vital role in their response to climate change (Rodrigues & Beldade, 2020). Insect pests show different responses to different stressors depending on the temperature, and it is necessary to define their effects on each stressor (Skendžić et al., 2021).

## 12.2 EFFECTS OF ANTHROPOGENICALLY INDUCED CLIMATIC CHANGE ON INSECT PESTS

Human-induced climate change has significant impacts on insect pests and their populations. Both plants and insects are directly affected by extreme weather events and human activities. Recent research shows that climate change and other human-caused pressures are putting enormous pressure on forests, which negatively impacts global exports. Understanding the taxonomy and immediate as well as long-term effects on insect populations is crucial for scientific and academic research. Clear and well-known evidence of climate change's effects on insects can support this research (Karthik et al., 2021; Mansoor et al., 2022; Harvey et al., 2020).

Hurricanes can have varying effects on different insect species. Before studying the impact of climate change on environmental factors that affect insects, it is crucial to examine the effects of changing climate conditions on hurricanes. Climate change can lead to economic losses for agricultural insects, and warming is a crucial factor that affects insect population dynamics (Lehmann et al., 2020). Therefore, it is essential to understand the impact of climate change on hurricanes and its indirect effects on insect populations.

## 12.3 ADAPTATION OF INSECTS TO DIVERSE HABITATS

Insects have been evolved for millions of years, possessing a variety of tools that they have evolved to adapt to different environments, such as aquatic and terrestrial environments (Table 12.2). Insects have developed tools, such as improvements in intracellular compartments, that allow them to adapt to different environments. In addition to these developments, insects have also acquired genetic and physiological developments to enable them to survive in aquatic environments. Ecology has also seen the activity of divergent growths in different environments, which increase the abundance of organisms (Wilson & Fox, 2021; Zhou et al., 2019; Jooste et al., 2020; Adams et al., 2020).

**TABLE 12.2**
**Exploring the Impact of Climate Change on Ecosystems: A Review of Previous Studies on the Interplay between Physical and Biological Components**

| Impact of Climate Change on Ecosystems | References |
|---|---|
| As ocean acidification increases, it reduces the protection of coastal biomes against the effects of other factors such as climate change. | Hall-Spencer & Harvey, 2019 |
| It has been proven that climate change is causing other problems around the world, one of which is increasing the acidity of seawater. As the amount of carbon dioxide ($CO_2$) and other toxic gases increases, so does the acidity, which is dangerous for marine animals and plants. | Ahmed et al., 2019 |
| Climate change refers to an increasing number of conditions that can lead to fisheries decline. That is, when there is a reduction in fish production or catch due to storms, sea level rise, change in seawater temperature, or increased water intensity, etc., fisheries decline may start. | Mendenhall et al., 2020 |
| One of the most important facts about environmental changes is that the pH of the world's oceans is decreasing due to the increase in $CO_2$ levels at critical levels. When $CO_2$ combines with oxygen to form carbonic acid, its active particles are present in the solution, which can have a negative impact on the life of marine animals in the future. In addition, it is possible that marine plants, cacti, rocks, and other key elements can adversely affect the performance and growth and survival of animals. | Mendenhall et al., 2020 |
| Due to climate change, the temperature of ocean pockets is increasing, which increases the productivity of oceans. This, in turn, increases fish populations and with it fisheries. Also, due to this fishing, more $CO_2$ enters the seawater which lowers the sea pH. Lowering ocean pH poses a threat to the life of algae, crustaceans, and other organisms. Without mud and crust, other marine animals cannot survive because too low a pH disrupts their physiological processes. | Lam et al., 2020 |
| The health of aquatic organisms, such as reefs, corals, plankton, and other micro- and macro-organisms, that obtain their nutrients from sunlight is affected by acidification. "UV radiation" and "oxygen depletion" are considered to be the effects of ocean acidification on marine ecosystems. | Gao et al., 2019 |
| With the help of coralline algae, the ability to protect against the effects of marine acidification can be achieved. When this algae is exposed to the effects of ocean acidification for several generations, it becomes more and more tolerant to ocean acidification. | Cornwall et al., 2020 |
| Seaweeds around the world are experiencing a decline in calcium carbonate production due to seawater erosion and warming. | Cornwall et al., 2021 |
| Climate change, a major problem facing the world, is an important topic that also affects microbiology. The relationship of microbes to climate change is considered, such as the emergence of traditional and new types of microbes, information about the future of disease outbreaks, and changes in the performance of microbes. | Hutchins et al., 2019 |
| Climate change poses threats to ecosystems. Changes in climate movement between large surfaces lead to changes in the movement of ecosystem materials that affect ecosystem interactions. In addition, the severity of drought is a threat to ecosystems. It affects the supply of food, water, and shelter for plants and animals which adversely affects their life. | Malhi et al., 2020 |
| Coastal areas are also affected by changing environmental conditions due to the effects of ocean acidification, which also affects these services. | Hall-Spencer & Harvey, 2019 |
| Plastic pollution makes an invisible contribution to greenhouse gases and climate change worldwide. That is, because of plastic, the rate of greenhouse gases increases in the world, and it also plays an important role in climate change. | Shen et al., 2020 |
| Mining causes huge amounts of direct and indirect greenhouse gases that are harmful to the environment. Overall, mining releases $CO_2$ which contributes to climate change. | Azadi et al., 2020 |

**TABLE 12.2 (*Continued*)**
**Exploring the Impact of Climate Change on Ecosystems: A Review of Previous Studies on the Interplay between Physical and Biological Components**

| Impact of Climate Change on Ecosystems | References |
|---|---|
| Hospitals in high-import countries cause environmental damage by masking greenhouse gases in the urban atmosphere. | Quitmann et al., 2023 |
| Economic growth in the BRICS (Brazil, Russia, China, India, South Africa) countries has exceeded the contribution of electricity intensity and the number of greenhouse gas reserves, and has included the dynamics of climate change. | Zheng et al., 2019 |
| The major source of greenhouse gas emissions arises from the extraction, production, and consumption of engineering materials. This statement clarifies that the predominant greenhouse gas emissions stem from the extraction, production, and consumption of energy resources like electricity, gas, diesel, or petroleum. | Mikhaylov et al., 2020 |
| All countries should compare the amount of greenhouse gases they emit from nuclear power, fossil fuels, and industrial activities within international climate change limits. In other words, when assessing height, we need to consider each country's overall electricity consumption without overly limiting other nations due to their higher consumption levels. | Ritchie et al., 2020 |
| We should create strategies to reduce greenhouse gas emissions by focusing on those with higher socioeconomic status. These people can be considered as the most important targets because on the one hand, they themselves emit large amounts of greenhouse gases, and on the other hand, they have the ability to discover ways to reduce these gases and reduce their consumption. | Nielsen et al., 2021 |
| Management of urban surface drainage systems plays an important role in climate change. Litter phoenixes and landfills have impacts on climate change due to the accumulation of greenhouse gases. | Maria et al., 2020 |
| The report assesses the dynamics of Earth's climate system and also looks at food systems, examining how both can be helped through both prevention and adaptation. | Ingram, 2011 |
| Climate and soil characteristics determine where no-till management may have the potential to reduce greenhouse gas emissions and store carbon in the soil. | Ogle et al., 2019 |

The adaptive biological strategies adopted by insects in different environments have contributed to their evolutionary success, and this success is threatened by environmental factors such as climate change, habitat loss, and the severe effects of pollution (Wilson & Fox, 2021). These factors reduce insect populations and their ability to adapt to new and changing environments, reducing insect diversity worldwide (Wilson & Fox, 2021). Conservation measures that focus on protecting the protected and diverse areas of different ecosystems can ensure the survival of these biodiverse organisms (Samways et al., 2020).

## 12.4 BEHAVIORAL ADAPTATIONS OF INSECTS TO TEMPERATURE CHANGES

Insects exhibit different behaviors to cope with temperature changes. These developments include physical, morphological, and behavioral changes. Insects regulate their body temperature through basking habits, selection of better microhabitats, and behavioral thermal enhancement. In addition, some insects have evolved oxygen-providing properties to cope with physiological challenges (Birrell et al., 2020).

Insects are prominent examples of thermal latches that allow them to adjust their physiology, lifespan, and geographic location to changes in temperature (Rodrigues & Beldade, 2020). For example, insect physiology has been shown to be critical for coping with winter, which explains the

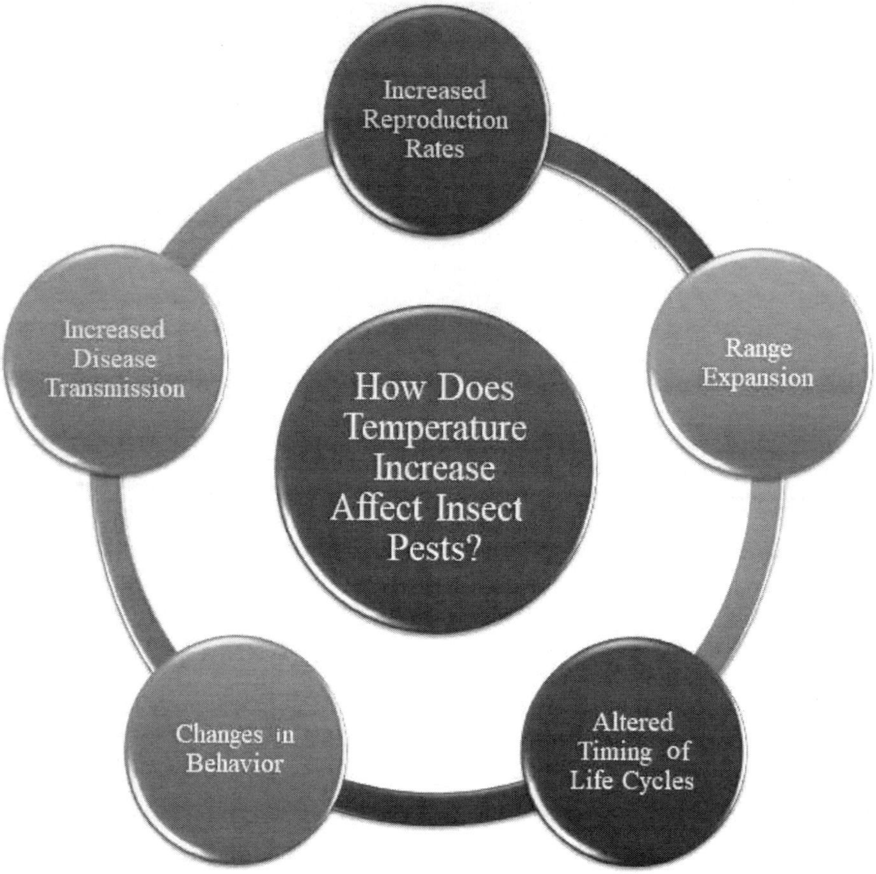


**FIGURE 12.1** Effects of Temperature Increase on Insect Pests.

importance of their physiology in terms of historical temperature and environmental adaptations (Lowe et al., 2021).

Insects use different strategies to cope with different temperatures such as changing their physiology, accommodation, and behavior (Figure 12.1). For example, bees have evolved different physiological and behavioral adaptations to cope with temperature and humidity dependencies (Ma et al., 2019; Abarca & Spahn, 2021).

### 12.4.1 Adjustment of Organisms to Challenging Environments

Organisms face various challenges to survive and reproduce, including extreme temperatures, salinity, and climate change. They adapt their behavior, function, and metabolism according to the conditions to cope with environmental changes (Bernhardt et al., 2020). For instance, organisms can adjust their metabolic rate to withstand temperature changes, change their gene expression to adapt to new environments, and in some cases, modify their physical structure and function to thrive in new surroundings.

Research has shown that organisms also have the ability to use predictive cues to alter their biological settings in response to impending environmental changes. For example, blackbirds adjust their metabolic rate settings according to changing temperature and light during seasonal movement (Figure 12.2). Also, some organisms regulate their internal conditions to maintain external

**FIGURE 12.2** Insect susceptibility to climate change.

environmental conditions such as temperature, salinity, and other parameters so that they can maintain their performance within a limited range (Bernhardt et al., 2020; Thurman et al., 2020; Gangloff & Greenberg, 2023).

## 12.5 EFFECTS OF INCREASED CARBON DIOXIDE LEVELS AND HIGHER TEMPERATURES ON INSECTS

Elevated carbon dioxide levels and increased temperatures are projected to exert substantial influences on insect populations, primarily through heightened rates of parasitism. These shifts may impact their physiological functions, behavior, metabolic processes, and reproduction rates. According to Jactel et al. (2019), elevated temperatures can cause changes in the insect life cycle, lifespan, and reproduction. In addition, thermal changes can reduce the intensity of tree hardening against pests, making them more vulnerable to insect attacks. Indirect effects may also occur due to changes in beetle physiology, as trees are less vigorous in summer and may become more susceptible to attack by these pests.

In addition to the effects on habitats, elevated $CO_2$ can also affect insect-plant interactions by altering the nutritional status of trees. According to Teawkul and Hwang (2019), increased $CO_2$ concentrations can alter the production of plant defense compounds, which can impact herbivore feeding. Insects that depend on certain compounds for food and detoxification can suffer from reduced nutritional status, which can lead to declines in health and population.

In addition to direct effects, increased heat and carbon dioxide concentrations may affect the distribution and numbers of insects, with effects on their pedogenesis and behavior. For example, Shrestha (2019) reported that increasing temperatures may lead to changes in insect activity patterns, including early hatching and increased activity in warmer seasons (Figure 12.3). This can affect the timing and severity of pest outbreaks and the effectiveness of pest control methods.

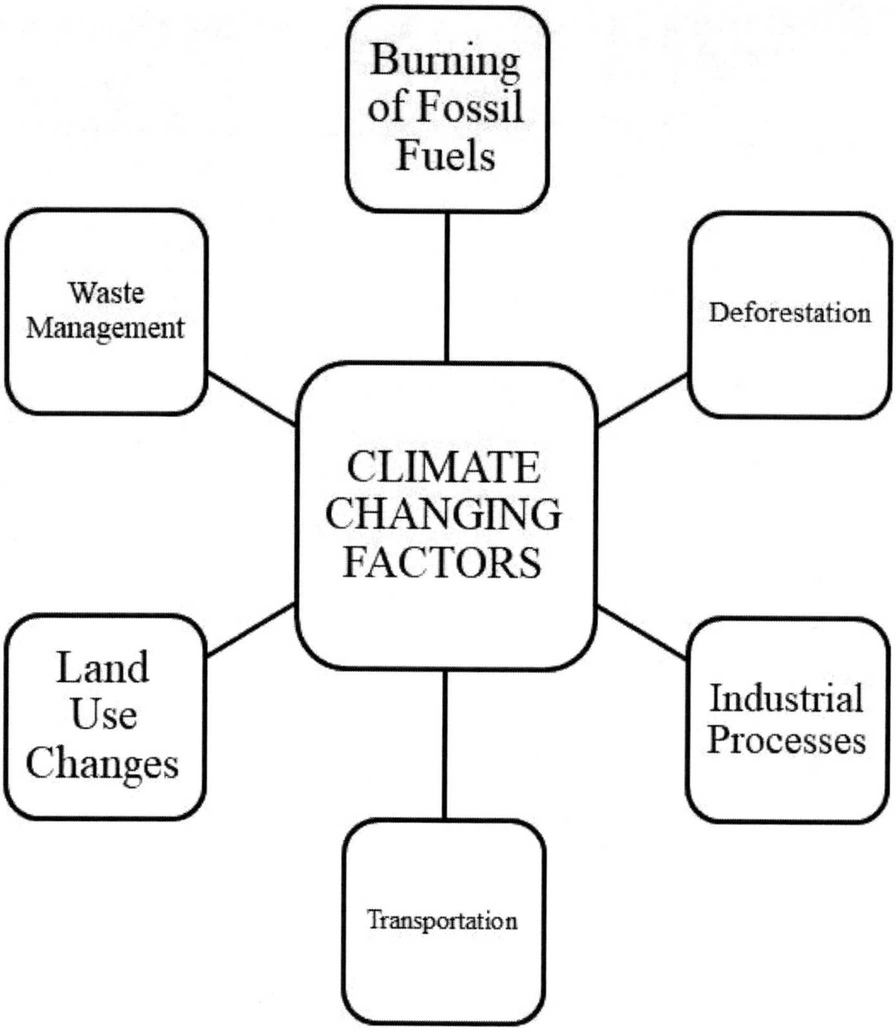


**FIGURE 12.3** Factors contributing to climate change.

### 12.5.1 Environmental Factors Affecting Insects

Insects are very intelligent organisms that are affected by different environmental factors such as temperature, humidity, and rainfall (Skendžić et al., 2021). Temperature is the most important environmental factor for insects (Skendžić et al., 2021). Environmental factors such as the illuminance, spectral quality, and wavelength of the insect's light actually affect the insect's phototactic behavior (Kim et al., 2019). Drought severity, food availability, and average annual rainfall are some of the environmental factors that influence the responses of ground beetles to environmental stressors (Gely et al., 2020).

Changes in environmental factors such as temperature, humidity, and precipitation, known as climate change, can affect dung beetles (Shrestha, 2019). The use of insects is increasingly being presented as a solution to reduce the environmental problems associated with livestock production and to enhance human food production (Sogari et al., 2019). Additionally, insect-fungal interactions have been implicated in causing insect damage because fungi can spread to contiguous areas under suitable environmental conditions (Islam et al., 2021).

## 12.6 COMPREHENSIVE ASSESSMENT OF THE IMPACTS OF CLIMATE CHANGE ON HOUSEHOLD INSECT PESTS

Pests on the floor can cause joint destruction and property damage, and their impact on minority populations must not be overlooked. Monitoring insect populations is essential to understand the complex responses of global insects, along with a thorough assessment of historical climate documentation (Lehmann et al., 2020). Skendžić et al. (2021) have pointed out that seasonal temperature changes are an important aspect of climate change for North American fauna, and insects respond differently to different causes of changes. However, Jactel et al. (2019) have observed that the effects of climate change on forest insects are not straightforward, and multiple factors such as temperature, precipitation, and extreme weather events interact to affect insect populations.

To mitigate future impacts on food production systems as a result of climate change, an understanding of the threats posed by pests and diseases to plants is crucial. According to Gomez-Zavaglia et al. (2020), assessing the impacts of climate change on crop productivity requires an estimate of the associated risks. Omerkhil et al. (2020) found that smallholder farmers in Takhar, Afghanistan's Yangi Qila district face challenges from pests, diseases, and natural disasters and are highly sensitive to the effects of climate change. Therefore, it is important to understand the impacts of climate change on plant productivity. According to Secretariat et al. (2021), understanding the effects of pests on plants is critical to predicting potential impacts on crop productivity.

Philanthropists use different means for their work in addressing climate change. Similarly, in order to increase remediation efficiency while controlling the spread of pests and maintaining crop production, various methods can be employed (Aryal et al., 2020). When discussing climate change mitigation programs, it is important to include measures to deal with the spread of pests in order to protect livestock from the adverse effects of crop production (Praveen & Sharma, 2019). In addition, it is important to protect insects as they play a critical role in mitigating the effects of climate change. Solutions should be implemented that are efficient, such as the use of thatched roofs that provide protection to insects (Samways et al., 2020).

## 12.7 POTENTIAL CONSEQUENCES OF CLIMATE CHANGE ON INSECT BEHAVIOR AND DAILY MOVEMENTS

The effects of climate change are expected to affect the movement and daily activities of insects, which may lead to loss of biodiversity. Changing temperature and precipitation patterns can cause changes in insect communities and development, which can lead to changes in their activity patterns, interests, and distribution. For example, warmer temperatures can cause pests to damage plants and other plants. In addition, light pollution caused by human activities can also affect insect activity, affecting their activity and movement. Future impacts of climate change may have severe impacts on insect populations, which may directly affect biodiversity and ecosystems. Insects play an important role in many biodiversity relationships, and changes in their behavior affect biodiversity and other services such as biodiversity. For example, changes in the temperature and distribution of mountain floodwaters due to climate change may affect the historical scale and behavior of aquatic insects, thereby affecting the composition and performance of these ecosystems. Pole-dwelling animals such as polar bears can also be affected. The effects may include changes in their emotional health, nutritional needs, and physical health. In addition, ecologically dependent insects such as mosquitoes, bed bugs, and polar insects may also be harmed by the effects of climate change. Changes in their behavior and populations have major implications for the health and survival of these insects. In addition, the effects of climate change may also affect the social behavior of insects, leading to changes in their morphology, reproduction, and evolutionary pathways. Changes in their relative abundance, sexual behavior and maturity, reproductive success, and population dynamics can all produce similar effects. Anticipated effects of environmental change reflect the fact that these birth and life functions are most influenced by the environment in which we were born. The foundation of human life is a chain that includes the rewards of all living beings. Unless we protect our environment, we cannot protect our own lives (Harvey et al., 2023; Birrell et al., 2020; Joly et al., 2020; Owens et al., 2020; Fisher et al., 2021).

## 12.8 STRATEGIES TO MANAGE HOUSEHOLD INSECT PESTS IN THE CONTEXT OF A CHANGING CLIMATE

Environmental changes affect domestic insects, influencing their behavior, distribution, and population statistics. With summer, their numbers increase, and their reproductive periods are extended for longer periods, which weakens the insect population. In addition, environmental changes cause changes in different methods of pest elimination, leading to the emergence of new pests in previously unaffected areas. To control domestic pests in a changing environment, several projects have been proposed.

One experiment aims to create potent insecticides capable of countering novel pests that exhibit adaptability to engineered insects in diverse conditions, including environmental changes. For example, changes in insect behavior may weaken the efficacy of synthetic insecticides, making it impossible to increase the use of effective insecticides and making it more beneficial to use animals with natural insect repellants. Another approach is to monitor insect populations and behavior using advanced technology and data analysis. This helps to identify new pest problems and develop more effective plans. This strategy is used to deal with various types of household pests such as bed bugs, flies, ants, and spiders that have a negative impact on human health and daily life. These insects can cause health problems such as allergies, asthma, and respiratory problems, and can increase the risk of foodborne illness. Through education and awareness programs, we strive to educate people about the dangers of household pests and provide safe measures against them. For example, we can train people to adopt peaceful measures such as upholstering and decorating household items, keeping food items in closed existing containers to rid the home of pests, and keeping open containers to prevent pests from entering. We also encourage closing doors and gaps. These measures can help people lead healthy lives and reduce the number of pests in their homes (Lehmann et al., 2020; Skendžić et al., 2021; Rehman et al., 2022).

## 12.9 CONCLUSION

The effects of climate change caused by human activities affect insects in a number of different ways. Along with this, various environmental factors like temperature, humidity, and carbon dioxide are also important for them. Insects exhibit amazing adaptations to different environments and environmental conditions, including motility and physical adaptations to cope with temperature changes and other stresses. But insect populations may face major threats due to climate change, which may facilitate changes in daily movements and speeds of insects, with far-reaching implications for public health and environmental management. To address these threats, there is a need to develop integrated domestic pest management strategies that take into account the effects of climate change. Effective management strategies have to be taken with regard to the impact on pests.

## REFERENCES

Abarca, M., & Spahn, R. (2021). Direct and indirect effects of altered temperature regimes and phenological mismatches on insect populations. Current Opinion in Insect Science, 47, 67–74.

Adams, B. J., Li, E., Bahlai, C. A., Meineke, E. K., McGlynn, T. P., & Brown, B. V. (2020). Local-and landscape-scale variables shape insect diversity in an urban biodiversity hot spot. Ecological Applications, 30(4), e02089.

Ahmed, N., Thompson, S., & Glaser, M. (2019). Global aquaculture productivity, environmental sustainability, and climate change adaptability. Environmental Management, 63, 159–172.

Aryal, J. P., Sapkota, T. B., Khurana, R., Khatri-Chhetri, A., Rahut, D. B., & Jat, M. L. (2020). Climate change and agriculture in South Asia: Adaptation options in smallholder production systems. Environment, Development and Sustainability, 22(6), 5045–5075.

Azadi, M., Northey, S. A., Ali, S. H., & Edraki, M. (2020). Transparency on greenhouse gas emissions from mining to enable climate change mitigation. Nature Geoscience, 13(2), 100–104.

Bernhardt, J. R., O'Connor, M. I., Sunday, J. M., & Gonzalez, A. (2020). Life in fluctuating environments. Philosophical Transactions of the Royal Society B, 375(1814), 20190454.

Birrell, J. H., Shah, A. A., Hotaling, S., Giersch, J. J., Williamson, C. E., Jacobsen, D., & Woods, H. A. (2020). Insects in high-elevation streams: Life in extreme environments imperiled by climate change. Global Change Biology, 26(12), 6667–6684.

Chidawanyika, F., Mudavanhu, P., & Nyamukondiwa, C. (2019). Global climate change as a driver of bottom-up and top-down factors in agricultural landscapes and the fate of host-parasitoid interactions. Frontiers in Ecology and Evolution, 7, 80.

Cianconi, P., Betrò, S., & Janiri, L. (2020). The impact of climate change on mental health: A systematic descriptive review. Frontiers in Psychiatry, 11, 74.

Cornwall, C. E., Comeau, S., DeCarlo, T. M., Larcombe, E., Moore, B., Giltrow, K., Puerzer, F., D'alexis, Q., & McCulloch, M. T. (2020). A coralline alga gains tolerance to ocean acidification over multiple generations of exposure. Nature Climate Change, 10(2), 143–146.

Cornwall, C. E., Comeau, S., Kornder, N. A., Perry, C. T., van Hooidonk, R., DeCarlo, T. M., Pratchett, M. S., Anderson, K. D., Browne, N., Carpenter, R., Diaz-Pulido, G., D'Olivo, Juan P., Doo, Steve S., Figueiredo, Joana, Fortunato, Sofia A. V., Kennedy, Emma, Lantz, Coulson A., McCulloch, Malcolm T., González-Rivero, Manuel, Schoepf, Verena, Smithers, Scott G., & Lowe, R. J. (2021). Global declines in coral reef calcium carbonate production under ocean acidification and warming. Proceedings of the National Academy of Sciences, 118(21), e2015265118.

Fisher, D. N., Kilgour, R. J., Siracusa, E. R., Foote, J. R., Hobson, E. A., Montiglio, P. O., Saltz, J. B., Wey, T. W., & Wice, E. W. (2021). Anticipated effects of abiotic environmental change on intraspecific social interactions. Biological Reviews, 96(6), 2661–2693.

Gangloff, E. J., & Greenberg, N. (2023). Biology of stress. In Health and welfare of captive reptiles (pp. 93–142). Springer.

Gao, K., Beardall, J., Häder, D. P., Hall-Spencer, J. M., Gao, G., & Hutchins, D. A. (2019). Effects of ocean acidification on marine photosynthetic organisms under the concurrent influences of warming, UV radiation, and deoxygenation. Frontiers in Marine Science, 6, 322.

Gely, C., Laurance, S. G., & Stork, N. E. (2020). How do herbivorous insects respond to drought stress in trees? Biological Reviews, 95(2), 434–448.

Gomez-Zavaglia, A., Mejuto, J. C., & Simal-Gandara, J. (2020). Mitigation of emerging implications of climate change on food production systems. Food Research International, 134, 109256.

González-Tokman, D., Córdoba-Aguilar, A., Dáttilo, W., Lira-Noriega, A., Sánchez-Guillén, R. A., & Villalobos, F. (2020). Insect responses to heat: Physiological mechanisms, evolution and ecological implications in a warming world. Biological Reviews, 95(3), 802–821.

Hall-Spencer, J. M., & Harvey, B. P. (2019). Ocean acidification impacts on coastal ecosystem services due to habitat degradation. Emerging Topics in Life Sciences, 3(2), 197–206.

Halsch, C. A., Shapiro, A. M., Fordyce, J. A., Nice, C. C., Thorne, J. H., Waetjen, D. P., & Forister, M. L. (2021). Insects and recent climate change. Proceedings of the National Academy of Sciences, 118(2), e2002543117.

Harvey, J. A., Heinen, R., Gols, R., & Thakur, M. P. (2020). Climate change-mediated temperature extremes and insects: From outbreaks to breakdowns. Global Change Biology, 26(12), 6685–6701.

Harvey, J. A., Tougeron, K., Gols, R., Heinen, R., Abarca, M., Abram, P. K., Basset, Y., Berg, M., Boggs, C., & Brodeur, J. (2023). Scientists' warning on climate change and insects. Ecological Monographs, 93(1), e1553.

Hutchins, D. A., Jansson, J. K., Remais, J. V., Rich, V. I., Singh, B. K., & Trivedi, P. (2019). Climate change microbiology—problems and perspectives. Nature Reviews Microbiology, 17(6), 391–396.

Ingram, J. (2011). A food systems approach to researching food security and its interactions with global environmental change. Food security, 3, 417–431.

Islam, W., Adnan, M., Shabbir, A., Naveed, H., Abubakar, Y. S., Qasim, M., Tayyab, M., Noman, A., Nisar, M. S., & Khan, K. A. (2021). Insect-fungal-interactions: A detailed review on entomopathogenic fungi pathogenicity to combat insect pests. Microbial Pathogenesis, 159, 105122.

Jactel, H., Koricheva, J., & Castagneyrol, B. (2019). Responses of forest insect pests to climate change: not so simple. Current Opinion in Insect Science, 35, 103–108.

Jansson, J. K., & Hofmockel, K. S. (2020). Soil microbiomes and climate change. Nature Reviews Microbiology, 18(1), 35–46.
Joly, K., Couriot, O., Cameron, M. D., & Gurarie, E. (2020). Behavioral, physiological, demographic and ecological impacts of hematophagous and endoparasitic insects on an arctic ungulate. Toxins, 12(5), 334.
Jooste, M. L., Samways, M. J., & Deacon, C. (2020). Fluctuating pond water levels and aquatic insect persistence in a drought-prone Mediterranean-type climate. Hydrobiologia, 847(5), 1315–1326.
Karthik, S., Reddy, M. S., & Yashaswini, G. (2021). Climate change and its potential impacts on insect-plant interactions. The Nature, Causes, Effects and Mitigation of Climate Change on the Environment, 10.
Kellermann, V., & van Heerwaarden, B. (2019). Terrestrial insects and climate change: Adaptive responses in key traits. Physiological Entomology, 44(2), 99–115.
Kim, K. N., Huang, Q. Y., & Lei, C. L. (2019). Advances in insect phototaxis and application to pest management: A review. Pest Management Science, 75(12), 3135–3143.
Lam, V. W., Allison, E. H., Bell, J. D., Blythe, J., Cheung, W. W., Frölicher, T. L., Gasalla, M. A., & Sumaila, U. R. (2020). Climate change, tropical fisheries and prospects for sustainable development. Nature Reviews Earth & Environment, 1(9), 440–454.
Larson, E. L., Tinghitella, R. M., & Taylor, S. A. (2019). Insect hybridization and climate change. Frontiers in Ecology and Evolution, 7, 348.
Lehmann, P., Ammunét, T., Barton, M., Battisti, A., Eigenbrode, S. D., Jepsen, J. U., Kalinkat, G., Neuvonen, S., Niemelä, P., & Terblanche, J. S. (2020). Complex responses of global insect pests to climate warming. Frontiers in Ecology and the Environment, 18(3), 141–150.
Lowe, W. H., Martin, T. E., Skelly, D. K., & Woods, H. A. (2021). Metamorphosis in an era of increasing climate variability. Trends in Ecology & Evolution, 36(4), 360–375.
Ma, W., Li, X., Shen, J., Du, Y., Xu, K., & Jiang, Y. (2019). Transcriptomic analysis reveals *Apis mellifera* adaptations to high temperature and high humidity. Ecotoxicology and Environmental Safety, 184, 109599.
Malhi, G. S., Kaur, M., & Kaushik, P. (2021). Impact of climate change on agriculture and its mitigation strategies: A review. Sustainability, 13(3), 1318.
Malhi, Y., Franklin, J., Seddon, N., Solan, M., Turner, M. G., Field, C. B., & Knowlton, N. (2020). Climate change and ecosystems: Threats, opportunities and solutions. Philosophical Transactions of the Royal Society B, 375(1794), 20190104.
Mansoor, S., Farooq, I., Kachroo, M. M., Mahmoud, A. E. D., Fawzy, M., Popescu, S. M., Alyemeni, M., Sonne, C., Rinklebe, J., & Ahmad, P. (2022). Elevation in wildfire frequencies with respect to the climate change. Journal of Environmental Management, 301, 113769.
Manzanedo, R. D., & Manning, P. (2020). COVID-19: Lessons for the climate change emergency. Science of the Total Environment, 742, 140563.
Maria, C., Góis, J., & Leitão, A. (2020). Challenges and perspectives of greenhouse gases emissions from municipal solid waste management in Angola. Energy Reports, 6, 364–369.
Mendenhall, E., Hendrix, C., Nyman, E., Roberts, P. M., Hoopes, J. R., Watson, J. R., Lam, V. W., & Sumaila, U. R. (2020). Climate change increases the risk of fisheries conflict. Marine Policy, 117, 103954.
Mikhaylov, A., Moiseev, N., Aleshin, K., & Burkhardt, T. (2020). Global climate change and greenhouse effect. Entrepreneurship and Sustainability Issues, 7(4), 2897.
Nielsen, K. S., Nicholas, K. A., Creutzig, F., Dietz, T., & Stern, P. C. (2021). The role of high-socioeconomic-status people in locking in or rapidly reducing energy-driven greenhouse gas emissions. Nature Energy, 6(11), 1011–1016.
Ogle, S. M., Alsaker, C., Baldock, J., Bernoux, M., Breidt, F. J., McConkey, B., Regina, K., Vazquez-Amabile, G. G. (2019). Climate and soil characteristics determine where no-till management can store carbon in soils and mitigate greenhouse gas emissions. Scientific Reports, 9(1), 11665.
Omerkhil, N., Chand, T., Valente, D., Alatalo, J. M., & Pandey, R. (2020). Climate change vulnerability and adaptation strategies for smallholder farmers in Yangi Qala District, Takhar, Afghanistan. Ecological Indicators, 110, 105863.
Owens, A. C., Cochard, P., Durrant, J., Farnworth, B., Perkin, E. K., & Seymoure, B. (2020). Light pollution is a driver of insect declines. Biological Conservation, 241, 108259.
Piao, S., Liu, Q., Chen, A., Janssens, I. A., Fu, Y., Dai, J., Liu, L., Lian, X., Shen, M., & Zhu, X. (2019). Plant phenology and global climate change: Current progresses and challenges. Global Change Biology, 25(6), 1922–1940.
Praveen, B., & Sharma, P. (2019). A review of literature on climate change and its impacts on agriculture productivity. Journal of Public Affairs, 19(4), e1960.

Quitmann, C., Sauerborn, R., Danquah, I., & Herrmann, A. (2023). 'Climate change mitigation is a hot topic, but not when it comes to hospitals': A qualitative study on hospital stakeholders' perception and sense of responsibility for greenhouse gas emissions. Journal of Medical Ethics, 49(3), 204–210.
Rehman, A., Saba, T., Kashif, M., Fati, S. M., Bahaj, S. A., & Chaudhry, H. (2022). A revisit of internet of things technologies for monitoring and control strategies in smart agriculture. Agronomy, 12(1), 127.
Ritchie, H., Roser, M., & Rosado, P. (2020). $CO_2$ and greenhouse gas emissions. Our World in Data.
Rocklöv, J., & Dubrow, R. (2020). Climate change: An enduring challenge for vector-borne disease prevention and control. Nature Immunology, 21(5), 479–483.
Rodrigues, Y. K., & Beldade, P. (2020). Thermal plasticity in insects' response to climate change and to multifactorial environments. Frontiers in Ecology and Evolution, 8, 271.
Samways, M. J., Barton, P. S., Birkhofer, K., Chichorro, F., Deacon, C., Fartmann, T., Fukushima, C. S., Gaigher, R., Habel, J. C., & Hallmann, C. A. (2020). Solutions for humanity on how to conserve insects. Biological Conservation, 242, 108427.
Santana, P. A., Kumar, L., Da Silva, R. S., & Picanço, M. C. (2019). Global geographic distribution of *Tuta absoluta* as affected by climate change. Journal of Pest Science, 92, 1373–1385.
Secretariat, I., Gullino, M., Albajes, R., Al-Jboory, I., Angelotti, F., Chakraborty, S., Garrett, K., Hurley, B., Juroszek, P., & Makkouk, K. (2021). Scientific review of the impact of climate change on plant pests. FAO on behalf of the IPPC Secretariat.
Shen, M., Huang, W., Chen, M., Song, B., Zeng, G., & Zhang, Y. (2020). (Micro) plastic crisis: un-ignorable contribution to global greenhouse gas emissions and climate change. Journal of Cleaner Production, 254, 120138.
Shrestha, S. (2019). Effects of climate change in agricultural insect pest. Acta Scientific Agriculture, 3(12), 74–80.
Skendžić, S., Zovko, M., Živković, I. P., Lešić, V., & Lemić, D. (2021). The impact of climate change on agricultural insect pests. Insects, 12(5), 440.
Sogari, G., Bogueva, D., & Marinova, D. (2019). Australian consumers' response to insects as food. Agriculture, 9(5), 108.
Teawkul, P., & Hwang, S. Y. (2019). Carbon dioxide-and temperature-mediated changes in plant defensive compounds alter food utilization of herbivores. Journal of Applied Entomology, 143(3), 289–298.
Thurman, L. L., Stein, B. A., Beever, E. A., Foden, W., Geange, S. R., Green, N., Gross, J. E., Lawrence, D. J., LeDee, O., & Olden, J. D. (2020). Persist in place or shift in space? Evaluating the adaptive capacity of species to climate change. Frontiers in Ecology and the Environment, 18(9), 520–528.
Tougeron, K. (2019). Diapause research in insects: Historical review and recent work perspectives. Entomologia Experimentalis et Applicata, 167(1), 27–36.
Tougeron, K., Brodeur, J., Le Lann, C., & van Baaren, J. (2020). How climate change affects the seasonal ecology of insect parasitoids. Ecological Entomology, 45(2), 167–181.
Tramblay, Y., Koutroulis, A., Samaniego, L., Vicente-Serrano, S. M., Volaire, F., Boone, A., Le Page, M., Llasat, M. C., Albergel, C., & Burak, S. (2020). Challenges for drought assessment in the Mediterranean region under future climate scenarios. Earth-Science Reviews, 210, 103348.
Trebicki, P. (2020). Climate change and plant virus epidemiology. Virus Research, 286, 198059.
Venäläinen, A., Lehtonen, I., Laapas, M., Ruosteenoja, K., Tikkanen, O. P., Viiri, H., . . . Peltola, H. (2020). Climate change induces multiple risks to boreal forests and forestry in Finland: A literature review. Global Change Biology, 26(8), 4178–4196.
Wilson, R. J., & Fox, R. (2021). Insect responses to global change offer signposts for biodiversity and conservation. Ecological Entomology, 46(4), 699–717.
Zheng, X., Streimikiene, D., Balezentis, T., Mardani, A., Cavallaro, F., & Liao, H. (2019). A review of greenhouse gas emission profiles, dynamics, and climate change mitigation efforts across the key climate change players. Journal of Cleaner Production, 234, 1113–1133.
Zhou, Y., Li, X., Katsuma, S., Xu, Y., Shi, L., Shimada, T., & Wang, H. (2019). Duplication and diversification of trehalase confers evolutionary advantages on lepidopteran insects. Molecular Ecology, 28(24), 5282–5298.

# 13 Climate Change and Agricultural Insect Pests

*Mohsin Ali, Ali Raza, Habib Ali, Aamir Riaz, Muhamamd Arshad, Yasir Ali, Muhammad Saleem, Saba, Ammara Riaz, and Muhammad Akmal*

## 13.1 INTRODUCTION

The only constant phenomenon that we have witnessed throughout earth's history is change. As the human population grows, so does the science, culture, economy, culture, and lifestyles of animals and production scenarios in agriculture. The agricultural sector has gone through many revolutionary changes that have influenced human advancements through technology. But this tremendously increasing population expansion threatens the global food supply by increasing food demand and food production. By 2050, the global crop must be doubled in order to encounter the blooming population's food demands (Amirova et al., 2022). Some studies have proposed that the most viable way to ensure food security is to increase crop productivity as opposed to clearing additional land surface for agricultural cultivation (Yang & Xu, 2023). Changing climate and associated phenomena, such as increasing temperatures worldwide and airborne carbon dioxide ($CO_2$) emissions, extreme temperature, droughts, floods, severe storms, and further unusual climate occurrences are the subject of current scientific investigation and agronomy (Kilag et al., 2023). Hence, when the propensity to limit yield loss owing to such circumstances develops, greater focus is placed on the abiotic elements stated earlier in agricultural research. Shifts in the pattern of precipitation may possibly be more significant for crop production than rising temperatures, particularly in areas where dry spells represent a constraint on crop productivity (AMIRI et al., 2023).

Pests constitute a significant biotic element that is also affected by global climate disruption and changes in the weather. The reproduction, sustenance, spread, distribution, and abundance of pests along with the relations between pest species, climate, and natural predators are all directly impacted by temperature fluctuations (Abbass et al., 2022). As a result, it is crucial to keep an eye on pests' presence and abundance because the circumstances around their existence might rapidly alter. In particular, this chapter discusses how increasing temperatures and global carbon dioxide densities, along with variable precipitation patterns, have an impact on the ecology and physiology of insect pests, particularly hostile pest species, that are likely to pose serious threats to crop production.

## 13.2 CHANGING CLIMATE AND ITS IMPACTS

Hydrology and aquatic ecosystems, glacial environments, freshwater and marine environments, terrestrial ecosystems, forests, and agriculture are just a few of the controlled and ecological resources that are heavily influenced by the weather patterns (Abbass et al., 2022). Climate change can be defined as a condition that prevails when ecological elements like humidity, temperature, and rainfall change over a long period of time. Total food supply is under grave risk as an outcome of increasing temperature, climatic extremes, elevated $CO_2$ and some other greenhouse gases, and altered rainfall patterns (Yang et al., 2023). Currently, an important issue that we are facing is global warming. As shown by the upsurge in atmospheric temperature as well as sea level, record-breaking values have been attained (Pörtner et al., 2022).

 DOI: 10.1201/9781003382089-13

According to the World Meteorological Organization (WMO), the world is around 1°C hotter in comparison to the time prior to extensive industrial development. The last three decades have gotten warmer overall: the 2000s were the warmest, as stated by the Intergovernmental Panel on Climate Change (IPCC). The earth is projected to warm by 1.4°C–5.8°C over the subsequent period, as per a wide range of global weather models and developmental projections (Pachauri et al., 2014). The level of greenhouse gases (GHGs) in the atmosphere have risen, which is the main factor contributing toward global warming. The most prevalent human-made events, for example the combustion of fossil fuels and altering land practices, give rise to the environmental gases nitrous oxide, $CO_2$, and methane (Yoro & Daramola, 2020). When contrasted to the pre-industrial age, the modern period of the past two centuries has seen a significant increase in GHG levels (Jiang et al.). $CO_2$ is significant and prevalent among the GHGs. One of the most noticeable global environmental shifts that has occurred over the past five decades has been the upsurge in levels of atmospheric $CO_2$ (Palazzesi et al., 2022). Equated to the 280 ppm reported for the pre-industrial time, its atmospheric amount has rapidly increased to 416 ppm, and it is predicted to nearly double by 2100 (Chowdhury et al., 2023). Due to its significant absorption of some infrared radiation emanated from the surface of earth, $CO_2$ is classified as a GHG. The proportion of radiation that is released from the atmosphere toward the surface of the earth increases with the number of gases in the atmosphere that absorb thermal infrared fallout of radiation from the planet's surface (Mol Lous et al., 2022).

From the middle of the 20th century, fluctuations in extreme climate and atmospheric events have been noted. Reduced cold temperature events, more frequent higher temperature extremes, faster degrees of rising sea levels, and a rise in the incidence of significant rainfall in many locations are only a few of the changes that have been linked to human-caused effects (Domeisen et al., 2023). Extreme weather phenomena, such as extreme weather and extended periods of excessive rainfall, are projected to increase in occurrence and intensity in some regions. The likelihood of fluctuating and uneven rainfall patterns is very high. The mean average rainfall appears to be rising at the higher latitudes and near the equatorial Pacific. Mean rainfall is predicted to decrease mostly in dry mid-latitude and subtropical areas of the world while increasing in the wet mid-latitude zones (So et al., 2022).

### 13.2.1 Crop Production under Climate Change

The growth of agriculture, which is generally alluded to as an "open-air industry," is mostly dependent on the climatic and some climatological circumstances so as to yield food along with numerous other substances essential to meet human needs. Furthermore, agriculture is a sector that is especially susceptible to weather change, and its effects are clouded by a variety of uncertainties (Deshar et al., 2020). Globally, it is projected that climate variation would mark the farming production in both advantageous and detrimental ways, with the latter outweighing the former (Carr et al., 2022). Ecosystems at all scales, from species to ecological levels, are significantly impacted by temperature rises, changing rainfall patterns, and elevated $CO_2$ (Siddha & Sahu, 2022). So, the impacts of climate variation on insect pests rely on the plant species that help these insects flourish and feed; thus, it is crucial to first discuss how it affects agricultural output.

#### 13.2.1.1 Potential Effect of Rising Temperature

Temperature is regarded as the most vital element regulating the distribution and community changes of plants because of the biological limitations of individual species. It is an element that affects the rate of maturation, development, and agricultural output as well as a factor that limits the regions in which certain crops may be grown (Rehman et al., 2015). In general, the effects of rising temperatures are correlated with other ecological factors, such as the incidence of wind gusts, the presence of water, and the quantity and duration of sunshine. Together with the direct detrimental effect on production, indirect effects of temperature on certain natural conditions might additionally have influence. For instance, a rising temperature increases the need for atmospheric water, which might lead to more severe water shortages because of higher water pressure, which is brought on by water imbalances, and which eventually results in a decrease in soil moisture and productivity (Zhao et al., 2017). Two other indirect effects of temperature rise include an upsurge in the occurrence of heat waves as well as negative effects of crop diseases, weeds, and insects (Krishna, 2023; Raza et al., 2022).

#### 13.2.1.2 Potential Effect of Rising Carbon Dioxide Concentrations

The important chemical component for photosynthesis, which employs sunlight to transform water and $CO_2$ into carbohydrates and sugars, is $CO_2$. $CO_2$ must enter via stomatal holes in order for photosynthesis to take place in plants. As carbon plays a crucial role in the plant's structure, higher $CO_2$ concentrations encourage quicker development because carbon is assimilated more quickly. Reduced evaporation, increased light and water usage efficacy, and an upsurge in the frequency of photosynthesis are the primary effects of increased $CO_2$ on plants. Hence, increased atmospheric $CO_2$ levels could directly affect ecosystems by endorsing plant growth and development (Drake et al., 1997).

#### 13.2.1.3 Potential Effect of Fluctuating Precipitation Patterns

Water availability has a significant impact on crop output. Precipitation patterns, moisture in the soil retention, evaporation, and outflow will all vary as a consequence of climate change. Rainfall is thought to be responsible for more than 80% of all agricultural output worldwide; hence, variations in seasonal total precipitation or its trends are crucial (Lachaud et al., 2022). The global water cycle is amplified, and temperature fluctuations have a significant impact on this process. Yet, because it relies on other climate variables including the severity and degree of severe weather, its effect on agricultural production remains extremely difficult to forecast (Gu et al., 2023).

## 13.3 DYNAMICS OF PEST POPULATIONS

The component of natural populations known as population dynamics deals with the variables influencing variations in population levels. Modifications in the propagation and growth of species of insects will be caused directly by weather pattern influences and continuous climate changes. The amount of host-plant exploitation as well as the extent, frequency, and biological characteristics of populations are all impacted by local temperature, which undoubtedly affects insect rates of growth, voltinism, and lifespan (Eisen & Brown, 2022). An environmental shift has an impact on the dynamics of the insect pest in two different ways: directly or indirectly by changing the physiology of the host. The frequency and kind of abiotic events also affect host accessibility and the likelihood of diseases and pests. Climate has an indirect impact on insect's herbivore growth because external conditions affect plant biology. Variations in temperature, rainfall, elevated $CO_2$ levels, and fluctuations in the nutrient supply all have complex interactions with plants and insects (Ziska, 2022).

### 13.3.1 Potential Effects of Climate Change on Insect Pests

Farming and agro-associated insect pests are mainly influenced by alterations to the worldwide climate. Climate alteration poses an effect on agronomic production as well as the pests that trouble them in all ways. The survival, reproduction, and distribution of pests are significantly affected, while the ecosystem and other species, including natural predators, rivals, carriers, and commensal, are indirectly impacted (Prakash et al., 2014). The temperature of the insect bodies is influenced by the ambient temperature. Consequently, the most significant environmental element influencing dispersal, breeding, growth, and insect behavior, is probably temperature (Kocmánková et al., 2009). The primary causes of climate change (more atmospheric $CO_2$, higher temperatures, and lower soil humidity) may therefore have a substantial impact on the distribution and abundance of insect pests and, consequently, on the proportion of yield reductions. As a consequence of climate change, unique ecological niches are created, giving insect pests the opportunity to thrive, proliferate, and migrate between different geographic areas. Rising temperatures and $CO_2$'s complicated physiological impacts can significantly impact how croplands and insect pests communicate (Caulfield & Bunce, 1994). Farmers should therefore prepare for new and severe pest issues in the upcoming years as a result of climate change. Food safety is jeopardized by the cross-border movement of crop pests, which is a worldwide issue affecting all nations and areas (Ristaino et al., 2021).

## 13.4 INFLUENCE OF CLIMATE CHANGE ON INSECT PESTS

Agriculture and agricultural insect pests are both significantly impacted by global climate change. Reproduction, development, survival, and dissemination of pests are directly impacted, whereas climate change indirectly influences the interactions between pests, the environment in which they live, and other insect species like mutualists, predators, and natural enemies (Harvey et al., 2023). Being poikilothermic organisms, insects' body temperatures are influenced by the ambient temperature. As temperature affects insect behavior, dispersal, development, and reproduction, it is likely the most significant environmental component (Kocmánková et al., 2009). The main roots of climate alteration may, consequently have a key effect on the population drifts of insect pests (Schneider et al., 2022). Climate change has created new niches that provide pest insects an opportunity to flourish, disseminate, and move from one place to another (Marchioro & Krechemer, 2021). The complex physiological effects of elevated temperatures and $CO_2$ can have a significant impact on the connection between agricultural production and insect pests. Because of climate change, farmers should be ready for serious new pest problems in the years to come (Ristaino et al., 2021).

### 13.4.1 Effects of Temperature on Insect Pests

Due to the sensitivity of insect bodies toward temperature, metabolic rates of insects characteristically double with a 10°C rise (Dukes et al., 2009). In light of this, according to research, increasing temperatures are likely to accelerate consumption, growth of insect, and movement. This can have an influence on population drifts by disturbing traits including fertility, mortality, growth stage, population size, and geographic distribution. While other species can thrive and spread swiftly in warmer environments, those that cannot adapt and evolve frequently have difficulty sustaining their generations. The probability of alteration in pest population and dynamics is influenced by temperature, which also affects host availability, movement, and metamorphosis (Shrestha, 2019) (Figure 13.1). It is possible to assume that increases in temperature should be associated with a rise in herbivory grounded on the distribution and behavior of current insects. It is anticipated that insect populations in tropical regions would face a decline in rate of growth as a result of global warming because the weather reported is already fairly near to the optimum for pest expansion and growth. In comparison, it is anticipated that predatory insects in temperate regions would develop more quickly (Deutsch et al., 2018). By modeling variations in insect population rise in the productivity of the three main grain crops of the world (wheat, maize, and rice) under several climate variation situations, the reporters were able to support this notion. As per the research, global warming will accelerate the growth of pests for wheat crops that are naturally harvested in temperate regions. They forecast a deterioration in the increase of pest populations for rice grown in tropical areas and inconsistent responses to the increase of pest populations for maize farmed in both moderate and tropical zones (Deutsch et al., 2018).

As soil is a highly isolating material that can shield temperature changes and so reduce their influence, the impacts of increased temperatures are more pronounced for flying insects compared with species that devote the majority of their life cycle in the soil. Parameters that regulate whitefly population include overall moisture, precipitation, and temperature. Whitefly population growth and high temperatures and humidity are positively correlated (Pathania et al., 2020).

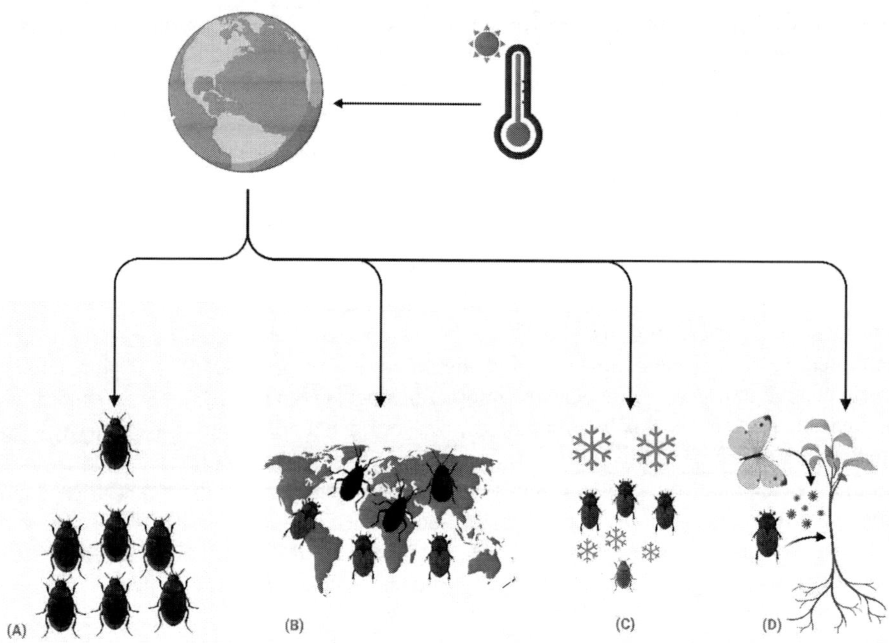


**FIGURE 13.1** Effect of increasing temperature on insect pests: (A) rise in generation numbers of insect pests; (B) increase in the geographical range of insect species over the globe; (C) increased survival in overwintering seasons; and (D) increased incidences of plant diseases transmitted by insect pests.

The frequency of the rise in global temperatures in the future will govern potential variations in population dynamics of insects. By the end of the present century, climate models project that the average global temperature will rise by 1.8°C–4°C (Deutsch et al., 2008). Under scenarios of global warming, the harshness of pest invasions is anticipated to worsen as ambient temperatures usually rise toward temperatures that are ideal for development and growth of several insect pest populations, possibly dropping thermal limitations on dynamics of population. Global warming might not necessarily result in an increase in pest profusion and consequent commercial crop losses due to the specific ecological niche needs, biological forbearances of insects, and varying influences of temperature on their phenology and life (Lehmann et al., 2020).

Several insect pest species responded differently to climate warming, according to the investigation by Lehmann et al. Their research findings suggested that in the majority of their insect case studies, rising temperatures increased the severity of pest problems. Nonetheless, 59% of all the species examined had reactions that could lessen their negative effects, primarily through decreased physiological ability and range limitation. Another study of over 1,100 insect species showed that 15%–37% of them would face extinction by 2050 as a result of climatic change brought on by global warming (Moore et al., 2022). The overall effects of global warming on dynamics of insect populations include terrestrial range extension, amplified overwintering population endurance rates, enhanced risk of invasive insect species introduction, and raised frequency of insect-transmitted plant ailments because of range extension.

### 13.4.2 Effects of Increased Carbon Dioxide Concentration on Insect Pests

Increased levels of $CO_2$ in the atmosphere can have an influence on the distribution, number, and efficiency of insects. Such intensifications may have an impression on an insect pest's population density, development rate, and fertility (Wang et al., 2023). The information now available indicates that the influence of amplified concentration of $CO_2$, in the atmosphere, on herbivory is tremendously unique to both insect-plant (pest-host) systems and specific insect species. The influence of increasing $CO_2$ levels on insect pests is mainly controlled by the plants. Higher $CO_2$ concentrations are probably going to have an impact on plant physiology by boosting photosynthetic activity, which will lead to better growth and higher plant output. As a result, this might pose an indirect effect on insects by changing the amount and quality of plants and foliage. A frequent trait of plants cultivated in high $CO_2$ environments is a variation in the chemical properties of leaves that may have an effect on the nutritive value of vegetation and the appeal of the plant to insects that feed on leaves (Ahammed et al., 2022) (Figure 13.2). Moreover, these plants frequently build up sugars and starches in the leaves, which alters the carbon-to-nitrogen ratio and decreases palatability. As nitrogen plays a crucial role in the growth of insects, some pest groups consume more plants when $CO_2$ concentrations are higher. Because pests need to feed on extra plant tissue to obtain the equal amount of energy, this might result in higher amounts of plant harm. As nitrogen levels fall, as projected by compensatory feeding and $CO_2$ fertilization, vegetation dependents like chewers, miners, and caterpillars often rise their rates of consumption (Hamilton et al., 2005). In an experiment (Hamilton et al., 2005), soybeans were cultivated at high ambient $CO_2$ levels. In comparison to soybeans cultivated in indoor atmospheric conditions, early season soybeans showed 57% more harm from insects like the *Empoasca fabae*, *Popilia japonica*, *Diabrotica virgifera*, and *Epilachna varivestis*. According to this study's findings, the detected increase in the number of simple sugars in soybean foliage may have prompted insects to feed in a supplemental manner. Insect herbivores have a tendency to eat extra plant material and hence harm more plants under these circumstances (Thomson et al., 2010). Several types of pests react differently to $CO_2$ fertilization. The number of thrips as well as other whole-cell consumers is growing.

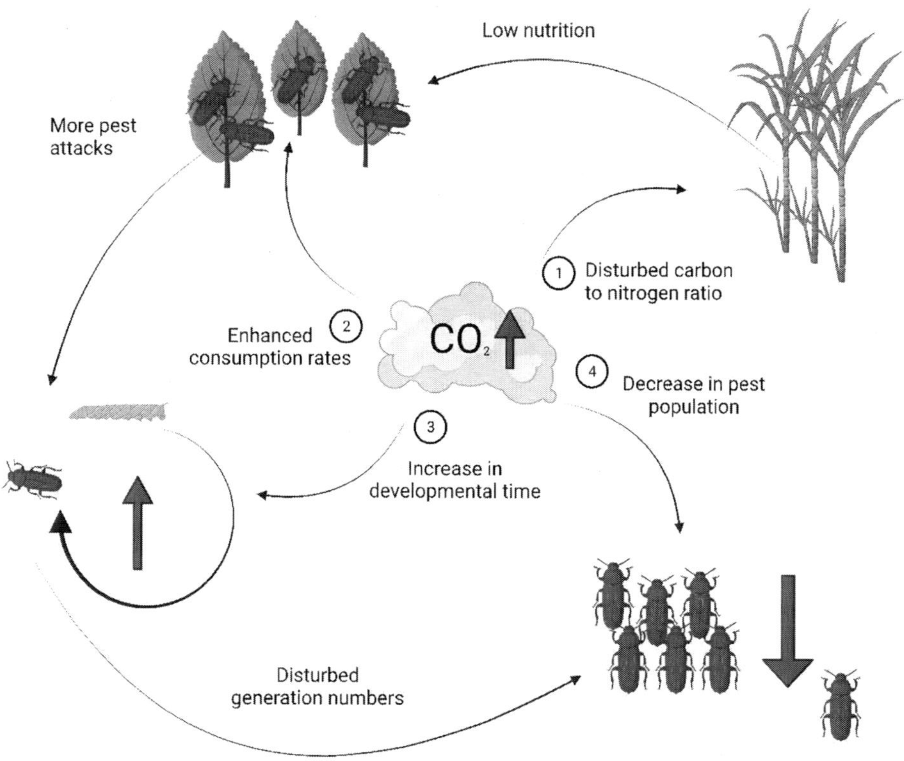


**FIGURE 13.2** Effect of increasing atmospheric carbon dioxide on insect pests of agriculture.

Whiteflies and aphids are examples of phloem-feeding insect pests that exhibit both an increase in rates of population growth and a drop in density of population. Although in some circumstances number and productivity may rise, findings on the impact of increasing $CO_2$ on sucking insects are conflicting. The indirect impacts of increased $CO_2$ on the life-cycle characteristics of herbivores were investigated by Stiling and Cornelissen using a meta-analytic research and evaluation of study documentation (Stiling & Cornelissen, 2007).

Their research revealed that insect pests had strong reactions to elevated $CO_2$ levels compared to ambient $CO_2$:

- Consumption rates rose by approximately 17%.
- Pest population decreased by approximately 22%.
- Growth times increased by approximately 4%.
- Relative growing rate decreased by 9%.

Moreover, chewers were shown to be more strongly affected by the rise in atmospheric $CO_2$ than other suckling associations, like sap-sucking herbivores. As a result, it has been demonstrated that despite the extensive research done to evaluate aphid reactions to rising levels of $CO_2$ in the atmosphere, it is still impossible to forecast future actions generally or to develop general guidelines for various aphid species to climate change (Sun & Ge, 2011).

### 13.4.3 Response of Insect Pests to Changeable Precipitation Pattern

Precipitation variations in terms of intensity, quantity, and occurrence are vital climate change indications. Rainfall has reduced in frequency but swelling in intensity, as has been observed in the majority of occurrences. Spells of droughts and overflows are more expected to happen when there is a pattern of this kind of precipitation. Overlapping rainfall patterns also have a direct effect on the kinds of insects that hibernate in the soil. In other ways, heavy rains may cause protracted water retention and flooding. This incidence puts the existence of insects at danger and affects the diapause (Figure 13.3). Furthermore, flooding and severe rains can wash away bug eggs and larvae. When it rains heavily, little pests like whiteflies, jassids, mite sand aphids, etc., can be swept away

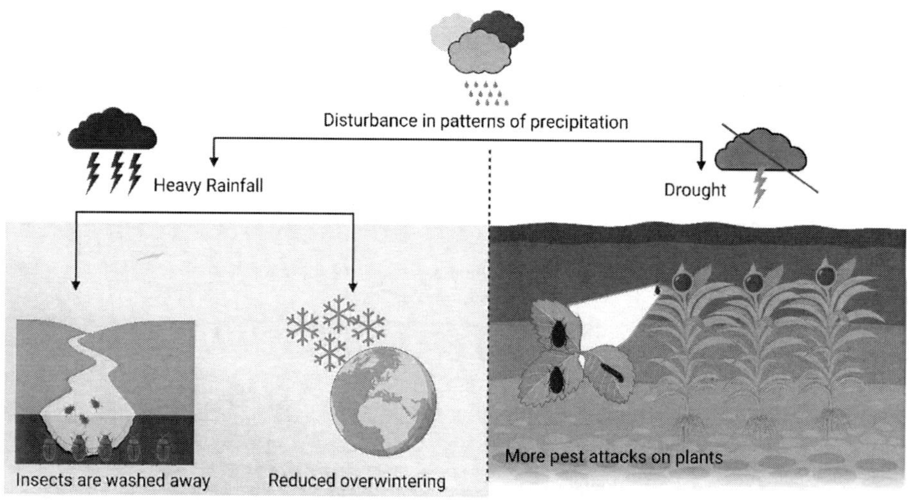


**FIGURE 13.3** Effects of disturbed rainfall patterns on insect pests of agriculture. Right side shows the effect of heavy rainfall on agricultural insect pests, while left side indicates how these insects respond to drought conditions and how their attacking incidences prevail in such xeric conditions.

(Pathak et al., 2012) (Figure 13.3). Rainfall that varies might have a significant effect on population of insects. For instance, Staley et al. 2007, studied how soil-dwelling wireworms (*Agriotes lineatus L.*) responded to drought and increased summer rainfall in grassland plots (Staley et al., 2007). Wireworms are extremely destructive pests of yields including sugar beetroot, corn, potatoes, and others, particularly when cultivated in grassland plans, and it is predicted that they would possibly develop a much bigger problematic condition as a result of climate change. In contrast to ambient and dry circumstances (Gregory et al., 2009), Staley et al. (2007) showed that enhanced summer rainfall sessions caused wireworm populations to grow quickly in the upper section of the soil.

Drought affects herbivorous insects in a number of ways:

- Dry regions may offer favorable climatic conditions for their growth.
- Drought-stressed plants may attract particular insect species. For instance, injurious bark beetles (*Scolytidae*) can perceive the ultrasonic acoustic discharge produced when, in the xylem, water columns separate or cavitate through transpiration.
- From drought, plants under pressure are more prone to insect outbreak because of decreases in their production of secondary compounds that play role defense functioning (Yihdego et al., 2019).

### 13.4.4 Increased Overwintering Survival

Being cold-blooded or poikilothermic, insects have a restricted ability to maintain homeostasis in response to changes in the surrounding temperature. To survive in thermally demanding environments, they have developed a range of survival mechanisms (González-Tokman et al., 2020). Winter

**TABLE 13.1**
**Effect of Climate Change on Population Dynamics of Some of the Important Insect Pests**

| Insect Species | Effecting Climate Parameter | Insect Response | References |
|---|---|---|---|
| *Ips typographus* | Increasing temperature | Early completion of life cycles (enhanced generation numbers) | Lange et al., 2006 |
| *Lymantria dispar* | Less frequent temperature extremes | Amplified persistence of hibernating eggs | Moore & Allard, 2008 |
| *Epirrita autumnata* | Less frequent temperature extremes | Amplified persistence of hibernating eggs | Moore & Allard, 2008 |
| *Operophthera brumata* | Less frequent temperature extremes | Amplified persistence of hibernating eggs | Veteli et al., 2005 |
| *Neodiprion sertifer* | Less frequent temperature extremes | Amplified persistence of hibernating eggs | Veteli et al., 2005 |
| *Thaumetopoea pityocampa* | Reduction in rate of snowfall | Population expansion | Kiritani, 2006 |
| *Helicoverpa armigera* | Increasing temp | Rise in abundance | Kiritani, 2006 |
| *Trichoplusia ni* | Increasing temp | Rise in abundance | Kiritani, 2006 |
| *Leptocorisa acuta* | Increasing temp | Increased generation time | Reji & Chander, 2008 |
| *Nephotettix cincticeps* | Increasing temp | Rise in abundance | Yamamura et al., 2006 |
| *C. suppressalis* | Increasing temp | Rise in abundance | Yamamura et al., 2006 |
| *Halyomorpha halys* | Increasing temp | Increased winter survival | Kiritani, 2006 |
| *Nezara viridula* | Increasing temp | Increased winter survival | Kiritani, 2006 |
| *Agriotes lineatus* | Increased precipitation | Rise in abundance | Staley et al., 2007 |
| *Schistocerca gregaria* | Disturbed precipitation pattern | Migration | Cheke & Tratalos, 2007 |
| *Diabrotica virgifera* | Elevated $CO_2$ concentration | Population expansion | Dermody et al., 2008 |
| *Aphis glycines* | Elevated $CO_2$ concentration | Population expansion | Dermody et al., 2008 |

is the most important time of year for many insect pests because the cold temperatures cause a significant rise in death, which lowers numbers the entire summer. Diapause confers increased cold hardiness in the absence of acclimatization to cold temperatures (Pullin, 1989). Diapause is an essential prerequisite for many species to successfully overwinter in colder and temperate climates.

Although the present environmental conditions may be favorable, the adaptive relevance of the periodic response to photoperiodism is to stop further growth and reproduction by making metabolic events for winter latency (Bradshaw & Holzapfel, 2010). The insect herbivores are highly adapted to their environments when they are in sync with their surroundings and host plants. However, a shift in spring development could result in a loss of synchronization with the atmosphere or the host plant because climatic warming can disturb the metabolic balance throughout diapause, which can greatly impact the time of emergence (Bale & Hayward, 2010).

The winter moth (*Operophtera brumata* L.), whose egg hatching happens far earlier than bud disruption on the host plant, the Pedunculate oak, is one of the effective examples (*Quercus robur* L.). The duration of bud burst is unlikely to be noticeably altered by a 2°C rise in temperature, but the duration of larval hatching is probably to be meaningly advanced, perhaps leading to larval hatching before bud burst, which is hazardous for the moth and may lessen this particular pest issue (Harrington et al., 2001). It is challenging to forecast the exact impacts of climate change on insect species by overwintering, because it appears that univoltine temperate species respond differentially to warmer winter temperatures (Tonnang et al., 2022). Warmer winters seem to increase survival rates for species that do not diapause, are frost-sensitive, and can overwinter in their active phases. Greater overwintering population and hence a greater number of insects on plants during the warmer season of the year could result from increased overwintering survival. A reliable biological indicator of global warming is the phenology of aphid flight (Akram et al., 2022). Several studies have demonstrated that a rise in temperature helps anholocyclic aphid species survive the winter in the United Kingdom and, in certain cases, delays the start of their departure by a maximum of one month. A hostile leaf miner (*Liriomyza huidobrensis*) may be more likely to overwinter outside of greenhouses in the United Kingdom if winters are milder there (Baker et al., 1996).

### 13.4.5 Increased Number of Generations

The most significant environmental component for insects, primarily determining their phenology, is temperature, as was already mentioned. According to the ambient energy concept, growth and reproduction are boosted by hot weather. As a result, greater temperatures or global warming are associated with larger population sizes, which can result in more species being in a state of dynamic equilibrium (Menéndez et al., 2007).

The development of species with annual life cycles is often faster than that of species with extended life cycles (Bale et al., 2002). Insects are predicted to respond to a change in temperature by advancing the time of adult emergence and lengthening their flight. The earlier start of the flight phase, which would enable the creation of an additional generation, is one explanation for the alterations in voltinism (Liu et al., 2023).

A 50-year analysis by Rothamsted Research, Harpenden, UK, of the time of the first drifting *M. persicae* individuals captured in a suction trap annually revealed a significant association with winter mean temperatures in January and February of every year. Another excellent example of phenological fluctuations is found in the members of the order Lepidoptera. Inside the United Kingdom, where 26 of the 35 identified species had advanced their first appearance, such alterations in butterflies have now been documented (Menéndez et al., 2007). In just 15 years, the earliest emergence of 17 species in Spain has moved by one to seven weeks (Pathania et al., 2020). The grapevine moth (*Lobesia botrana*) in Spain exhibited higher voltinism because of its early emergence. Several central European Lepidoptera species have seen an increase in the number of populations each year during the 1980s, and some univoltine or bivoltine species have changed to bivoltine or multivoltine life cycles (Altermatt, 2010). The frequency of second or later generations is anticipated to rise in partially bivoltine or multivoltine species (Altermatt, 2010). Insects are among the creatures that react to global warming because of their strong adaptation to environmental change, which is supported by the observed alterations in voltinism (Yamamura et al., 1998).

## 13.5 CONCLUSION

It is commonly acknowledged that the production of agricultural crops and the insect pests linked with them are significantly impacted by changing climate, despite the fact there remain many unanswered questions related to it. Narrow climate variability, including temperature rise, rise in $CO_2$ concentration, shifting patterns of precipitation, moisture content, and other variables constitute a few of the uncertainties surrounding various elements of climate variation that are pertinent to insect pests. It is projected that various insect species will react differently to climate change in various regions worldwide due to the huge diversity of species of insects, their intermediate hosts, and earth's climate fluctuation. Climate change has complicated consequences on insects since it affects their range, variety, abundance, growth, development, and phenotypic expression while favoring some and hindering others. Also, it is usually anticipated that there would be a rise in pest outbreaks as a whole. It is conceivable that insects would widen their geographic range. Certain pests will become more numerous as a result of higher overwintering endurance rates and the capacity to produce additional generations. There would most certainly be an increase in plant illnesses spread by insects as well as invasive pest species that may colonize new places more easily. We run a high danger of suffering substantial financial losses and threats to human food and nutrition security if changing climate variables create optimal conditions for pest infestation and crop failure. Dealing with this issue will require systematic methods. As a result, planning and developing solutions for mitigation as well as adaptation are extremely important.

## REFERENCES

Abbass, K., Qasim, M. Z., Song, H., Murshed, M., Mahmood, H., Younis, I. J. E. S., & Research, P. (2022). A review of the global climate change impacts, adaptation, and sustainable mitigation measures. *Environmental Science and Pollution Research*, *29*(28), 42539–42559. https://doi.org/10.1007/s11356-022-19718-6

Ahammed, G. J., Li, X. J. E., & Botany, E. (2022). Elevated carbon dioxide-induced regulation of ethylene in plants. *Environmental and Experimental Botany*, 105025. https://doi.org/10.1016/j.envexpbot.2022.105025

Akram, R., Amanet, K., Iqbal, J., Fatima, M., Mubeen, M., Hussain, S., Ali, M., Nasim, W., Ahmad, A., Farid, H. U., El Sabagh, A., & Fahad, S. (2022). Climate change, insects and global food production. In *Climate change and ecosystems* (pp. 47–60). CRC Press, Boca Raton, Florida, United States.

Altermatt, F. (2010). Climatic warming increases voltinism in European butterflies and moths. *Proceedings of the Royal Society B: Biological Sciences*, *277*(1685), 1281–1287. https://doi.org/10.1098/rspb.2009.1910

Amiri, N., Yacoubi, M., & Messouli, M. (2023). Population projections, food consumption, and agricultural production are used to optimize agriculture under climatic constraints. In *Intelligent solutions for optimizing agriculture and tackling climate change: Current and future dimensions* (pp. 169–192). IGI Global. https://doi.org/10.4018/978-1-6684-4649-2.ch007

Amirova, E. F., Gavrilyeva, N. K., Romanishina, T. S., & Asfandiarova, R. A. (2022). On the problem of the development of 'sustainable' agriculture in modern economic realities. *Siberian Journal of Life Sciences and Agriculture*, *14*(3), 392–406. https://doi.org/10.12731/2658-6649-2022-14-3-392-406

Baker, R., Cannon, R., & Walters, K. J. (1996). An assessment of the risks posed by selected non-indigenous pests to UK crops under climate change. *Aspects of Applied Biology*, (45), 323–330.

Bale, J., & Hayward, S. (2010). Insect overwintering in a changing climate. *Journal of Experimental Biology*, *213*(6), 980–994. https://doi.org/10.1242/jeb.037911

Bale, J. S., Masters, G. J., Hodkinson, I. D., Awmack, C., Bezemer, T. M., Brown, V. K., . . . Farrar, J. (2002). Herbivory in global climate change research: Direct effects of rising temperature on insect herbivores. *Global Change Biology*, *8*(1), 1–16. https://doi.org/10.1046/j.1365-2486.2002.00451.x

Bradshaw, W. E., & Holzapfel, C. M. (2010). Light, time, and the physiology of biotic response to rapid climate change in animals. *Annual Review of Physiology*, *72*, 147–166. https://doi.org/10.1146/annurev-physiol-021909-135837

Carr, T. W., Mkuhlani, S., Segnon, A. C., Ali, Z., Zougmoré, R., Dangour, A. D., . . . Scheelbeek, P. (2022). Climate change impacts and adaptation strategies for crops in West Africa: A systematic review. *Environmental Research Letters*, *17*(5), 053001. https://doi.org/10.1088/1748-9326/ac61c8

Caulfield, F., & Bunce, J. A. (1994). Elevated atmospheric carbon dioxide concentration affects interactions between *Spodoptera exigua* (Lepidoptera: Noctuidae) larvae and two host plant species outdoors. *Environmental Entomology*, *23*(4), 999–1005. https://doi.org/10.1093/ee/23.4.999

Cheke, R. A., & Tratalos, J. A. (2007). Migration, patchiness, and population processes illustrated by two migrant pests. *Bioscience*, *57*(2), 145–154. https://doi.org/10.1641/B570209

Chowdhury, S., Dubey, V. K., Choudhury, S., Das, A., Jeengar, D., Sujatha, B., . . . Kumar, V. J. (2023). Insects as bioindicator: A hidden gem for environmental monitoring. *Frontiers in Environmental Science*, *11*. https://doi.org/10.3389/fenvs.2023.1146052

Dermody, O., O'Neill, B. F., Zangerl, A. R., Berenbaum, M. R., & DeLucia, E. H. (2008). Effects of elevated $CO_2$ and $O_3$ on leaf damage and insect abundance in a soybean agroecosystem. *Arthropod-Plant Interactions*, *2*, 125–135. https://doi.org/10.1007/s11829-008-9045-4

Deshar, R., & Koirala, M. (2020). Climate change and gender policy. In *Global climate change and environmental policy* (pp. 411–422). https://doi.org/10.1007/978-981-13-9570-3_14

Deutsch, C. A., Tewksbury, J. J., Huey, R. B., Sheldon, K. S., Ghalambor, C. K., Haak, D. C., & Martin, P. R. (2008). Impacts of climate warming on terrestrial ectotherms across latitude. *Proceedings of the National Academy of Sciences*, *105*(18), 6668–6672. https://doi.org/10.1073/pnas.0709472105

Deutsch, C. A., Tewksbury, J. J., Tigchelaar, M., Battisti, D. S., Merrill, S. C., Huey, R. B., & Naylor, R. L. (2018). Increase in crop losses to insect pests in a warming climate. *Science*, *361*(6405), 916–919. https://doi.org/10.1126/science.aat34

Domeisen, D. I., Eltahir, E. A., Fischer, E. M., Knutti, R., Perkins-Kirkpatrick, S. E., Schär, C., . . . Wernli, H. (2023). Prediction and projection of heatwaves. *Nature Reviews Earth & Environment*, *4*(1), 36–50. https://doi.org/10.1038/s43017-022-00371-z

Drake, B. G., Gonzàlez-Meler, M. A., & Long, S. P. (1997). More efficient plants: A consequence of rising atmospheric $CO_2$? *Annual Review of Plant Niology*, *48*(1), 609–639. https://doi.org/10.1146/annurev.arplant.48.1.609

Dukes, J. S., Pontius, J., Orwig, D., Garnas, J. R., Rodgers, V. L., Brazee, N., . . . Harrington, R. (2009). Responses of insect pests, pathogens, and invasive plant species to climate change in the forests of northeastern North America: What can we predict? *Canadian Journal of Forest Research*, *39*(2), 231–248. https://doi.org/10.1139/X08-171

Eisen, M. B., & Brown, P. O. (2022). Rapid global phaseout of animal agriculture has the potential to stabilize greenhouse gas levels for 30 years and offset 68 percent of $CO_2$ emissions this century. *PLoS Climate*, *1*(2), e0000010. https://doi.org/10.1371/journal.pclm.0000010

González-Tokman, D., Córdoba-Aguilar, A., Dáttilo, W., Lira-Noriega, A., Sánchez-Guillén, R. A., & Villalobos, F. (2020). Insect responses to heat: Physiological mechanisms, evolution and ecological implications in a warming world. *Biological Reviews*, *95*(3), 802–821. https://doi.org/10.1111/brv.12588

Gregory, P. J., Johnson, S. N., Newton, A. C., & Ingram, J. S. (2009). Integrating pests and pathogens into the climate change/food security debate. *Journal of Experimental Botany*, *60*(10), 2827–2838. https://doi.org/10.1093/jxb/erp080

Gu, L., Yin, J., Slater, L. J., Chen, J., Do, H. X., Wang, H. M., . . . Zhao, T. (2023). Intensification of global hydrological droughts under anthropogenic climate warming. *Water Resources Research*, e2022WR032997. https://doi.org/10.1029/2022WR032997

Hamilton, J. G., Dermody, O., Aldea, M., Zangerl, A. R., Rogers, A., Berenbaum, M. R., & Delucia, E. H. (2005). Anthropogenic changes in tropospheric composition increase susceptibility of soybean to insect herbivory. *Environmental Entomology*, *34*(2), 479–485. https://doi.org/10.1603/0046-225X-34.2.479

Harrington, R., Fleming, R. A., & Woiwod, I. P. (2001). Climate change impacts on insect management and conservation in temperate regions: Can they be predicted? *Agricultural and Forest Entomology*, *3*(4), 233–240. https://doi.org/10.1046/j.1461-9555.2001.00120.x

Harvey, J. A., Tougeron, K., Gols, R., Heinen, R., Abarca, M., Abram, P. K., . . . Brodeur, J. (2023). Scientists' warning on climate change and insects. *Ecological Monographs*, *93*(1), e1553. https://doi.org/10.1002/ecm.1553

Kilag, O. K. T., Miñoza, J. L., Ledesma, E. N. B., Dosdos, B. V. C., Poloyapoy, B. N. M., & Lisao, C. Y. (2023). Shepherding nature: The environmental degradation. *Science and Education*, *4*(1), 129–147.

Kiritani, K. (2006). Predicting impacts of global warming on population dynamics and distribution of arthropods in Japan. *Population Ecology*, *48*(1), 5–12. https://doi.org/10.1007/s10144-005-0225-0

Kocmánková, E., Trnka, M., Juroch, J., Dubrovský, M., Semerádová, D., Možný, M., & Žalud, Z. (2009). Impact of climate change on the occurrence and activity of harmful organisms. *Plant Protection Science*, *45*(Special Issue), S48. https://doi.org/10.17221/2835-PPS

Krishna, G. V. (2023). Impact of climate change on plant viral diseases. In *Enhancing resilience of dryland agriculture under changing climate: Interdisciplinary and convergence approaches* (pp. 475–486). Springer. https://doi.org/10.1007/978-981-19-9159-2_23

Lachaud, M. A., Bravo-Ureta, B. E., & Ludena, C. E. (2022). Economic effects of climate change on agricultural production and productivity in Latin America and the Caribbean (LAC). *Agricultural Economics*, *53*(2), 321–332. https://doi.org/10.1111/agec.12682

Lange, H., Økland, B., & Krokene, P. (2006). Thresholds in the life cycle of the spruce bark beetle under climate change. *Complex Systems*, *1648*, 1–10.

Lehmann, P., Ammunét, T., Barton, M., Battisti, A., Eigenbrode, S. D., Jepsen, J. U., . . . Björkman, C. (2020). Complex responses of global insect pests to climate warming. *Frontiers in Ecology and the Environment*, *18*(3), 141–150. https://doi.org/10.1002/fee.2160

Liu, K., Wang, Z., Zhang, T., & He, K. (2023). Intra-population alteration on voltinism of Asian corn borer in response to climate warming. *Biology*, *12*(2), 187. https://doi.org/10.3390/biology12020187

Marchioro, C. A., & Krechemer, F. S. (2021). Prevention is better than cure: Integrating habitat suitability and invasion threat to assess global biological invasion risk by insect pests under climate change. *Pest Management Science*, *77*(10), 4510–4520. https://doi.org/10.1002/ps.6486

Menéndez, R., González-Megías, A., Collingham, Y., Fox, R., Roy, D. B., Ohlemüller, R., & Thomas, C. D. (2007). Direct and indirect effects of climate and habitat factors on butterfly diversity. *Ecology*, *88*(3), 605–611. https://doi.org/10.1890/06-0539

Mol Lous, M., Helled, R., & Mordasini, C. (2022). Potential long-term habitable conditions on planets with primordial H-He atmospheres. *Nature Astronomy*, *6*(7), 819–827. https://doi.org/10.48550/arXiv.2206.13859

Moore, B., & Allard, G. (2008). Climate change impacts on forest health. In *Forest health and biosecurity working paper FBS/34E*. Rome: Food and Agriculture Organization of the United Nations.

Moore, M. E., Hill, C. A., & Kingsolver, J. G. (2022). Developmental timing of extreme temperature events (heat waves) disrupts host-parasitoid interactions. *Ecology and Evolution*, *12*(3), e8618. https://doi.org/10.1002/ece3.8618

Pachauri, R. K., Allen, M. R., Barros, V. R., Broome, J., Cramer, W., Christ, R., . . . Dasgupta, P. (2014). *Climate change 2014: Synthesis report. Contribution of working groups I, II and III to the fifth assessment report of the intergovernmental panel on climate change*. Geneva: IPCC.

Palazzesi, L., Hidalgo, O., Barreda, V. D., Forest, F., & Höhna, S. (2022). The rise of grasslands is linked to atmospheric $CO_2$ decline in the late Palaeogene. *Nature Communications*, *13*(1), 293. https://doi.org/10.1038/s41467-021-27897-y

Pathak, H., Aggarwal, P. K., & Singh, S. D. (2012). Climate change impact, adaptation and mitigation in agriculture: Methodology for assessment and applications. *Agricultural Research Institute*, *302*.

Pathania, M., Verma, A., Singh, M., Arora, P. K., & Kaur, N. (2020). Influence of abiotic factors on the infestation dynamics of whitefly, *Bemisia tabaci* (Gennadius 1889) in cotton and its management strategies in North-Western India. *International Journal of Tropical Insect Science*, *40*, 969–981. https://doi.org/10.1007/s42690-020-00155-2

Pörtner, H.-O., Roberts, D. C., Adams, H., Adler, C., Aldunce, P., Ali, E., . . . Biesbroek, R. (2022). *Climate change 2022: Impacts, adaptation and vulnerability*. Geneva: IPCC.

Prakash, A., Rao, J., Mukherjee, A. K., Berliner, J., Pokhare, S. S., Adak, T., Munda, S., & Shashank, P. R. (2014). *Climate change: Impact on crop pests*. Applied Zoologists Research Association (AZRA), Central Rice Research Institute, Cuttack, India.

Pullin, A., & Bale, J. S. (1989). Influence of diapause and temperature on cryoprotectant synthesis and cold hardiness in pupae of *Pieris brassicae*. Comparative Biochemistry and Physiology Part A: Physiology, *94*(3), 499–503.

Raza, A., Hairs, M., Riaz, A., Rafiq, R., Basharat, M., Shaheen, M., . . . Genetics, M. (2022). Trending killers of plants; An overview of the most prevailing pathogenic fungi. *Journal of Microbiology and Molecular Genetics*, *3*(3), 97–147. https://doi.org/10.52700/jmmg.v3i3.73

Rehman, M. U., Rather, G. H., Gull, Y., Mir, M. R., Mir, M. M., Waida, U. I., & Hakeem, K. R. (2015). Effect of climate change on horticultural crops. In *Crop production and global environmental issues* (pp. 211–239). https://doi.org/10.1007/978-3-319-23162-4_9

Reji, G., & Chander, S. J. (2008). A degree-day simulation model for the population dynamics of the rice bug, *Leptocorisa acuta* (Thunb.). *Journal of Applied Entomology*, *132*(8), 646–653. https://doi.org/10.1111/j.1439-0418.2007.01178.x

Ristaino, J. B., Anderson, P. K., Bebber, D. P., Brauman, K. A., Cunniffe, N. J., Fedoroff, N. V., . . . Jones, C. M. (2021). The persistent threat of emerging plant disease pandemics to global food security. *Proceedings of the National Academy of Sciences*, *118*(23), e2022239118. https://doi.org/10.1073/pnas.2022239118

Schneider, L., Rebetez, M., & Rasmann, S. (2022). The effect of climate change on invasive crop pests across biomes. *Current Opinion in Insect Science*, 100895. https://doi.org/10.1016/j.cois.2022.100895

Shrestha, S. (2019). Effects of climate change in agricultural insect pest. *Acta Scientific Agriculture*, *3*(12), 74–80. https://doi.org/10.31080/ASAG.2019.03.0727

Siddha, S., & Sahu, P. (2022). Impact of climate change on the river ecosystem. In *Ecological significance of river ecosystems* (pp. 79–104). Elsevier. https://doi.org/10.1016/B978-0-323-85045-2.00014-5

So, K. W., Tam, C.-Y., & Lau, N.-C. (2022). Impacts of global warming on Meiyu-Baiu extreme rainfall and associated mid-latitude synoptic-scale systems as inferred from 20km AGCM simulations. *Climate Dynamics*, 1–13. https://doi.org/10.21203/rs.3.rs-383898/v1

Staley, J. T., Hodgson, C. J., Mortimer, S. R., Morecroft, M. D., Masters, G. J., Brown, V. K., & Taylor, M. E. (2007). Effects of summer rainfall manipulations on the abundance and vertical distribution of herbivorous soil macro-invertebrates. *European Journal of Soil Biology*, *43*(3), 189–198. https://doi.org/10.1016/j.ejsobi.2007.02.010

Stiling, P., & Cornelissen, T. (2007). How does elevated carbon dioxide ($CO_2$) affect plant-herbivore interactions? A field experiment and meta-analysis of $CO_2$-mediated changes on plant chemistry and herbivore performance. *Global Change Biology*, *13*(9), 1823–1842. https://doi.org/10.1111/j.1365-2486.2007.01392.x

Sun, Y., & Ge, F. (2011). How do aphids respond to elevated $CO_2$? *Journal of Asia-Pacific Entomology*, *14*(2), 217–220. https://doi.org/10.1016/j.aspen.2010.08.001

Thomson, L. J., Macfadyen, S., & Hoffmann, A. A. (2010). Predicting the effects of climate change on natural enemies of agricultural pests. *Biological Control*, *52*(3), 296–306. https://doi.org/10.1016/j.biocontrol.2009.01.022

Tonnang, H. E., Sokame, B. M., Abdel-Rahman, E. M., & Dubois, T. (2022). Measuring and modelling crop yield losses due to invasive insect pests under climate change. *Current Opinion in Insect Science*, 100873. https://doi.org/10.1016/j.cois.2022.100873

Veteli, T., Lahtinen, A., Repo, T., Niemelä, P., & Varama, M. (2005). Geographic variation in winter freezing susceptibility in the eggs of the European pine sawfly (*Neodiprion sertifer*). *Agricultural and Forest Entomology*, *7*(2), 115–120. https://doi.org/10.1111/j.1461-9555.2005.00259.x

Wang, J., Shi, X., Lucas-Borja, M. E., Guo, Q., Mao, J., Tan, Y., & Zhang, G. (2023). Soil nematode abundances drive agroecosystem multifunctionality under short-term elevated $CO_2$ and $O_3$. *Global Change Biology*, *29*(6), 1618–1627. https://doi.org/10.1111/gcb.16546

Yamamura, K., & Kiritani, K. (1998). A simple method to estimate the potential increase in the number of generations under global warming in temperate zones. *Applied Entomology and Zoology*, *33*(2), 289–298. https://doi.org/10.1303/aez.33.289

Yamamura, K., Yokozawa, M., Nishimori, M., Ueda, Y., & Yokosuka, T. (2006). How to analyze long-term insect population dynamics under climate change: 50-year data of three insect pests in paddy fields. *Population Ecology*, *48*, 31–48. https://doi.org/10.1007/s10144-005-0239-7

Yang, Q., Du, T., Li, N., Liang, J., Javed, T., Wang, H., . . . Liu, Y. (2023). Bibliometric analysis on the impact of climate change on crop pest and disease. *Agronomy*, *13*(3), 920. https://doi.org/10.3390/agronomy13030920

Yang, R., & Xu, H. (2023). Does agricultural water-saving policy improve food security? Evidence from the Yellow River Basin in China. *Water Policy*. https://doi.org/10.2166/wp.2023.217

Yihdego, Y., Salem, H. S., & Muhammed, H. H. (2019). Agricultural pest management policies during drought: Case studies in Australia and the state of Palestine. *Natural Hazards Review*, *20*(1), 05018010. https://doi.org/10.1061/(ASCE)NH.1527-6996.0000312

Yoro, K. O., & Daramola, M. O. (2020). $CO_2$ emission sources, greenhouse gases, and the global warming effect. In *Advances in carbon capture* (pp. 3–28). Elsevier. https://doi.org/10.1016/B978-0-12-819657-1.00001-3

Zhao, C., Liu, B., Piao, S., Wang, X., Lobell, D. B., Huang, Y., . . . Ciais, P. (2017). Temperature increase reduces global yields of major crops in four independent estimates. *Proceedings of the National Academy of Sciences*, *114*(35), 9326–9331. https://doi.org/10.1073/pnas.1701762114

Ziska, L. H. (2022). Rising carbon dioxide and global nutrition: Evidence and action needed. *Plants*, *11*(7), 1000. https://doi.org/10.3390/plants11071000

# 14 Climate-Smart Agriculture
## *An Integrated Approach to Address Climate Change and Food Security*

*Prasanta Kumar Majhi, Ipsita Samal, Tanmaya Kumar Bhoi, Prachi Pattnaik, Chandini Pradhan, Akhilesh Gupta, Deepak Kumar Mahanta, and Subrat Kumar Senapati*

### 14.1 INTRODUCTION

According to the United Nations climate change refers to long-term changes in temperature and weather events. Climate change is characterized by rising global temperatures, extreme weather events such as severe droughts, floods, heat waves, cyclones, and shifting precipitation patterns. The temperature rise (global warming) is the most important of these components since it begins the story of climate change and impacts the other climatic factors of the earth's ecosystem. The combustion of fossil fuels—coal, oil, and gas—is primarily responsible for the warming of the earth's biosphere, accounting for more than 75% of global greenhouse gas (GHG) emissions and almost 90% of total carbon dioxide ($CO_2$) emissions. GHGs function as a blanket around the earth, trapping heat from the sun and boosting temperatures. The warmer temperatures of the earth's ecosystem over time cause melting glaciers and rising sea levels, change weather patterns, and disrupt the usual balance of nature. Climate change possesses many great challenges to human life directly or indirectly with its far-reaching and potentially devastating impacts on society and the natural world. One of the most significant areas of concern is agriculture and food security. Agricultural production heavily relies on weather variables like precipitation, temperature, relative humidity, and sunshine hours which affect crop establishment, growth and development, and ultimately crop production. Climate change disrupts a well-established, long-standing ecosystem of crop-weather relationship through temperature rise, drought, changing precipitation patterns, hailstorms, soil degradation, and pests and disease outbreaks, etc. Climate change has an impact not just on crops and food production but also on food security through direct or indirect effects on other components of agricultural production systems (Birthal et al., 2014; Panda et al., 2021, 2023). In India, around 36 mha of the agricultural area was reported to be affected due to hydro-meteorological weather events like heavy rain and floods which account for 6.65 mha in 2016 to 5.04 mha in 2021.

The Intergovernmental Panel on Climate Change's (IPCC) assessment report (IPCC, 2021) says that by the year 2100, the surface temperature of earth will be 1.0°C to 1.8°C higher in the very low GHG emissions scenario (2.1°C to 3.5°C) higher in the intermediate GHG emissions scenario, and 3.3°C to 5.7°C higher in the very high GHG emissions scenario. Changes in temperature and rainfall that are expected will cause the yield of all major crops to drop by 1%–10% by 2035, 3%–18% by 2065, and 4%–26% by 2100. Climate change could cut crop yields by up to 2% per decade for the rest of the century. This may not seem like a lot, but when you consider that the global population is projected to reach 9.7 billion by 2050, it becomes clear that even small reductions in crop yields could have a major impact on food security. In a nutshell, climate change poses a significant threat to agriculture and food security, and strong strategic steps are required to be taken to mitigate its impacts. The investment in adaptation strategies and strengthening of farmers' support system are necessary to ensure that they continue producing food for a growing global population, even in the face of a changing climate.

DOI: 10.1201/9781003382089-14

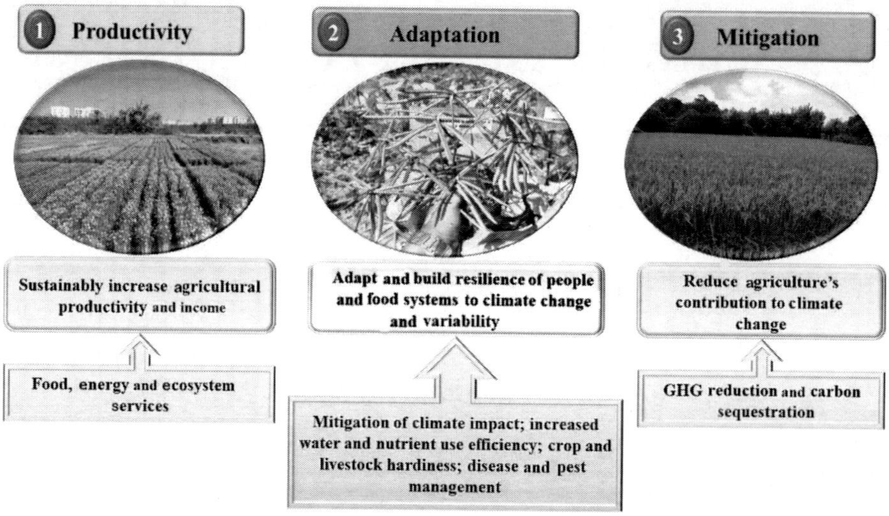


**FIGURE 14.1** The three pillars of climate-smart agriculture: (1) productivity, (2) adaptation, and (3) mitigation.

Climate-smart agriculture (CSA) is a sustainable way of mitigating the threat of climate change through farming that aims to increase productivity, improve food security, and reduce GHGs. It involves integrating three main objectives: (a) sustainably increasing agricultural productivity and incomes, (b) adapting and building resilience to climate change, and (c) reducing and/or removing GHG emissions, where possible. These three broad objectives are called the three pillars of CSA (Figure 14.1). The role of CSA is to provide farmers with the tools and practices they need to adapt to a changing climate, while also reducing their carbon footprint. This includes practices such as conservation agriculture, crop diversification, agroforestry, water management, and the use of climate-resilient crops. The needs for CSA include access to information and training on CSA practices, investment in research and development to identify and promote effective CSA practices, access to financing and support for farmers to implement CSA practices, and policies and regulations that incentivize and promote the adoption of CSA practices. CSA has a significant opportunity to make it simpler to accomplish the 17 sustainable development goals (SDGs). Finding possible synergies and trade-offs between the SDGs' three pillars and five implementation steps and CSA's three pillars and five implementation steps is a key step towards achieving this scope. Therefore, the present chapter focusses on soil and water management strategies for CSA, pest and disease dynamics under climate change scenario, climate-resilient variety development through genomics approaches, and CSA in the Indian context including government initiatives and policies.

## 14.2 SOIL AND WATER MANAGEMENT STRATEGIES FOR CLIMATE-SMART AGRICULTURE

Water and nutrients are the basic requirement for any viable agricultural system. Success or productivity of any agricultural system depends on how efficiently the inputs like water and nutrients are being used by the crop. Modern-day agriculture is being threatened under the current scenario of climate change where water scarcity and environmental pollution due to excessive fertilizer use are a rising concern. Identifying methods which lead to efficient management of both water and nutrients in order to have higher productivity, more profitability, and reduced losses is mandatory to have a CSA.

### 14.2.1 Improving Water and Nutrient Use Efficiency

#### 14.2.1.1 Water-Use Efficiency

Water-use efficiency (WUE) is the term which means the amount of grain or biomass produced per unit of water used by the plant. Currently, WUE is being considerably impacted by the climate change phenomenon which can be attributed to four reasons: rise in $CO_2$ levels, elevating temperature, highly erratic precipitation, and alteration in humidity. Under such circumstances, enhancing the WUE is crucial in order to be resilient to the changing climate, and some methods are discussed in Table 14.1.

Nutrient use efficiency (NUE) basically indicates how efficiently the nutrient applied is used by the plant with minimal losses. It is the amount of yield or biomass produced by the plant per unit of nutrient applied. Excessive or inefficient utilization of nutrients causes loss of nutrients into the environment leading to pollution and also increases financial burden on the farmers to buy more fertilizers. It also leads to soil degradation and imbalanced nutrition to plants. NUE of any cropping system is based on the 4Rs of nutrient management—right source, right amount, right time, and right place (Johnston and Bruulsema, 2014):

- *Right source*: Right source of fertilizer depends on the crop, properties of the soil, and method of application (Stewart et al., 2009). The best source for any nutrient is the fertilizer which has high nutrient content, good physical properties, compatibility with the soil pH, positive nutrient interactions, and low price. For example, in acidic soils, rock phosphate is considered as the best fertilizer, but in low land rice conditions, soluble P sources are better than rock phosphates.

**TABLE 14.1**
**Different Methods to Improve the Water-Use Efficiency in Crops' Nutrient Use Efficiency**

| Sl. No. | Method | Cause | Example | References |
|---|---|---|---|---|
| 1. | Application of mulch | Reduces soil water evaporation, reduces water losses from weeds, maintains low soil temperature | Water-use efficiency (WUE) enhanced by 61% along with rise in productivity in maize. | Zhang et al., 2017 |
| 2. | Micro-irrigation | Precise application of water, low soil evaporation loss, lower soil percolation and leaching loss | Micro-irrigation increased yield by 37% while reducing water use for wheat by 23%. | Fan et al., 2018 |
| 3. | Narrow spacing between crop rows | Partitioning between the soil's water evaporation and the canopy's transpiration | When maize crop had a nitrogen and water shortage, the close row spacing raised WUE by 17% with considerable impacts. | Barbieri et al., 2012 |
| 4. | Crop rotation and intercropping | Diversified crop leads to varying rooting depth and better utilization of soil water | When peanut was planted as an intercrop with watermelon and okra, it had higher WUE as opposed to a monoculture. | Franco et al., 2018 |
| 5. | Breeding of improved varieties | Determine the genotypes that are most tolerant to extreme temperatures and water shortages. With relatively consistent net photosynthetic rate, stomatal conductance, and transpiration rate under water stress, the drought-tolerant genotype showed a higher WUE | Based on both economic yield and carbon dioxide fixation, plant breeders have created and chosen crop varieties with improved WUE. | Krupnik et al., 2012; Shardadevi et al., 2017; Ul-Allah et al., 2018 |

- *Right amount*: The rate or amount of nutrient required to be applied through fertilizer can be determined through soil and plant analysis. Factors like soil's native nutrient supply, the target yield required, and growth stage of the plant determine the rate of nutrient required. Modern techniques like soil plant analysis development (SPAD) meter, leaf colour charts, and optical sensors also aid in determining how much nutrient is required by the plant. Thus, application of just as much required nutrients by the crop leads to enhanced NUE.
- *Right time*: Fertilizers or nutrients for their effective utilization must be applied at the time of peak uptake and critical stages (Murrell et al., 2009). Moreover, application of fertilizers should be avoided during the time when chances of losses are maximum. Therefore, the application of fertilizers in three to two splits as per requirement of crop leads to increased NUE and minimal losses. Use of control and slow-release fertilizers like neem-coated urea, sulphur-coated urea, polymer-coated urea, isobutylidene-diurea, crotonaldehydediure, nitrification inhibitors, and urease inhibitors facilitate release of nutrients at the right time, thus leading to increased nutrient efficiency (Trenkel, 2021).
- *Right place*: Application or placement of the fertilizer at the place that is conducive and accessible to the roots or the plant parts enhances NUE. Placement of fertilizers largely depends on properties of the fertilizer itself. For instance, ammoniacal fertilizers should be placed at the reduced zone of submerged rice, whereas nitrate fertilizers should be placed at the oxidized layer to avoid denitrification losses. Fertilizers can be applied through broadcast, band application, deep banding, or foliar application depending on their properties.

### 14.2.2 Role of Conservation Tillage in Restoring Soil Health

In modern agriculture, the term "conservation tillage" refers to agronomic practices that ensure that at least 30% of the surface soil is covered in crop residues after the following crop is sown. It involves direct planting, which implies growing crops without mechanically preparing the seedbed and with the least degree of soil disturbance possible after the harvest of the previous crop. Conservation tillage has a significant positive impact on all the properties of soil which affect soil health. With the increasing concern of climate change, conservation tillage acts as an effective tool for CSA. The impact of conservation tillage on soil health is described in Figure 14.2.

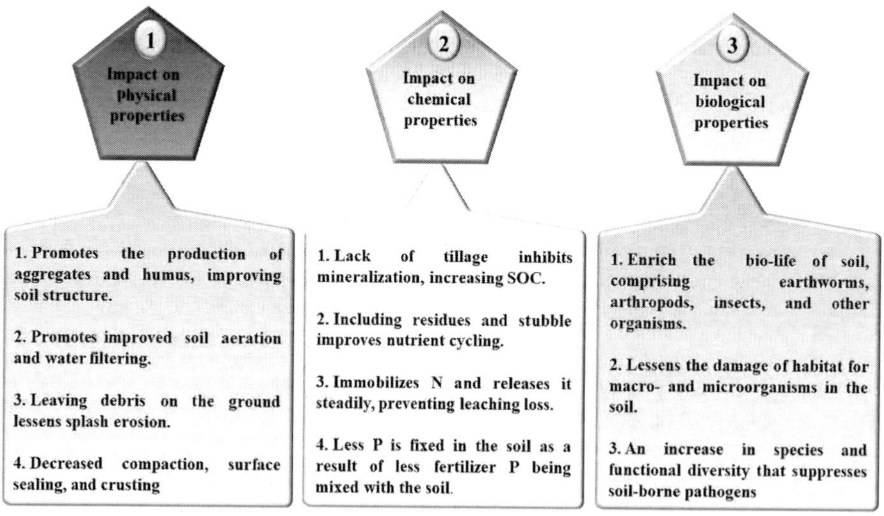


**FIGURE 14.2** The impact of conservation tillage on soil health.

### 14.2.3 Good Agricultural Practices for Soil Health and Carbon Sequestration

Both soil health and carbon sequestration invariably depend on soil organic matter. Soil organic matter or organic carbon in soil is a key factor which influences physical, chemical, and biological properties of soil. By using improved management techniques to remediate degraded soils and ecosystems, the soil's capacity to capture atmospheric $CO_2$ can be significantly increased. Such techniques enhance the soil organic carbon (SOC) pool, promote soil health, and preserve soil water. Some of the practices are enlisted below:

1. *Switching from convectional tillage to conservation tillage*: Soil can store more carbon through inclusion of cover crops in the rotational cycle, reduction of soil disturbance, and shortening of the fallow period. No-till management and use of residue as mulch improves infiltration capacity, decreases bulk density, and enriches the soil structure which leads to better soil health (Shaver et al., 2002).
2. *Inclusion of cover crop in the cropping cycle*: Use of leguminous cover crops boosts biodiversity, the quality of residue input, and SOC pool (Blanco-Canqui and Ruis, 2020). Soils with richer biodiversity have better soil health than soils with narrow diversity.
3. *Balanced nutrition and inclusion of organic nutrient sources*: Adequate supply of nutrition from both organic and inorganic sources as per site requirement leads to greater utilization of nutrients and results in increased biomass. Increased biomass eventually leads to greater organic matter returned to soil, thereby ensuring both carbon sequestration and good soil health (Yadav et al., 2019).
4. *Judicious irrigation*: A drought-prone soil can yield greater biomass, restore higher aboveground and root biomass to the soil, and have a richer SOC content with the judicious management of irrigation water (Chatterjee et al., 2018).
5. *Restoration of degraded soil and ecosystem*: Most of the degraded soils have lost a huge amount of carbon ranging from 40 to 60 Mg C/ha. These soils can sequester more carbon since they have lost a large portion of their initial SOC pool, by using restorative methods like conservation tillage, incorporation of organic residues, replacing monoculture with crop rotation, integrated nutrient management, etc. (Yang et. al. 2019).
6. *Conversion of agricultural land into forest*: There is a significant potential for SOC sequestration through the reforestation of marginal agricultural soils or degraded soils. The increased biomass and deeper root system of forest cover has a positive impact on soil health too; moreover, it also reduces erosion loss of soil (Favero et al., 2020).

### 14.2.4 Geospatial Application for Water Resources Management

The rising population has led to a rise in food demand, which can be met only if the right amount of necessary inputs like water is present and provided for agriculture in adequate amount. On the other hand, the impact of climate change has made rainfall, a source of water, more erratic. Under such conditions, irrigated agriculture is the only option to ensure adequate food production in spite of climate change. But unfortunately, the irrigated agriculture itself has its negative repercussions on the environment like water logging and salinization due to injudicious management of the irrigation sources or the water resources. Efficient management of the water resources will lead to better utilization of water, reduce wastage, and decrease the chances of water logging and salinization. Geospatial tools such as geographic information systems (GIS) and remote sensing techniques can be used to monitor and manage the water resources (Singh, 2016; Panda et al., 2022; Sahoo and Panda, 2020). GIS is a system for efficiently acquiring, storing, processing, evaluating, and presenting spatial information, whereas remote sensing is detecting or analyzing an area without being in contact with the target, with the help of sensors. With the emergence of high-speed electronic computers in recent decades, the application of remote sensing and GIS techniques in the management of water resource issues has expanded significantly (Schmid et al., 2005; Leblanc et al., 2007; Xu et al., 2011; Muller and van Niekerk, 2016).

Uses of geospatial techniques in water resource management are listed below:

1. Remote sensing has been used extensively as a tool in groundwater recharge-based studies, mostly to support conventional evaluation and modelling methods as well as basic geophysical procedures (Schultz and Engman, 2012). The GIS technique also has significant use in groundwater resource numerical modellings (Chenini and Ben Mammou, 2010).
2. Waterlogged areas have been mapped and monitored using GIS and remote sensing techniques in many areas (Huang et al., 2008; Dwivedi et al., 2001)). Geospatial techniques can be applied to investigate how different topography, hydrological, and anthropogenic factors affect waterlogging (Singh and Pandey, 2014).
3. Remote sensing and GIS can be used together to analyze large amounts of hydrogeologic data and simulate complex subsurface flow and transport processes under both saturated and unsaturated conditions.
4. GIS is a relatively easy to use, quick, and affordable tool to evaluate the biophysical resources of a watershed. This approach provides a good substitute for the manual method when relatively level terrain is unavailable in watersheds.
5. GIS can be used to create a quality evaluation information management model, which can be helpful to evaluate the hazards associated with the use of irrigation water (Peragón et al., 2015).
6. Soil evaporation and transpiration needs can be simulated using remote sensing estimations in combination with daily weather data, GIS information, and live and dead crop cover.
7. GIS can handle complex issues of managing water resources by assessing geographic heterogeneity in crop, soil, and water supply in a canal command.

## 14.3 PEST AND DISEASE DYNAMICS UNDER CLIMATE CHANGE SCENARIO

### 14.3.1 Climate Change Impact on Disease-Crop Interactions

Plant diseases are critical to agricultural output (Agrios, 2005), and climate change has produced changes in distribution patterns, outbreaks, and the introduction of new pest/disease pathotypes. Rising temperatures, unpredictable rainfall, and diminishing irrigated soil water tables could cost South Asian countries such as India 40% of their agricultural productivity by 2100 (IPCC, 2007). Some pests and diseases have become more harmful as a result of environmental changes. Due to sporadic late rain, *Ascochyta* blight and chickpea pod infection impacted productivity and quality (Abang and Malhotra, 2008). Climate has an impact on pathogenic microbes and host plants, resulting in sickness. Climate, diseases, and controls all have an impact on the relationship. Pathogen reproduction is influenced by temperature and season length. Temperature increases lengthen seasons and cause infections to evolve. Temperature changes allow viruses and vectors to hybridize and transmit disease, posing a hazard to food security (Brasier, 2001). Garrett et al. (2006) proposed researching plant disease and climate change in order to understand trait inheritance and pathogen evolution. Pfender and Vollmer (1999) discovered that winter temperature variations have a significant impact on *Puccinia graminis* survival. Climate change exacerbated the illness of biotrophic and necrotrophic fungi (Chakraborty et al., 2002). Necrotrophy is caused by biomass and host crop replacements (Melloy et al., 2010). Climate change accelerates gene allele recombination and the emergence of antagonistic pathotypes (Sturrock et al., 2011). Temperature variation, according to Eastburn et al. (2010), affects stomata opening and closing, enhancing leaf metabolism and decreasing pathogen invasion (Butterworth et al., 2010). The detailed effects of impact of climate dynamics on crop-disease interactions are discussed next.

#### 14.3.1.1 Effect of Increased $CO_2$

Rising $CO_2$ levels and temperatures have a negative impact on metabolic performance, leaf physiology, morphology, and crop output (Ainsworth and Rogers, 2007). $CO_2$ has an impact on both the host and the pathogen. Researchers have found an increase in leaf quantity, leaf area, branches, leaf area per plant, plant height, and agricultural output when $CO_2$ levels rise (Eastburn et al., 2011). McElrone et al. (2005) discovered that host metabolic alterations raised $CO_2$ and decreased the occurrence and severity of *Phyllosticta minima*–caused red maple disease. Further, warming and increased $CO_2$ levels may promote the proliferation of harmful microorganisms (Yáñez-López et al., 2012). As $CO_2$ levels rise, so does biomass output. Plant biomass, slower litter decomposition, and higher temperatures may favour pathogen survival and the development of microclimates for mildews, leaf spots, rusts, and blights (Lambers et al., 2008). *Erysiphe graminis* lowers barley leaf infection, and high $CO_2$ extends the latent period of *Maravalia cryptostegiae* (Mina and Sinha, 2008). These microclimates promote the growth of biotrophic fungus and infection (Chakraborty et al., 2002). High-temperature and $CO_2$ plants boosted photosynthesis, water, and fertilizer utilization (Chakraborty and Datta, 2003). As $CO_2$ levels rise, roots sprout and discharge chemicals into the rhizosphere (Pickles et al., 2012). High atmospheric $CO_2$ levels, according to Kobayashi et al. (2006), enhanced rice blast and sheath blight infection. Barley powdery mildew and *Colletotrichum gloeosporioides* anthracnose disease bacteria multiply more quickly at higher CO concentrations. Further, it was anticipated that rising $CO_2$ levels in Finland will increase the risk of potato blight. Climate change may have an impact on *Fusarium* species disease (Chitarra et al., 2015).

#### 14.3.1.2 Impact of Increased Temperature on Host and Pathogens

The large cumulative temperature increase may affect crop physiology and disease resistance. Climate change alters host-pathogen distribution and plant-pathogen interactions (Mina and Sinha, 2008). Growing temperature, exposure period, and climate change affect plant disease severity (Evans et al., 2008) and disease occurrence (Ashoub et al., 2015). Temperature affects pathogen penetration, survival, development, reproduction, spread, and plant growth. Pathogen reproduction may depend on temperature (Ladanyi and Horvath, 2010). Host plants become rust-sensitive but pathogenic fungus–resistant as temperature rises (Coakley et al., 1999). Due to temperature changes, inactive pathogen races can trigger rapid epidemics. Richerzhagen (2011) found that sugar beet leaf spot disease shifted due to 0.8°C–1°C annual mean temperature increase. Pg3 and Pg4 of oat stem and Lr210 and Lr217 of wheat leaf are temperature-sensitive rust-resistance genes (Das et al., 2016). Bacterial pathogens including *Burkholderia glumea, Acidovorax avenae*, and *Ralstonia solanacearum* grow pathogenic at higher temperatures and spread to new areas (Ahanger et al., 2013). Temperature changes also affected viral, vector-borne illnesses, and aphid survival (van Munster et al., 2017). Host tissue accumulates phytoalexins or protective pigments in sunlight. Mboup et al. (2012) investigated temperature-specific adaptation of European wheat yellow rust populations and found that stripe rust isolates were more aggressive at higher temperatures. In addition, a model simulation conducted by Salinari et al. (2006) demonstrated that temperature and humidity had an impact on infections. More dangerous viruses thrive in damp environments (McElrone et al., 2005). In addition, *Phytophthora infestans*, the fungus responsible for the late blight of solanaceous plants (tomatoes and potatoes), thrives in humid environments with temperatures between 7.2°C and 26.8°C (Fry et al., 2013). Wet conditions favour the spread of foliar diseases and several soil-borne infections Some fungal infections may become more prevalent as a result of dryness brought on by climate change. Dry root rot and powdery mildew are two diseases that can be brought on by a lack of rain in a legume crop (Gautam et al., 2013).

### 14.3.2 Impact of Climate Change on Dynamics of Insect Pests

#### 14.3.2.1 Anthropogenic Climate Change and Its Resultant Effects on Insects

Long-term warming has an impact on insect population, reproduction, survival, and performance thresholds, as well as population dynamics, dispersal, phenology, and abundance change (Harvey et al., 2020). The tropics and subtropics are warming quicker than high-latitude temperate, boreal, and arctic regions (Post et al., 2018). Ecosystems shift upslope as the tropical highlands warm by 0.017°C per year (Feeley et al., 2013). Winter nights are getting warmer (IPCC, 2014). Ectothermic organisms are concerned about resource depletion since their metabolic rates increase with temperature (Williams et al., 2012). Warming promotes growth, development, and reproduction at optimal temperatures, but above this point, metabolic costs rise disproportionately and thermal damage accumulate, resulting in developmental failure, decreased fecundity, diminished dispersion capability, worse fitness, and higher mortality (Sweeney et al., 2018). Climate change can produce bottlenecks, unrelated genetic variety loss, and outbreeding or inbreeding depression, all of which can impair population persistence (Halsch et al., 2021). Some insect populations have delayed local adaptation due to a lack of heat-tolerance genetic variation (García-Robledo and Baer, 2021). Lighter insects may better manage their body temperature in hot conditions. Thus, the most major impact of climate change is the destabilization of insect communities, which increases the danger of population decreases and potentially species extinction if geographic distributions, behavioural activity patterns, and seasonal phenologies are not adapted to the new environment (Halsch et al., 2021). Ecological community reorganization affects species relationships, ecosystem health, and services.

#### 14.3.2.2 Influence on the Spatial Distribution of Species

Climate change has an impact on species distribution (Colwell et al., 2008). To live, organisms at lower latitudes or elevations must migrate to cooler environments. Ecological and evolutionary factors influence thermocline monitoring in insects. According to bioclimatic models, low, moderate, and severe warming will result in the extinction of 34,000 insects by 2100. Fifty percent of insect species lose 50% or more of their range at +3.2°C, compared to 18% at +2°C and 6% at +1.5°C (Warren et al., 2021). Insects and other species are being pushed poleward by global warming. Lower latitudes and altitudes experience slower population loss. More butterfly species will become extinct when trailing edge populations lose climate appropriateness. Long-term temperature data and demographic models demonstrate that residents evacuate the trailing edge and seek shelter even with a 2°C global average temperature increase. Expanding ranges may help many animals survive climate change. Warming near range limits and genetic variation for thermal tolerance suggest some species can adapt swiftly to shifting thermoclines. In uneven terrain with altitude gradients, human land use can separate and isolate insect habitats. If old and new niches overlap, climate change may kill natural competition.

#### 14.3.2.3 Resultant Impact on Interspecies Interactions

Insects, like other animals, have complex social networks that influence their life. Predation and other multitrophic interactions are caused by long-term coevolutionary processes in stable climates. Temperature fluctuations, on the other hand, can have an impact on the many species in a system. This is supported by research on multitrophic systems, including plants, herbivores, natural enemies (parasitoids, predators, and diseases), and hyperparasitoids (Bhoi et al., 2022; Pardikes et al., 2022). Herbivore fitness and biocontrol services may suffer if plants adapted to hotter, carbon monoxide–rich, and/or drier circumstances provide insufficient nutritional resources (Han et al., 2022). Warmer temperatures and more herbivorous insects may prolong the damaging instar and feeding period for host plants (Wang et al., 2021). Climate change may reduce plant-insect pollination (Tylianakis et al., 2008). Since lower trophic levels are scarcer and their effects depend on host specialization and adaptability, higher levels should have bigger effects (Monticelli et al., 2021). Chemically altered agro-landscapes and island communities may be vulnerable (Raven and Wagner, 2021). Climate change affects floral biogenic volatile organic compounds (BVOCs), which attract pollinators and facilitate plant-pollinator mutualism. Temperature affects global warming BVOC emissions in *Petunia hybrida* flowers. The right amount of ozone destroyed *Brassica nigra* flowers' BVOCs in a compound-specific manner, changing their scent and decreasing *Bombus terrestris'* attraction. Insect distribution and phenology may also induce regional and temporal incompatibility among interacting species. Interacting species may differ generationally due to trophic level reactivity (Gérard et al., 2020). Specialized species, those limited by hosts or prey, or those in constrained ecological niches may be more vulnerable (Damien and Tougeron, 2019). Whiteflies and damselflies, closely related yet with varied thermal tolerances, can experience sexual conflicts due to temperature differences (Abarca & Spahn, 2021).

#### 14.3.2.4 Consequence for Phenology

Climate change also affects insect phenology. Voltinism, which takes advantage of the extended growth season and faster development at higher temperatures, can improve insect reproduction. Despite modest temperature variations, seasonal or annual generation increases (Parmesan and Yohe, 2003). Higher summer temperatures allow more generations in Lepidoptera, although voltinism may produce smaller adults (Verberk et al., 2021). Overwintering strategy and photoperiod (beginning of diapause and fewer days) affect later-season generation augmentation for many temperate-zone insects (Marshall et al., 2020). For example, aphid parasitoids are active year-round, suggesting repeated mild winters may cause winter diapause to disappear (Tougeron et al., 2017). Active insects need physiological and behavioural thermotolerance, including fast cold hardening or brief, reversible rest intervals, to survive winter (Diniz et al., 2017). Warmer winters change honeybee colony phenology, affecting nectar and pollen (Nürnberger et al., 2019). Warmer winters help honeybee colonies generate brood, but they also attract aggressive *Varroa destructor* mites (Vercelli et al., 2021). Winter conditions vary more locally and geographically than summer ones, hence latitudinal change may affect biotic activity more (Bonan, 2004). Further, asynchrony in time and diapause duration have influenced the vulnerability of insects to winter predators (Cooper, 2014).

#### 14.3.2.5 Impact of Climate Change on Plant and Insect Defensive Traits

Temperature affects plant secondary chemicals and herbivore-protective structural characteristics. Temperature can affect insects and plants. Temperature changes host-plant phytochemistry and morphology. At night temperatures of 17°C, tomato catecholic phenolics (chlorogenic acid and rutin) were much greater than at other temperatures (Bradfield and Stamp, 2004). Rivero et al. (2003) found inadequate polyphenol oxidase (PPO) and peroxidase (POX) activity in tomatoes at 35°C. Below 22°C, tomatoes lose protease inhibitor action (Green and Ryan, 1973). High temperatures thicken leaf trichomes (Bickford, 2016). BVOCs mediate antagonistic and mutualistic multitrophic interactions and influence the troposphere's oxidative capacity (Peñuelas and Llusià, 2003). Herbivores and illnesses cause BVOC mixtures that disclose plant location, quality, infestation history, and insect presence. At higher temperatures, alfalfa (*Medicago sativa*) secondary metabolites (sapogenins and saponins) increased, limiting caterpillar growth (*Spodoptera exigua*). The green-veined butterfly, *Pieris napi*, ingested far more plant tissue in response to warming-mediated poor-quality Brassicaceae leaves (Bauerfeind and Fischer, 2013). Aphids (*Myzus persicae* and *Brevicoryne brassicae*) fed on oilseed rape plants at different temperatures with nutritional quality discrepancies survived and developed normally (Himanen et al., 2008). At higher temperatures, tobacco hornworm, *Manduca sexta*, on *Nicotiana tabacum* was larger (Diamond and Kingsolver, 2010).

### 14.3.3 Approaches Used to Investigate the Effects of Climate Change on Plant Pests

Temperature, $CO_2$, ozone, ultraviolet-B radiation, water and humidity patterns, and insect pests have all been examined in relation to climate change over the last 30–40 years. Pests have been investigated in tropical and plantation crops, forest trees, wheat, barley, rice, soybeans, and potatoes (Ghini et al., 2011). Some research manipulated weather to investigate its influence on ecosystems, while another looked at temperature change based on species distributions along latitudinal or elevational gradients. Experts or simulation models have anticipated how future temperature or atmospheric composition changes may affect pest and other species distribution, prevalence, severity, and control, as well as meta-analysis of published results and long-term data sets (Pautasso et al., 2012).

Climate change research in open-topped chambers and free-air $CO_2$ enrichment (FACE) facility systems has revealed the elements that spread pests over a wide range of crops (Eastburn et al., 2011). Weeds and insects have been studied in similar circumstances (Delucia et al., 2012). According to Ainsworth and Long, elevated $CO_2$ levels in FACE systems worsened the majority of pest and disease issues (2021). Short-term $CO_2$ and temperature increases on host-insect interactions were also researched using phytotrons, or environmental chambers (Betz et al., 2020). Furthermore, researchers conducted meta-analyses of previously published data sets in order to find broad pest responses to various environmental circumstances (Vilà et al., 2021). Predators are more vulnerable to climate change than plants or herbivorous insects, according to a meta-analysis of laboratory studies (Fussmann et al., 2014). Due to a scarcity of field data, this is critical when examining the function of natural enemies in insect pest dynamics and biological management (Thomson et al., 2010). Aside from these, simulation models can predict the effects of climate change on pests and aid pest management tactics (Hill and Thomson, 2015). "Climate matching" is a modelling technique that forecasts pest dynamics by analyzing a site with an environment similar to the future climate of the area of interest and extrapolating the results (Angelotti et al., 2017).

#### 14.3.3.1 Forecasting the Climate and the Development of Models

Climate models and a pest species' biological needs can produce accurate worldwide change estimates. Thus, modelling pest risk and plant host reactions to climate change improves insect infestation predictions (Raza et al., 2015). Ecological niche models (ENMs) including mechanistic and simulation models are used for predicting pest infestation. Using past occurrence data and environmental element values, a correlative model can predict the species' best habitats. MaxEnt, Bioclim, and Random Forest are three current and popular correlational models (Kumar et al., 2014). Climate change policy mainly relies on correlative species distribution models to evaluate biodiversity implications of climate change (Gillson et al., 2013). Correlative models can predict species distributions, assess extinction risks, and prioritize conservation (Porfirio et al., 2014). These models first establish a statistical relationship between a species' current geographic distribution and climatic parameters to anticipate where it may live in a warming world (Pacifici et al., 2015). Comparing the predicted area of climate-suitable habitats with the species' actual distribution can estimate the risk of introduction and establishment (Evans et al., 2015).

Mechanistic models can foresee outcomes by integrating data about a species' environmental conditions and its capacity to endure them (Kumar et al., 2014). Correlative and mechanistic species dispersion models differ in their focus on environmental influences on physiologic function. To estimate future species distributions, an exclusion technique excludes prospective habitats that impede physiological performance to the point that they hinder persistence, development, or reproduction (Kearney et al., 2010). Mechanistic models are the best way to solve a variety of management issues since they can generalize beyond particular examples and separate biogeography-affecting factors (Cuddington et al., 2013). Based on physiological, behavioural, and climatic factors, CLIMEX and other semi-mechanistic models predict a species' optimum habitats (Kriticos et al., 2015). Creating such models and analyzing meteorological and historical weather data will also help predict pest threats. As the climate changes, new pest control methods may emerge (Yonow et al., 2018).

Crop disease and pest modelling efforts have so far primarily concentrated on short-term, tactical problems, analyzing individual crop-pest systems in specific conditions utilizing empirical connections based on multiseasonal observations (Donatelli et al., 2017). They normally speak about pathogens and the chain of events that goes from infection to injury to damage, which is then described as one to many damage functions that affect crop health. Examples of these model frameworks include NAPPFAST (Magarey et al., 2007), which includes a template for arthropod pest models based on degree days and a framework for infectious phytopathogens, and can be used to garner pest risk maps; iPiPE (Hong et al., 2015), which is based on NAPPFAST and specifically intended to forecast the timeframe of pest incidences to enable countermeasures; and the APSIM-DYMEX complex. DYMEX is a framework for making process-based models of insect, weed, and disease populations. APSIM is a model of an agricultural system (Keating et al., 2003). The linked models have been used to study, among other things, how rust affects the size of wheat leaves and how nematodes grow in non-host fallows. Furthermore, the EPIC-GILSYM linked crop-insect model realistically replicates two trophic levels (insects and plants), their interactions, and the impact of crop management on both levels. It has been successfully tested against field trial data collected on five plots in different locations of the United States over the course of eight years. It may become a valuable tool in the assessment of insect pest impacts and appropriate control methods with thorough calibration and validation to various sites and climates (Rasche and Taylor, 2019). There are already many separate models for how insect populations change, such as matrix models, phenology models, differential equations, competition models, and regression models (Tonnang et al., 2017). Other models, like MaxEnt (Phillips et al., 2006), CLIMEX, and ILCYM, are mostly used to find places where insect pests are likely to spread (Sporleder et al., 2013). Using presence/absence data and other biological and climate variables, these models can predict whether a particular place is a decent location for a species to live. They can also be used to figure out the best way to control insect pests as the climate changes.

#### 14.3.3.2 Monitoring Distribution and Abundance

Long-term data are needed to assess how climate change affects insect pest populations (Yamamura et al., 2006). Without this knowledge, it is difficult to accurately analyze pest population changes in response to changing climatic regimes or predict future population dynamics. Long-term pest population and behaviour monitoring in climate change–vulnerable locations may reveal early biological responses to climate change (Heeb et al., 2019). Tracking vectors, diseases, host populations, and distribution changes is crucial. Climate change helps alien species expand worldwide. Invasive species must be monitored and controlled before they become a financial burden in unknown areas (Hellmann et al., 2008). Innovative biosecurity and pest control are needed. Climate change may need pest control efforts to be strengthened. Monitoring, forecasting, chemical manipulation, and biological control are among examples (Heeb et al., 2019). For reliable monitoring and risk assessment, a global strategy is needed since many insect pests can cross borders. A global infrastructure is needed to convey insect, invasive alien species, ecological, and meteorological data. Thus, nations and regional, national, and international organizations must cooperate more (Perrings et al., 2010). The U.S. Department of Agriculture's early warning and fast response programme and the European and Mediterranean Plant Protection Organizations (EPPO's) Early Warning and Information System for invasive alien species will prioritize entry point monitoring and speedy eradication (Ziska et al., 2011). Farmers can also prevent pest problems by monitoring weather, pests, and climate and pest risk prediction data (Heeb et al., 2019).

## 14.4 GENOMICS PERSPECTIVES TO COPE WITH CLIMATE CHANGE: A WAY FOR CLIMATE-RESILIENT VARIETY DEVELOPMENT

Drought, heavy metal stress, nutrient stress, hunger, salinity, new diseases, and insect pests are all examples of extreme events made worse by the way climate change has altered the pattern of rainfall, temperature, $CO_2$ levels, and production of greenhouse gases. This has serious repercussions for agricultural output, food security, economic well-being, and dietary intake. By increasing the frequency and severity of extreme climatic events that cause population relocation and decline in agricultural output, climate change is having a devastating impact on millions of people around the world, particularly on the most vulnerable (Majhi et al., 2023). The ability of agriculture to adjust to climate variability depends on the research, development, identification, and implementation of climate-resilient crop varieties and other adaptation and mitigation methods. To improve the lives of the millions of people who rely on agriculture for a living, it is crucial to increase farming communities' production and profitability while reducing agricultural risk. Plants respond to biotic and abiotic challenges by altering their gene expression, signal transduction pathways, metabolism, and molecular responses according to source and sink relations. The abiotic pressures that plants can withstand vary depending on the crop and the variety. In order to find high-yielding short-duration cultivars that can survive the terminal drought and other abiotic and biotic challenges, it is important to learn about photoperiod sensitivity and genetic diversity in transpiration efficiency. To breed plants that can withstand adverse conditions, it is necessary to learn which characteristics let them continue growing and developing normally even while under pressure. Tolerance to abiotic stresses shows up as the ability to adapt to diminished resources when under stress and then flourish again with increased productivity after the stress has abated. Genetic control of stress resistance and resource use efficiency is quantitatively inherited in various crops, involving multiple loci across the genome.

Exploring the vast genetic variety of the soil microbiome, which lies just below our feet, will be just as difficult. Very little is understood about the organisms responsible for processing these natural cycles, despite the fact that soils cycle five times as much carbon as humans emit in a year. The advancement of (meta)genomics will allow for more precise geographical, temporal, and experimental replication, as well as functional modification, to begin unravelling this complexity. To combat the widespread use of transcriptional regulatory networks and the resulting destruction of previously productive landscapes, it may be possible to distil microbial coevolutionary networks that transform soil from dirt. Spontaneous genomics experiments are now taking place in every living system, and future research will illuminate the recurrent evolutionary processes that yield useful biological improvements (Borevitz, 2021). Several biotic and abiotic stresses have been successfully bred out of crops with the help of molecular breeding techniques. These include drought tolerance in rice (Song et al., 2011; Hoang et al., 2019; Majhi et al., 2022a; Singh et al., 2022), salinity tolerance in rice (Ibrahimova et al., 2021), bruchid tolerance in mungbean (Majhi et al., 2022b; Manmathan et al., 2013). Lower-throughput, more expensive, and labor-intensive, conventional crop phenotyping presents a significant "phenotyping bottleneck" for dynamic drought research. To better understand the genetic basis of complex quantitative traits under abiotic stresses, such as phosphate deficiency tolerance of *Brassica napus* (Gepts, 2014), drought response of wheat (Shi et al., 2016), salinity tolerance of rice (Ibrahimova et al., 2021), and drought resistance of rice (Ibrahimova et al., 2021), high-throughput non-destructive plant phenotyping techniques have been rapidly developing and popular. The majority of these analyses have concentrated on the visible effects, such as changes in morphology, biomass, and greenness. Little is known about how plants react on the inside during drought. Genetic dissection of several agronomic aspects of crops has made extensive use of genome-wide association studies based on linkage disequilibrium (LD) (Wu et al., 2021).

## 14.5 CSA IN INDIAN CONTEXT: GOVERNMENT INITIATIVES AND POLICIES

The expanding global population and changing diets are driving up the demand for more food. This challenge is further intensified by the proneness of agriculture system to extreme vagaries of climate. The intensification of unambiguous warming of the climate system since the 1950s has been stated by the fifth assessment report (AR5) of the IPCC. Globally, numerous obstacles are being witnessed to growth and development of crops due to changes in climatic pattern. The potential crop yields and investment returns are threatened by the frequent occurrence of floods and droughts. The biggest limiting aspect is the surging scarcity of natural resources, particularly water. Other significant factors that lower agricultural output include the increasing frequency of extreme climate uncertainties like hot and cold waves. CSA is an integrated approach that include actions needed for transformation and reorientation of agricultural systems so as to address the interlinked hurdles of food security and increased productivity, reduced emissions, as well as enhanced resilience to changing climate. It includes acclimatizing to extreme climate changes in some regions and responding to possibilities offered by uncertain climate in other regions. Land and water management techniques plays a key component in CSA. Studies conducted in India revealed considerable harmful effects of climate change, which, depending on the intensity and distribution of warming, may result in diminishing yields by 4.5%–9.0%. In India, it has been observed that a moderate rise in temperature has a substantial effect on yield of wheat, rice, and maize (Aryal et al., 2020). In crops like wheat, shortening of growing period, improper grain filling, water scarcity, and poor vernalization may be the reasons for decline in yield caused by changing climate. The increasing occurrence of surface soil erosion caused widespread deficiencies of soil nutrients thus resulting in inadequate nutrient supply to plants. Indian farmers have been the immediate victim of systemic neglect who face challenging situations like crop loss brought on by natural disasters and unfavourable meteorological conditions, which result in hopelessness and suicides. Weather shocks have resulted in 60% fluctuation in yield as reported by Tiwari et al. (2020). Several government initiatives (Table 14.2) are in action to fulfil the goals of the National Action Plan on Climate Change launched during the year 2008 which identified agriculture as a priority area and includes various schemes and missions under its umbrella.

**TABLE 14.2**
**Government Initiatives and Policies to Tackle the Climate Crisis in India**

| Sl. No. | Policies | Function | Outcome | References |
|---|---|---|---|---|
| 1. | National Innovation on Climate Resilient Agriculture (NICRA) | • Established by Indian Council of Agricultural Research (ICAR) in 2011.<br>• To conduct research on climate change covering crops and livestock | • Development of seven climate-resilient varieties and 650 district agricultural contingency plans<br>• The first Vulnerability Atlas of India was created at district level<br>• Adoption of Natural resource management (NRM) technologies like water foot prints, Biochar, Conservation Agriculture (CA), and emission reduction through efficient energy management<br>• Maximization of rainwater storage and conservation which made the beneficiaries in NICRA villages to attain higher yield by constructing farm ponds and percolation ponds<br>• After the implementation of NICRA, adoption of water-saving technologies like intercropping and multicropping which increased the cropping intensity | Pise et al., 2018 |
| 2. | National Water Mission | • Approved by cabinet in year 2011<br>• To create a comprehensive water database in the public domain<br>• Increasing water-use efficiency by 20%<br>• To formulate river-linking projects | • Establishment of India-WRIS WebGIS, or 'Generation of database and implementation of web-enabled water resources information system in the country', is a joint venture of the Central Water Commission, Ministry of Water Resources, Central government and Indian Space Research Organization (ISRO)<br>• About 702 Hydrological Observation Stations were established until 2016–2017<br>• Establishing 6,376 new groundwater monitoring wells | Kumar, 2018; Harikrishna et al., 2019 |

*(Continued)*

**TABLE 14.2 (*Continued*)**
**Government Initiatives and Policies to Tackle the Climate Crisis in India**

| Sl. No. | Policies | Function | Outcome | References |
|---|---|---|---|---|
| 3. | National Mission on Sustainable Agriculture (NMSA) | • Operational from 2014 to 2015<br>• Strengthening the stability of farming communities<br>• Mainly emphasis on conservation of soil and water, efficient use of water resources, soil health management and rain-fed area development, Climate Change and Sustainable Agriculture Monitoring, Modelling and Networking (CCSAMMN)<br>• To focus on appropriate soil and moisture conservation techniques<br>• Work towards economic stability at the national level<br>• To expand Automatic Weather Stations networks to the Panchayat level | • At present NMSA has been designed for large- and medium-sized farms and landholdings in different states | Patel, 2016 https://nmsa.dac.gov.in/frmObjectives.aspx |
| 4. | The National Adaptation Fund for Climate Change (NAFCC) | • Established in August 2015<br>• NABARD is the implementing entity which carries out different projects under NAFCC<br>• To meet the cost of adaptation for Indian State and UTs, those particularly vulnerable to the negative impacts of climate change | • 30 projects were sanctioned (Rs. 847.48 cr/USD 113 million) covering 25 states and 2 UTs | |
| 5. | Climate-Smart Village (CSV) | • To enhance farmers' abilities to cope with the impacts of future climate change<br>• To test, implement, modify, and promote CSA at the local level<br>• To understand the socioeconomic, gender, and biophysical constraints to adopt new practice<br>• It adopts a wide range of interventions covering various farm activities such as "Water-Smart" technologies (e.g. raised-bed planting, micro-irrigation, direct seeded rice, bunding, precision land levelling); "Nutrition Smart" (e.g. inclusion of legume, nutrient expert decision support tool); "Carbon and Energy Smart" (e.g. conservation agriculture, no-tillage system, management of residues); and "Weather Smart" (e.g. e-agriculture-based agroservices, weather forecast, index-based insurance) | • CSVs were piloted in Karnal district of Haryana and Vaishali district of Bihar state in 2011<br>• Studies showed that crop yields are higher in laser-levelled fields compared to traditionally levelled ones by adopting such climate-resilient practice | Hariharan et al., 2020; Aryal et al., 2020 |

**TABLE 14.2 (*Continued*)**
**Government Initiatives and Policies to Tackle the Climate Crisis in India**

| Sl. No. | Policies | Function | Outcome | References |
|---|---|---|---|---|
| 6. | National Solar Mission (NSM) | • Established on 11 January 2010<br>• To promote sustainable growth while addressing India's energy security challenges<br>• To install 100 GW grid-connected solar power plants by the year 2022<br>• To lessen the intensity emission of its GDP by 33%–35% by 2030.<br>• The mission has three phases: phase-I (2010–2013), phase-II (2013–2017) and phase-III (2017–2022)<br>• Various schemes are being operated by the government to encourage generation of solar power in the country like Solar Park Scheme, CPSU Scheme, Defence Scheme, VGF Schemes, Canal bank and Grid Connected Solar Rooftop Scheme, Bundling Scheme, etc. | • The installed capacity of solar power has reached around 61.97 GW as on 30 November 2022<br>• Green energy corridors have been created from renewable energy sources with dedicated transmission system for power<br>• 36 solar parks in 21 states have been sanctioned with a cumulative capacity of 20,700 MW<br>• A mobile app "Surya Mitra," has been launched that can handle thousands of calls simultaneously and efficiently monitor all activities on the app | https://mnre.gov.in/solar/current-status/<br>Kumar, 2018; Akoijam and Krishna, 2017 |
| 7. | Pradhan Mantri Krishi Sinchayee Yojna (PMSKY) | • Conceived on 1 July 2015<br>• Vision to extend the acreage of irrigation "Har Khet ko pani" and improving water-use efficiency "More crop per drop" in a focused manner<br>• To undertake practices for water conservation such as rainwater conservation, water harvesting structures, small check dams, contour bunding, and construction of farm pond, etc. | • Micro-irrigation area covered under drip irrigation is 3.58 Lakh Ha, and under sprinkler irrigation is 6.57 Lakh Ha<br>• Potential created in 0.06 Lakh Ha for protective irrigation<br>• 2.25 lakh hectares of sugarcane area has been brought under drip technology in Maharashtra by implementing PMKSY | https://pmksy.gov.in/<br>Gulati and Mohan, 2018 |
| 8. | National Mission for Green India | • Adopted in 2014<br>• Main focus on protecting, restoring, and increasing India's forest cover and responding to climate change<br>• To increase forest cover as well as quality of forest cover<br>• To escalate forest-based livelihood incomes | • The actions taken under Green India Mission are implemented by converging it with National Afforestation Program, Mahatma Gandhi National Rural Employment Guarantee Act and Compensatory Afforestation Fund Management and Planning Authority (CAMPA)<br>• The Compensatory Afforestation Bill was passed in the Parliament in 2016 to address the issue of decreasing forest cover | Kumar, 2018 |

*(Continued)*

**TABLE 14.2 (*Continued*)**
**Government Initiatives and Policies to Tackle the Climate Crisis in India**

| Sl. No. | Policies | Function | Outcome | References |
|---|---|---|---|---|
| 9. | Pradhan Mantri Fasal Bima Yojna (PMFBY) | • Launched on 18 February 2016.<br>• Helps stabilizing the income of farmers by providing a comprehensive insurance cover against crop failure<br>• The scheme covers all food and oilseeds crops and annual commercial/horticultural crops for which past yield data are available | • The cropped area under insurance increased from 23% in 2013–2014 to 29% in 2016–2017, and then declined to 26% in 2017–2018 after the launch of PMFBY | Tiwari et al., 2020 |
| 10. | Soil Health Card Scheme | • Launched by the Government of India on 19 February 2015<br>• This card can help the farmers for judicious use of inputs for increasing output by providing crop-wise recommendations of nutrients and fertilizers required for the individual farms<br>• Contain the status of soil in accordance with 12 parameters, namely N, P, K, S, Zn, Fe, Cu, Mn, Bo, and pH, EC, OC | • This has been implemented in 32 states and UTs<br>• Savings were up to 8%–10% because of application of fertilizer and micronutrients based on recommendations of Soil Health Card (SHC)<br>• Increase in yield of crops was reported to the tune of 5%–6% after adopting the SHC recommendations<br>• The farmers even received SHC at their doorsteps or in campaigns by agricultural officers (Mittal and Mehar, 2016)<br>• As per studies, awareness of the SHC scheme was greater in developed states (92.4% of farmers) than in less developed states (55.1%)<br>• The main advantage from using SHC was the reduction in use of fertilizers. Cost of production was reduced by about 10% due to lower fertilizer use in cotton, paddy, and soybean, and 4% in wheat | Naya and Nishad (2022) |
| 11. | National Mission for Enhanced Energy Efficiency (NMEEE) | • Approved in June 2010<br>• Frame novel policy and regulatory frameworks, business models and funding schemes that transparently build and maintain markets for energy efficiency across many economic sectors<br>• NMEEE comprises four initiatives: Perform, Achieve and Trade, Market Transformation for Energy Efficiency, Energy Efficiency Financing Platform and Framework for Energy Efficient Economic Development (FEEED) | • Distribution of energy-efficient appliances, such as 231 million LED lights were distributed by the government under the Pradhan Mantri Ujjwala Yojana to provide free clean cooking gas connections to women below the poverty line | Gupta and Sengupta, (2012) |

## 14.6 CONCLUSION AND FUTURE PROSPECTIVES

Even though CSA has many positive impacts on society, economy, and the environment, its adoption on farms in underdeveloped nations is minimal. Climate change is just one of many factors jeopardizing the long-term viability of the world's food supply. Adapting to such innovativeness presents both a difficulty and an opening. The difficulty is in thinking of ways to mitigate its effects, while the possibility is found in the discovery of novel approaches, competencies, and treatments that foster originality. The development of sustainable agri-food systems and enhanced rural livelihoods are threatened by climate change, and a greater knowledge of these threats can be attained by accumulating more evidence of them. It provides a framework for tracking progress towards the SDGs and reviewing current policies and procedures. Evidence-based decision-making in policy-making and implementation aids in making the most of opportunities while minimizing risks. Efforts to increase the body of evidence supporting the use of CSAs are most effective when they are informed by a thorough analysis of the specific problems, possibilities, and obstacles at a given location, and when they welcome and encourage participation from all relevant parties and types of knowledge. The only way to increase crop yields through the development of climate-resilient cultivars is to quantify and understand these genetic linkages. There are a number of agricultural enhancement initiatives that aim to increase yield by making crops more resistant to abiotic stresses including drought, heat, cold, salt, flooding, etc. Technology advancement such as green technology for agriculture and livestock production; climate and hydrological modelling for cropping systems; artificial intelligence; remote sensing and GIS application; smart mobile-based applications for disease and crop and soil management has imparted a great change in agricultural production. Much progress has been made in our capacity to analyze the physiology, biochemistry, and molecular genetics of plant responses to abiotic stressors. Early vigour, short duration, osmotic adjustment, leaf senescence, green habitats, etc. are just some of the adaptive features that have been investigated and exploited to improve drought resistance (Figure 14.3). The term "stay green" is commonly used to describe a plant's resistance to the post-flowering senescence brought on by dryness. The roots of plants are also crucial in their ability to adapt to drought. Despite the wealth of data at breeders' disposal, increasing crop productivity in water-limited environments continues to be a formidable obstacle. Precision phenotyping is essential for identifying drought-resistant genotypes and understanding the quantitative genetic features that influence crop adaptation in the face of

**FIGURE 14.3** Emerging technologies in advancing climate-smart agriculture.

drought. Evidence-gathering initiatives should also compare the outcomes of different CSA policies and practises to a baseline state and to the outcomes of other interventions to determine the synergies and trade-offs between them. An in-depth familiarity with the human, material, and financial resources needed to implement CSA would tremendously aid ongoing efforts to increase the body of evidence supporting the practise.

## REFERENCES

Abang, M., & Malhotra, R. (2008). Chickpea and climate change. *ICARDA Caravan*, *25*, 48–50.

Abarca, M., & Spahn, R. (2021). Direct and indirect effects of altered temperature regimes and phenological mismatches on insect populations. *Current Opinion in Insect Science*, *47*, 67–74. https://doi.org/10.1016/j.cois.2021.04.008

Agrios, G. N. (2005). *Plant pathology* (5th ed., pp. 249–263). Elsevier.

Ahanger, R. A., Bhat, H. A., Bhat, T. A., Ganie, S. A., Lone, A. A., Wani, I. A., et al. (2013). Impact of climate change on plant diseases. *International Journal of Modern Plant and Animal Sciences*, *1*(3), 105–115.

Ainsworth, E. A., & Long, S. P. (2021). 30 years of free-air carbon enrichment (FACE): What have we learned about future crop productivity and its potential for adaptation? *Global Change Biology*, *27*(1), 27–49. https://doi.org/10.1111/gcb.15375

Ainsworth, E. A., & Rogers, A. (2007). The response of photosynthesis and stomatal conductance to rising [CO2]: Mechanisms and environmental interactions. *Plant, Cell and Environment*, *30*(3), 258–270. https://doi.org/10.1111/j.1365-3040.2007.01641.x

Akoijam, A. S., & Krishna, V. V. (2017). Exploring the Jawaharlal Nehru national solar mission (JNNSM): Impact on innovation ecosystem in India. *African Journal of Science, Technology, Innovation and Development*, *9*(5), 573–585. https://doi.org/10.1080/20421338.2017.1359466

Angelotti, F., Hamada, E., Magalhães, E. E., Ghini, R., Garrido, L. D. R., & Pedro Júnior, M. J. P. (2017). Climate change and the occurrence of downy mildew in Brazilian grapevines. *Pesquisa Agropecuaria* Brasileira, *52*(6), 426–434. https://doi.org/10.1590/s0100-204x2017000600006

Aryal, J. P., Sapkota, T. B., Rahut, D. B., & Jat, M. L. (2020). Agricultural sustainability under emerging climatic variability: The role of climate-smart agriculture and relevant policies in India. *International Journal of Innovation and Sustainable Development*, *14*(2), 219–245. https://doi.org/10.1504/IJISD.2020.106243

Ashoub, A., Baeumlisberger, M., Neupaertl, M., Karas, M., & Brüggemann, W. (2015). Characterization of common and distinctive adjustments of wild barley leaf proteome under drought acclimation, heat stress and their combination. *Plant Molecular Biology*, *87*(4–5), 459–471. https://doi.org/10.1007/s11103-015-0291-4

Barbieri, P., Echarte, L., Della Maggiora, A., Sadras, V. O., Echeverria, H., & Andrade, F. H. (2012). Maize evapotranspiration and water-use efficiency in response to row spacing. *Agronomy Journal*, *104*(4), 939–944. https://doi.org/10.2134/agronj2012.0014

Bauerfeind, S. S., & Fischer, K. (2013). Increased temperature reduces herbivore host-plant quality. *Global Change Biology*, *19*(11), 3272–3282. https://doi.org/10.1111/gcb.12297

Betz, O., Srisuka, W., & Puthz, V. (2020). Elevational gradients of species richness, community structure, and niche occupation of tropical roove beetles (Coleoptera: Staphylinidae: Steninae) across mountain slopes in Northern Thailand. *Evolutionary Ecology*, *34*(2), 193–216. https://doi.org/10.1007/s10682-020-10036-2

Bhoi, T. K., Samal, I., Majhi, P. K., Komal, J., Mahanta, D. K., Pradhan, A. K., Saini, V., Nikhil Raj, M., Ahmad, M. A., Behera, P. P., & Ashwini, M. (2022). Insight into aphid mediated potato virus Y transmission: A molecular to bioinformatics prospective. *Frontiers in Microbiology*, *13*, 1001454. https://doi.org/10.3389/fmicb.2022.1001454

Bickford, C. P. (2016). Ecophysiology of leaf trichomes. *Functional Plant Biology*, *43*(9), 807–814. https://doi.org/10.1071/FP16095

Birthal, P. S., Khan, T., Negi, D. S., & Agarwal, S. (2014). Impact of climate change on yields of major food crops in India: Implications for food security. *Agricultural Economics Research Review*, *27*(2), 145–155. https://doi.org/10.5958/0974-0279.2014.00019.6

Blanco-Canqui, H., & Ruis, S. J. (2020). Cover crop impacts on soil physical properties: A review. *Soil Science Society of America Journal*, *84*(5), 1527–1576. https://doi.org/10.1002/saj2.20129

Bonan, G. B. (2004). Biogeophysical feedbacks between land cover and climate. *Washington DC American Geophysical Union Geophysical Monograph Series*, *153*, 61–72.

Borevitz, J. (2021). Utilizing genomics to understand and respond to global climate change. *Genome Biology*, *22*(1), 91. https://doi.org/10.1186/s13059-021-02317-y

Bradfield, M., & Stamp, N. (2004). Effect of night time temperature on tomato plant defensive chemistry. *Journal of Chemical Ecology*, *30*(9), 1713–1721. https://doi.org/10.1023/b:joec.0000042397.42061.9f

Brasier, C. M. (2001). Rapid evolution of introduced plant pathogens via interspecific hybridization. *BioScience*, *51*, 123–133.

Butterworth, M. H., Semenov, M. A., Barnes, A., Moran, D., West, J. S., & Fitt, B. D. L. (2010). North-South divide; contrasting impacts of climate change on crop yields in Scotland and England. *Journal of the Royal Society, Interface*, *7*(42), 123–130. https://doi.org/10.1098/rsif.2009.0111

Chakraborty, S., & Datta, S. (2003). How will plant pathogens adapt to host plant resistance at elevated $CO_2$ under a changing climate? *New Phytologist*, *159*(3), 733–742. https://doi.org/10.1046/j.1469-8137.2003.00842.x

Chakraborty, S., Murray, G., & White, N. (2002). Potential impact of climate change on plant diseases of economic significance to Australia. *Australasian Plant Pathology*, *27*, 15–35.

Chatterjee, S., Bandyopadhyay, K. K., Pradhan, S., Singh, R., & Datta, S. P. (2018). Effects of irrigation, crop residue mulch and nitrogen management in maize (*Zea mays* L.) on soil carbon pools in a sandy loam soil of Indo-gangetic plain region. *CATENA*, *165*, 207–216. https://doi.org/10.1016/j.catena.2018.02.005

Chenini, I., & Ben Mammou, A. B. (2010). Groundwater recharge study in arid region: An approach using GIS techniques and numerical modeling. *Computers and Geosciences*, *36*(6), 801–817. https://doi.org/10.1016/j.cageo.2009.06.014

Chitarra, W., Siciliano, I., Ferrocino, I., Gullino, M. L., & Garibaldi, A. (2015). Effect of elevated atmospheric $CO_2$ and temperature on the disease severity of rocket plants caused by fusarium wilt under phytotron conditions. *PLoS ONE*, *10*(10), e0140769. https://doi.org/10.1371/journal.pone.0140769

Coakley, S. M., Scherm, H., & Chakraborty, S. (1999). Climate change and plant disease management. *Annual Review of Phytopathology*, *37*, 399–426. https://doi.org/10.1146/annurev.phyto.37.1.399

Colwell, R. K., Brehm, G., Cardelús, C. L., Gilman, A. C., & Longino, J. T. (2008). Global warming, elevational range shifts, and lowland biotic attrition in the wet tropics. *Science*, *322*(5899), 258–261. https://doi.org/10.1126/science.1162547

Cooper, E. J. (2014). Warmer shorter winters disrupt arctic terrestrial ecosystems. *Annual Review of Ecology, Evolution, and Systematics*, *45*(1), 271–295. https://doi.org/10.1146/annurev-ecolsys-120213-091620

Cuddington, K., Fortin, M.-J., Gerber, L. R., Hastings, A., Liebhold, A., O'Connor, M., & Ray, C. (2013). Process-based models are required to manage ecological systems in a changing world. *Ecosphere*, *4*(2), 1–12. https://doi.org/10.1890/ES12-00178.1

Damien, M., & Tougeron, K. (2019). Prey-predator phenological mismatch under climate change. *Current Opinion in Insect Science*, *35*, 60–68. https://doi.org/10.1016/j.cois.2019.07.002

Das, T., Majumdar, M. H. D., Devi, R. T., & Rajesh, T. (2016). Climate change impacts on plant diseases. *SAARC Journal of Agriculture*, *14*(2), 200–209. https://doi.org/10.3329/sja.v14i2.31259

Delucia, E. H., Nabity, P. D., Zavala, J. A., & Berenbaum, M. R. (2012). Climate change: Resetting plant-insect interactions. *Plant Physiology*, *160*(4), 1677–1685. https://doi.org/10.1104/pp.112.204750

Diamond, S. E., & Kingsolver, J. G. (2010). Fitness consequences of host plant choice: A field experiment. *Oikos*, *119*(3), 542–550. https://doi.org/10.1111/j.1600-0706.2009.17242.x

Diniz, D. F. A., de Albuquerque, C. M. R., Oliva, L. O., de Melo-Santos, M. A. V., & Ayres, C. F. J. (2017). Diapause and quiescence: Dormancy mechanisms that contribute to the geographical expansion of mosquitoes and their evolutionary success. *Parasites and Vectors*, *10*(1), 310. https://doi.org/10.1186/s13071-017-2235-0

Donatelli, M., Magarey, R. D., Bregaglio, S., Willocquet, L., Whish, J. P. M., & Savary, S. (2017). Modelling the impacts of pests and diseases on agricultural systems. *Agricultural Systems*, *155*, 213–224. https://doi.org/10.1016/j.agsy.2017.01.019

Dwivedi, R. S., Ramana, K. V., Thammappa, S. S., & Singh, A. N. (2001). The utility of IRS-1C LISS-III and PAN-merged data for mapping salt-affected soils. *Photogrammetric Engineering and Remote Sensing*, *67*(10), 1167–1175.

Eastburn, D. M., Degennaro, M. M., Delucia, E. H., Dermody, O., & Mcelrone, A. J. (2010). Elevated atmospheric carbon dioxide and ozone alter soybean diseases at Soy FACE. *Global Change Biology*, *16*(1), 320–330. https://doi.org/10.1111/j.1365-2486.2009.01978.x

Eastburn, D. M., McElrone, A. J., & Bilgin, D. D. (2011). Influence of atmospheric and climatic change on plant-pathogen interactions. *Plant Pathology*, *60*(1), 54–69. https://doi.org/10.1111/j.1365-3059.2010.02402.x

Evans, N., Baierl, A., Semenov, M. A., Gladders, P., & Fitt, B. D. L. (2008). Range and severity of a plant disease increased by global warming. *Journal of the Royal Society, Interface*, *5*(22), 525–531. https://doi.org/10.1098/rsif.2007.1136

Evans, T. G., Diamond, S. E., & Kelly, M. W. X. (2015). Mechanistic species distribution modelling as a link between physiology and conservation. *Conservation Physiology*, *3*(1), cov056. https://doi.org/10.1093/conphys/cov056

Fan, Y., Park, S., & Nan, Z. (2018). Participatory water management and adoption of micro-irrigation systems: Smallholder farmers in arid north-western China. *International Journal of Water Resources Development*, *34*(3), 434–452. https://doi.org/10.1080/07900627.2017.1400424

Favero, A., Daigneault, A., & Sohngen, B. (2020). Forests: Carbon sequestration, biomass energy, or both? *Science Advances*, *6*(13), eaay6792. https://doi.org/10.1126/sciadv.aay6792

Feeley, K. J., Hurtado, J., Saatchi, S., Silman, M. R., & Clark, D. B. (2013). Compositional shifts in Costa Rican forests due to climate-driven species migrations. *Global Change Biology*, *19*(11), 3472–3480. https://doi.org/10.1111/gcb.12300

Franco, J. G., King, S. R., & Volder, A. (2018). Component crop physiology and water use efficiency in response to intercropping. *European Journal of Agronomy*, *93*, 27–39. https://doi.org/10.1016/j.eja.2017.11.005

Fry, W. E., McGrath, M. T., Seaman, A., Zitter, T. A., McLeod, A., Danies, G., Small, I. M., Myers, K., Everts, K., Gevens, A. J., Gugino, B. K., Johnson, S. B., Judelson, H., Ristaino, J., Roberts, P., Secor, G., Seebold, K., Snover-Clift, K., Wyenandt, A.,. . . Smart, C. D. (2013). The 2009 late blight pandemic in the Eastern United States—causes and results. *Plant Disease*, *97*(3), 296–306. https://doi.org/10.1094/PDIS-08-12-0791-FE

Fussmann, K. E., Schwarzmüller, F., Brose, U., Jousset, A., & Rall, B. C. (2014). Ecological stability in response to warming. *Nature Climate Change*, *4*(3), 206–210. https://doi.org/10.1038/nclimate2134

García-Robledo, C., & Baer, C. S. (2021). Positive genetic covariance and limited thermal tolerance constrain tropical insect responses to global warming. *Journal of Evolutionary Biology*, *34*(9), 1432–1446. https://doi.org/10.1111/jeb.13905

Garrett, K. A., Dendy, S. P., Frank, E. E., Rouse, M. N., & Travers, S. E. (2006). Climate change effects on plant disease: Genomes to ecosystems. *Annual Review of Phytopathology*, *44*, 489–509. https://doi.org/10.1146/annurev.phyto.44.070505.143420

Gautam, H. R., Bhardwaj, M. L., & Kumar, R. (2013). Climate change and its impact on plant diseases. *Current Science*, *105*(12), 1685–1691.

Gepts, P. (2014). The contribution of genetic and genomic approaches to plant domestication studies. *Current Opinion in Plant Biology*, *18*, 51–59. https://doi.org/10.1016/j.pbi.2014.02.001

Gérard, M., Vanderplanck, M., Wood, T., & Michez, D. (2020). Global warming and plant-pollinator mismatches. *Emerging Topics in Life Sciences*, *4*(1), 77–86. https://doi.org/10.1042/ETLS20190139

Ghini, R., Hamada, E., & Bettiol, W. (2011). *Impactos das mudanças climáticas sobre doenças de importantes culturas no Brasil*. Embrapa. Meio ambiente.

Gillson, L., Dawson, T. P., Jack, S., & McGeoch, M. A. (2013). Accommodating climate change contingencies in conservation strategy. *Trends in Ecology and Evolution*, *28*(3), 135–142. https://doi.org/10.1016/j.tree.2012.10.008

Green, T. R., & Ryan, C. A. (1973). Wound-induced proteinase inhibitor in tomato leaves: Some effects of light and temperature on the wound response. *Plant Physiology*, *51*(1), 19–21. https://doi.org/10.1104/pp.51.1.19

Gulati, A., & Mohan, G. (2018). *Towards sustainable, productive and profitable agriculture*, *358*.

Gupta, M., & SenGupta, R. (2012). Fiscal and monetary policy aspects under Framework for Energy Efficient Economic Development (FEED) mechanism of National Mission for Enhanced Energy Efficiency (NMEEE). *Journal of Economic Policy & Research*, *14*(1).

Halsch, C. A., Shapiro, A. M., Fordyce, J. A., Nice, C. C., Thorne, J. H., Waetjen, D. P., & Forister, M. L. (2021). Insects and recent climate change. *Proceedings of the National Academy of Sciences of the United States of America*, *118*(2), e2002543117. https://doi.org/10.1073/pnas.2002543117

Han, P., Lavoir, A. V., Rodriguez-Saona, C., & Desneux, N. (2022). Bottom-up forces in agroecosystems and their potential impact on arthropod pest management. *Annual Review of Entomology*, *67*, 239–259. https://doi.org/10.1146/annurev-ento-060121-060505

Hariharan, V. K., Mittal, S., Rai, M., Agarwal, T., Kalvaniya, K. C., Stirling, C. M., & Jat, M. L. (2020). Does climate-smart village approach influence gender equality in farming households? A case of two contrasting ecologies in India. *Climatic Change*, *158*(1), 77–90. https://doi.org/10.1007/s10584-018-2321-0

Harikrishna, Y. V., Naberia, S., Pradhan, S., & Hansdah, P. (2019). Agro-economic impact of climate resilient practices on farmers in Anantapur district of Andhra Pradesh. *Indian Journal of Extension Education*, *55*(4), 91–95.

Harvey, J. A., Heinen, R., Gols, R., & Thakur, M. P. (2020). Climate change-mediated temperature extremes and insects: From outbreaks to breakdowns. *Global Change Biology*, *26*(12), 6685–6701. https://doi.org/10.1111/gcb.15377

Heeb, L., Jenner, E., & Cock, M. J. W. (2019). Climate-smart pest management: Building resilience of farms and landscapes to changing pest threats. *Journal of Pest Science*, *92*(3), 951–969. https://doi.org/10.1007/s10340-019-01083-y

Hellmann, J. J., Byers, J. E., Bierwagen, B. G., & Dukes, J. S. (2008). Five potential consequences of climate change for invasive species. *Conservation Biology*, *22*(3), 534–543. https://doi.org/10.1111/j.1523-1739.2008.00951.x

Hill, M. P., & Thomson, L. J. (2015). Species distribution modelling in predicting response to climate change. In C. Björkman & P. Niemelä (Eds.), *Climate change and insect pests* (pp. 16–37). CABI Publishing.

Himanen, S. J., Nissinen, A., Dong, W. X., Nerg, A. M., Stewart, C. N., Poppy, G. M., & Holopainen, J. K. (2008). Interactions of elevated carbon dioxide and temperature with aphid feeding on transgenic oilseed rape: Are *Bacillus thuringiensis* (Bt) plants more susceptible to nontarget herbivores in future climate? *Global Change Biology*, *14*(6), 1437–1454. https://doi.org/10.1111/j.1365-2486.2008.01574.x

Hoang, G. T., Van Dinh, L., Nguyen, T. T., Ta, N. K., Gathignol, F., Mai, C. D., Jouannic, S., Tran, K. D., Khuat, T. H., Do, V. N., Lebrun, M., Courtois, B., & Gantet, P. (2019). Genome-wide association study of a panel of Vietnamese rice landraces reveals new QTLs for tolerance to water deficit during the vegetative phase. *Rice*, *12*(1), 4. https://doi.org/10.1186/s12284-018-0258-6

Hong, S. C., Magarey, R., Borchert, D. M., Vargas, R. I., & Souder, S. (2015). Site-specific temporal and spatial validation of a generic plant pest forecast system with observations of *Bactrocera dorsalis* (oriental fruit fly). *NeoBiota*, *27*, 37–67. https://doi.org/10.3897/neobiota.27.5177

Huang, D., Liu, C., Fang, H., & Peng, S. (2008). Assessment of waterlogging risk in Lixiahe region of Jiangsu Province based on AVHRR and MODIS image. *Chinese Geographical Science*, *18*(2), 178–183. https://doi.org/10.1007/s11769-008-0178-2

Ibrahimova, U., Zivcak, M., Gasparovic, K., Rastogi, A., Allakhverdiev, S. I., Yang, X., & Brestic, M. (2021). Electron and proton transport in wheat exposed to salt stress: Is the increase of the thylakoid membrane proton conductivity responsible for decreasing the photosynthetic activity in sensitive genotypes? *Photosynthesis Research*, *150*(1–3), 195–211. https://doi.org/10.1007/s11120-021-00853-z

IPCC. (2007). *Contribution of working group I, II and III to the fourth assessment report of the Intergovernmental Panel on Climate Change (IPCC)* (R. K. Pachauri & A. Reisinger, Eds., Climate Change. 2007: Synthesis report). IPCC.

IPCC. (2014). Climate change 2014 the physical science basis: Final draft underlying scientific-technical assessment: Synthesis report. In T. F. Stocker (Ed.), *Contribution of working groups I, II and III to the fifth assessment report of the Intergovernmental Panel on Climate Change* (p. 1535). Cambridge University Press.

IPCC CC. (2021). The physical science basis. Contribution of working group I to the fourth assessment report of the Intergovernmental Panel on Climate Change (pp. 113–119, p. 996). Cambridge University Press.

Johnston, A. M., & Bruulsema, T. W. (2014). 4R nutrient stewardship for improved nutrient use efficiency. *Procedia Engineering*, *83*, 365–370. https://doi.org/10.1016/j.proeng.2014.09.029

Kearney, M. R., Wintle, B. A., & Porter, W. P. (2010). Correlative and mechanistic models of species distribution provide congruent forecasts under climate change. *Conservation Letters*, *3*(3), 203–213. https://doi.org/10.1111/j.1755-263X.2010.00097.x

Keating, B. A., Carberry, P. S., Hammer, G. L., Probert, M. E., Robertson, M. J., Holzworth, D., Huth, N. I., Hargreaves, J. N. G., Meinke, H., Hochman, Z., McLean, G., Verburg, K., Snow, V., Dimes, J. P., Silburn, M., Wang, E., Brown, S., Bristow, K. L., Asseng, S.,. . . Smith, C. J. (2003). An overview of APSIM, a model designed for farming systems simulation. *European Journal of Agronomy*, *18*(3–4), 267–288. https://doi.org/10.1016/S1161-0301(02)00108-9

Kobayashi, T., Ishiguro, K., Nakajima, T., Kim, H. Y., Okada, M., & Kobayashi, K. (2006). Effects of elevated atmospheric $CO_2$ concentration on the infection of rice blast and sheath blight. *Phytopathology*, *96*(4), 425–431. https://doi.org/10.1094/PHYTO-96-0425

Kriticos, D. J., Maywald, G. F., Yonow, T., Zurcher, E. J., Herrmann, N. I., & Sutherst, R. (2015). Exploring the effects of climate on plants, animals and diseases. *CLIMEX Version*, *4*, 184.

Krupnik, T. J., Rodenburg, J., Haden, V. R., Mbaye, D., & Shennan, C. (2012). Genotypic trade-offs between water productivity and weed competition under the system of rice intensification in the Sahel. *Agricultural Water Management*, *115*, 156–166. https://doi.org/10.1016/j.agwat.2012.08.016

Kumar, S., Neven, L. G., & Yee, W. L. (2014). Evaluating correlative and mechanistic niche models for assessing the risk of pest establishment. *Ecosphere*, *5*, 1–23.

Kumar, V. (2018). *Coping with climate change: An analysis of India's state action plans on climate change*. Centre for Science and Environment.

Ladanyi, M., & Horvath, L. (2010). A review of the potential climate change impact on insect populations—General and agricultural aspects. *Applied Ecology and Environmental Research*, *8*(2), 143–151. https://doi.org/10.15666/aeer/0802_143151

Lambers, H., Chapin, F. S., & Pons, T. L. (2008). *Plant physiological ecology* (2nd ed., p. 604). Springer Science+Business Media LLC.

Leblanc, M., Favreau, G., Tweed, S., Leduc, C., Razack, M., & Mofor, L. (2007). Remote sensing for groundwater modelling in large semiarid areas: Lake Chad Basin, Africa. *Hydrogeology Journal*, *15*(1), 97–100. https://doi.org/10.1007/s10040-006-0126-0

Magarey, R. D., Fowler, G. A., Borchert, D. M., Sutton, T. B., Colunga-Garcia, M., & Simpson, J. A. (2007). NAPPFAST: An internet system for the weather-based mapping of plant pathogens. *Plant Disease*, *91*(4), 336–345. https://doi.org/10.1094/PDIS-91-4-0336

Majhi, P. K., Bhoi, T. K., Mogali, S. C., Shiv, A., Sahoo, K. C., & Saini, V. (2022b). Advances in molecular breeding for bruchid (*Callosobruchus* spp.) resistance in mungbean [*Vigna radiata* (L.) Wilczek]: A review. *Legume Research*, *45*, 933–941. https://doi.org/10.18805/LR-4695

Majhi, P. K., Raza, B., Behera, P. P., Singh, S. K., Shiv, A., Mogali, S. C., Bhoi, T. K., Patra, B., & Behera, B. (2023). Future-proofing plants against climate change: A path to ensure sustainable food systems. In *Biodiversity, functional ecosystems and sustainable food production* (pp. 73–116). Springer International Publishing.

Majhi, P. K., Singh, S. K., Anandan, A., Khaire, A. R., Korada, M., Habde, S. V., Singh, D. K., Bhagvan, A. P., Jayasudha, S., & Singh, A. (2022a). Parental polymorphism survey for evaluation and selection of contrasting parents for drought tolerance in rice (*Oryza sativa* L.) by using SSR markers. *International Journal of Environment and Climate Change*, *12*(11), 2507–2519. https://doi.org/10.9734/ijecc/2022/v12i1131244

Manmathan, H., Shaner, D., Snelling, J., Tisserat, N., & Lapitan, N. (2013). Virus-induced gene silencing of *Arabidopsis thaliana* gene homologues in wheat identifies genes conferring improved drought tolerance. *Journal of Experimental Botany*, *64*(5), 1381–1392. https://doi.org/10.1093/jxb/ert003

Marshall, K. E., Gotthard, K., & Williams, C. M. (2020). Evolutionary impacts of winter climate change on insects. *Current Opinion in Insect Science*, *41*, 54–62. https://doi.org/10.1016/j.cois.2020.06.003

Mboup, M., Bahri, B., Leconte, M., De Vallavieille-Pope, C., Kaltz, O., & Enjalbert, J. (2012). Genetic structure and local adaptation of European wheat yellow rust populations: The role of temperature-specific adaptation. *Evolutionary Applications*, *5*(4), 341–352. https://doi.org/10.1111/j.1752-4571.2011.00228.x

McElrone, A. J., Reid, C. D., Hoye, K. A., Hart, E., & Jackson, R. B. (2005). Elevated $CO_2$ reduces disease incidence and severity of a red maple fungal pathogen via changes in host physiology and leaf chemistry. *Global Change Biology*, *11*(10), 1828–1836. https://doi.org/10.1111/j.1365-2486.2005.001015.x

Melloy, P., Hollaway, G., Luck, J., Norton, R., Aitken, E., & Chakraborty, S. (2010). Production and fitness of *Fusarium pseudograminearum* inoculum at elevated carbon dioxide in FACE. *Global Change Biology*, *16*(12), 3363–3373. https://doi.org/10.1111/j.1365-2486.2010.02178.x

Mina, U., & Sinha, P. (2008). Effects of climate change on plant pathogens. *EnviroNews*, *14*(4), 6–10.

Mittal, S., & Mehar, M. (2016). Socio-economic factors affecting adoption of modern information and communication technology by farmers in India: Analysis using multivariate probit model. *Journal of Agricultural Education and Extension*, *22*(2), 199–212. https://doi.org/10.1080/1389224X.2014.997255

Monticelli, L. S., Bishop, J., Desneux, N., Gurr, G. M., Jaworski, C. C., McLean, A. H. C., Thomine, E., & Vanbergen, A. J. (2021). Multiple global change impacts on parasitism and biocontrol services in future agricultural landscapes. *Advances in Ecological Research*, *65*, 245–304. https://doi.org/10.1016/bs.aecr.2021.10.002

Muller, S. J., & Van Niekerk, A. (2016). An evaluation of supervised classifiers for indirectly detecting salt-affected areas at irrigation scheme level. *International Journal of Applied Earth Observation and Geoinformation*, *49*, 138–150. https://doi.org/10.1016/j.jag.2016.02.005

Murrell, T. S., Lafond, G. P., & Vyn, T. J. (2009, March-April). Know your fertilizer rights: Right place. *Crops and Soils* (American Society of Agronomy. Madison, WI).

Naya, N., & Nishad, S. (2022). Awareness and demonstration of the soil health card scheme benefits for rural development. *3-D Printing in Food Product Development*, *29*.

Nürnberger, F., Härtel, S., & Steffan-Dewenter, I. (2019). Seasonal timing in honey bee colonies: Phenology shifts affect honey stores and *Varroa* infestation levels. *Oecologia*, *189*(4), 1121–1131. https://doi.org/10.1007/s00442-019-04377-1

Pacifici, M., Foden, W. B., Visconti, P., Watson, J. E. M., Butchart, S. H. M., Kovacs, K. M., Scheffers, B. R., Hole, D. G., Martin, T. G., Akçakaya, H. R., Corlett, R. T., Huntley, B., Bickford, D., Carr, J. A., Hoffmann, A. A., Midgley, G. F., Pearce-Kelly, P., Pearson, R. G., Williams, S. E.,. . . Rondinini, C. (2015). Assessing species vulnerability to climate change. *Nature Climate Change*, *5*(3), 215–224. https://doi.org/10.1038/nclimate2448

Panda, K. C., Kumar, A., Pradhan, S. N., De, N., & Meena, V. S. (2021). Impact of soil moisture stress on rice productivity in warming climate over Indian mid-Indo-gangetic plain. *Climate Change and Environmental Sustainability*, *9*(1), 21–31. https://doi.org/10.5958/2320-642X.2021.00003.X

Panda, K. C., Singh, R. M., Singh, V. K., Singla, S., & Paramaguru, P. K. (2023). Impact of climate change induced future rainfall variation on dynamics of arid-humid zone transition in the western province of India. *Journal of Environmental Management*, *325*(B), 116646. https://doi.org/10.1016/j.jenvman.2022.116646

Panda, K. C., Singh, R. M., Thakural, L. N., & Sahoo, D. P. (2022). Representative grid location-multivariate adaptive regression spline (RGL-MARS) algorithm for downscaling dry and wet season rainfall. *Journal of Hydrology*, *605*, 127381. https://doi.org/10.1016/j.jhydrol.2021.127381

Pardikes, N. A., Revilla, T. A., Lue, C. H., Thierry, M., Souto-Vilarós, D., & Hrcek, J. (2022). Effects of phenological mismatch under warming are modified by community context. *Global Change Biology*, *28*(13), 4013–4026. https://doi.org/10.1111/gcb.16195

Parmesan, C., & Yohe, G. (2003). A globally coherent fingerprint of climate change impacts across natural systems. *Nature*, *421*(6918), 37–42. https://doi.org/10.1038/nature01286

Patel, A. (2016). Climate change and agriculture in India: Effective implementation of national mission for sustainable agriculture. *International Journal of Agricultural Research, Innovation and Technology*, *5*(3), 2319–1473.

Pautasso, M., Döring, T. F., Garbelotto, M., Pellis, L., & Jeger, M. J. (2012). Impacts of climate change on plant diseases—Opinions and trends. *European Journal of Plant Pathology*, *133*(1), 295–313. https://doi.org/10.1007/s10658-012-9936-1

Peñuelas, J., & Llusià, J. (2003). BVOCs: Plant defense against climate warming? *Trends in Plant Science*, *8*(3), 105–109. https://doi.org/10.1016/S1360-1385(03)00008-6

Peragón, J. M., Delgado, A., & Pérez-Latorre, F. J. (2015). A GIS-based quality assessment model for olive tree irrigation water in southern Spain. *Agricultural Water Management*, *148*, 232–240. https://doi.org/10.1016/j.agwat.2014.10.009

Perrings, C., Burgiel, S., Lonsdale, M., Mooney, H., & Williamson, M. (2010). International cooperation in the solution to trade-related invasive species risks. *Annals of the New York Academy of Sciences*, *1195*, 198–212. https://doi.org/10.1111/j.1749-6632.2010.05453.x

Pfender, W. F., & Vollmer, S. S. (1999). Freezing temperature effect on survival of *Puccinia graminis* sub sp. Graminicola in *Festuca arundinacea* and *Lolium perenne*. *Plant Disease*, *83*(11), 1058–1062. https://doi.org/10.1094/PDIS.1999.83.11.1058

Phillips, S. J., Anderson, R. P., & Schapire, R. E. (2006). Maximum entropy modeling of species geographic distributions. *Ecological Modelling*, *190*(3–4), 231–259. https://doi.org/10.1016/j.ecolmodel.2005.03.026

Pickles, B. J., Genney, D. R., Anderson, I. C., & Alexander, I. J. (2012). Spatial analysis of ectomycorrhizal fungi reveals that root tip communities are structured by competitive interactions. *Molecular Ecology*, *21*(20), 5110–5123. https://doi.org/10.1111/j.1365-294X.2012.05739.x

Pise, G. K., Ahire, R. D., & Kale, N. D. (2018). Impact of national innovations on climate resilient agriculture (NICRA) project on its beneficiaries. *International Journal of Current Microbiology and Applied Sciences*, *6*, 2928–2935.

Porfirio, L. L., Harris, R. M., Lefroy, E. C., Hugh, S., Gould, S. F., Lee, G., Bindoff, N. L., & Mackey, B. (2014). Improving the use of species distribution models in conservation planning and management under climate change. *PLoS ONE*, *9*(11), e113749. https://doi.org/10.1371/journal.pone.0113749

Post, E., Steinman, B. A., & Mann, M. E. (2018). Acceleration of phenological advance and warming with latitude over the past century. *Scientific Reports*, *8*(1), 3927. https://doi.org/10.1038/s41598-018-22258-0

Rasche, L., & Taylor, R. A. J. (2019). EPIC-GILSYM: Modelling crop-pest insect interactions and management with a novel coupled crop-insect model. *Journal of Applied Ecology*, *56*(8), 2045–2056. https://doi.org/10.1111/1365-2664.13426

Raven, P. H., & Wagner, D. L. (2021). Agricultural intensification and climate change are rapidly decreasing insect biodiversity. *Proceedings of the National Academy of Sciences of the United States of America*, *118*(2), e2002548117. https://doi.org/10.1073/pnas.2002548117

Raza, M. M., Khan, M. A., Arshad, M., Sagheer, M., Sattar, Z., Shafi, J., Haq, Eu., Ali, A., Aslam, U., Mushtaq, A., Ishfaq, I., Sabir, Z., & Sattar, A. (2015). Impact of global warming on insects. *Archives of Phytopathology and Plant Protection*, *48*(1), 84–94. https://doi.org/10.1080/03235408.2014.882132

Rivero, R. M., Sánchez, E., Ruiz, J. M., & Romero, L. (2003). Influence of temperature on biomass, iron metabolism and some related bioindicators in tomato and watermelon plants. *Journal of Plant Physiology*, *160*(9), 1065–1071. https://doi.org/10.1078/0176-1617-00907

Richerzhagen, C. (2011). Effective governance of access and benefit-sharing under the convention on biological diversity. *Biodiversity and Conservation*, *20*, 2243–2261.

Sahoo, S. P., Panda, K. C. (2020). Prediction of Climate Change Using Statistical Downscaling Techniques. In: Rakshit, A., Singh, H., Singh, A., Singh, U., Fraceto, L. (eds). *New Frontiers in Stress Management for Durable Agriculture*. Springer, Singapore. https://doi.org/10.1007/978-981-15-1322-0_19

Saradadevi, R., Palta, J. A., & Siddique, K. H. M. (2017). ABA-mediated stomatal response in regulating water use during the development of terminal drought in wheat. *Frontiers in Plant Science*, *8*, 1251. https://doi.org/10.3389/fpls.2017.01251

Salinari, S., Bertuzzi, A., Gandolfi, A., Greco, A. V., Scarfone, A., Manco, M., & Mingrone, G. (2006). Dodecanedioic acid overcomes metabolic inflexibility in type 2 diabetic subjects. *American Journal of Physiology-Endocrinology and Metabolism*, *291*(5), E1051-8.

Schmid, T., Koch, M., & Gumuzzio, J. (2005). Multisensor approach to determine changes of wetland characteristics in semiarid environments (Central Spain). *IEEE Transactions on Geoscience and Remote Sensing*, *43*(11), 2516–2525. https://doi.org/10.1109/TGRS.2005.852082

Schultz, G. A., & Engman, E. T. (Eds.). (2012). *Remote sensing in hydrology and water management*. Springer Science+Business Media.

Shaver, T. M., Peterson, G. A., Ahuja, L. R., Westfall, D. G., Sherrod, L. A., & Dunn, G. (2002). Surface soil physical properties after twelve years of dryland no till management. *Soil Science Society of America Journal*, *66*(4), 1296–1303. https://doi.org/10.2136/sssaj2002.1296

Shi, J., Gao, H., Wang, H., Lafitte, H. R., Archibald, R. L., Yang, M. (2016). ARGOS8 variants generated by CRISPR-Cas9 improve maize grain yield under field drought stress conditions. *Plant Biotechnology Journal*, *15*, 207–216.

Singh, A. (2016). Managing the water resources problems of irrigated agriculture through geospatial techniques: An overview. *Agricultural Water Management*, *174*, 2–10. https://doi.org/10.1016/j.agwat.2016.04.021

Singh, S. K., & Pandey, A. C. (2014). Geomorphology and the controls of geohydrology on waterlogging in Gangetic Plains, North Bihar, India. *Environmental Earth Sciences*, *71*(4), 1561–1579. https://doi.org/10.1007/s12665-013-2562-1

Singh, S. K., Majhi, P. K., Anandan, A., Korada, M., Habde, S. V., Khaire, A. R. Singh D. K, Jayasudha S, Singh A, Bhagvan A. P. (2022). Microsatellites based parental polymorphism survey for moisture stress in Rice (*Oryza sativa* L.) between the parental genotypes HUR-1309 and CR Dhan 801. *Emergent Life Science Research*, *8*(2), 248–258.

Song, S. Y., Chen, Y., Chen, J., Dai, X. Y., & Zhang, W. H. (2011). Physiological mechanisms underlying OsNAC5-dependent tolerance of rice plants to abiotic stress. *Planta*, *234*(2), 331–345. https://doi.org/10.1007/s00425-011-1403-2

Sporleder, M., Tonnang, H. E., Carhuapoma, P., Gonzales, J. C., Juarez, H., & Kroschel, J. (2013). Insect Life Cycle Modelling (ILCYM) software-a new tool for regional and global insect pest risk assessments under current and future climate change scenarios. In *Potential invasive pests of agricultural crops* (pp. 412–427). Wallingford UK: CABI.

Stewart, W. M., Sawyer, J. E., & Alley, M. M. (2009). The four fertilizer rights: Timing. *Crops Soils*, *42*(5), 24–28.

Sturrock, R. N., Frankel, S. J., Brown, A. V., Hennon, P. E., Kliejunas, J. T., Lewis, K. J., Worrall, J. J., & Woods, A. J. (2011). Climate change and forest diseases. *Plant Pathology*, *60*(1), 133–149. https://doi.org/10.1111/j.1365-3059.2010.02406.x

Sweeney, B. W., Funk, D. H., Camp, A. A., Buchwalter, D. B., & Jackson, J. K. (2018). Why adult mayflies of *Cloeon dipterum* (Ephemeroptera: Baetidae) become smaller as temperature warms. *Freshwater Science*, *37*(1), 64–81. https://doi.org/10.1086/696611

Thomson, L. J., MacFadyen, S., & Hoffmann, A. A. (2010). Predicting the effects of climate change on natural enemies of agricultural pests. *Biological Control*, *52*(3), 296–306. https://doi.org/10.1016/j.biocontrol.2009.01.022

Tiwari, R., Chand, K., & Anjum, B. (2020). Crop insurance in India: A review of Pradhan mantri fasal bima yojana (PMFBY). *FIIB Business Review*, *9*(4), 249–255. https://doi.org/10.1177/2319714520966084

Tonnang, H. E. Z., Hervé, B. D. B., Biber-Freudenberger, L., Salifu, D., Subramanian, S., Ngowi, V. B., Guimapi, R. Y. A., Anani, B., Kakmeni, F. M. M., Affognon, H., Niassy, S., Landmann, T., Ndjomatchoua, F. T., Pedro, S. A., Johansson, T., Tanga, C. M., Nana, P., Fiaboe, K. M., Mohamed, S. F.,. . . Borgemeister, C. (2017). Advances in crop insect modelling methods—Towards a whole system approach. *Ecological Modelling*, *354*, 88–103. https://doi.org/10.1016/j.ecolmodel.2017.03.015

Tougeron, K., Le Lann, C., Brodeur, J., & van Baaren, J. (2017). Are aphid parasitoids from mild winter climates losing their winter diapause? *Oecologia*, *183*(3), 619–629. https://doi.org/10.1007/s00442-016-3770-7

Trenkel, M. E. (2021). *Slow- and controlled-release and stabilized fertilizers: An option for enhancing nutrient use Efficiency in agriculture*. International Fertilizer Industry Association (IFA).

Tylianakis, J. M., Didham, R. K., Bascompte, J., & Wardle, D. A. (2008). Global change and species interactions in terrestrial ecosystems. *Ecology Letters*, *11*(12), 1351–1363. https://doi.org/10.1111/j.1461-0248.2008.01250.x

Ul-Allah, S., Iqbal, M., Maqsood, S., Naeem, M., Ijaz, M., Ashfaq, W., & Hussain, M. (2018). Improving the performance of bread wheat genotypes by managing irrigation and nitrogen under semi-arid conditions. *Archives of Agronomy and Soil Science*, *64*(12), 1678–1689. https://doi.org/10.1080/03650340.2018.1450974

van Munster, M., Yvon, M., Vile, D., Dader, B., Fereres, A., & Blanc, S. (2017). Water deficit enhances the transmission of plant viruses by insect vectors. *PLoS ONE*, *12*(5), e0174398. https://doi.org/10.1371/journal.pone.0174398

Verberk, W. C. E. P., Atkinson, D., Hoefnagel, K. N., Hirst, A. G., Horne, C. R., & Siepel, H. (2021). Shrinking body sizes in response to warming: Explanations for the temperature-size rule with special emphasis on the role of oxygen. *Biological Reviews of the Cambridge Philosophical Society*, *96*(1), 247–268. https://doi.org/10.1111/brv.12653

Vercelli, M., Novelli, S., Ferrazzi, P., Lentini, G., & Ferracini, C. (2021). A qualitative analysis of beekeepers' perceptions and farm management adaptations to the impact of climate change on honey bees. *Insects*, *12*(3), 228. https://doi.org/10.3390/insects12030228

Vilà, M., Beaury, E. M., Blumenthal, D. M., Bradley, B. A., Early, R., Laginhas, B. B., Trillo, A., Dukes, J. S., Sorte, C. J. B., & Ibáñez, I. (2021). Understanding the combined impacts of weeds and climate change on crops. *Environmental Research Letters*, *16*(3), 034043. https://doi.org/10.1088/1748-9326/abe14b

Wang, Y., Yan, J., Sun, J., Shi, W., Harwood, J. D., Monticelli, L. S., Tan, X., & Chen, J. (2021). Effects of field simulated warming on feeding behavior of *Sitobion avenae* (Fabricius) and host defense systems. *Entomologia Generalis*, *41*(6), 567–578. https://doi.org/10.1127/entomologia/2021/1271

Warren, M. S., Maes, D., van Swaay, C. A. M., Goffart, P., Van Dyck, H., Bourn, N. A. D., Wynhoff, I., Hoare, D., & Ellis, S. (2021). The decline of butterflies in Europe: Problems, significance, and possible solutions. *Proceedings of the National Academy of Sciences of the United States of America*, *118*(2), e2002551117. https://doi.org/10.1073/pnas.2002551117

Williams, C. M., Marshall, K. E., MacMillan, H. A., Dzurisin, J. D. K., Hellmann, J. J., & Sinclair, B. J. (2012). Thermal variability increases the impact of autumnal warming and drives metabolic depression in an overwintering butterfly. *PLoS ONE*, *7*(3), e34470. https://doi.org/10.1371/journal.pone.0034470

Wu, X., Feng, H., Wu, D., Yan, S., Zhang, P., Wang, W., Zhang, J., Ye, J., Dai, G., Fan, Y., Li, W., Song, B., Geng, Z., Yang, W., Chen, G., Qin, F., Terzaghi, W., Stitzer, M., Li, L.,. . . Dai, M. (2021). Using high-throughput multiple optical phenotyping to decipher the genetic architecture of maize drought tolerance. *Genome Biology*, *22*(1), 185. https://doi.org/10.1186/s13059-021-02377-0

Xu, X., Huang, G., Qu, Z., & Pereira, L. S. (2011). Using MODFLOW and GIS to assess changes in groundwater dynamics in response to water saving measures in irrigation districts of the upper Yellow River Basin. *Water Resources Management*, *25*(8), 2035–2059. https://doi.org/10.1007/s11269-011-9793-2

Yadav, G. S., Lal, R., Meena, R. S., Babu, S., Das, A., Bhowmik, S. N., Datta, M., Layak, J., & Saha, P. (2019). Conservation tillage and nutrient management effects on productivity and soil carbon sequestration under double cropping of rice in north eastern region of India. *Ecological Indicators*, *105*, 303–315. https://doi.org/10.1016/j.ecolind.2017.08.071

Yamamura, K., Yokozawa, M., Nishimori, M., Ueda, Y., & Yokosuka, T. (2006). How to analyze long-term insect population dynamics under climate change: 50-year data of three insect pests in paddy fields. *Population Ecology*, *48*(1), 31–48. https://doi.org/10.1007/s10144-005-0239-7

Yáñez-López, R., Torres-Pacheco, I., Guevara-Gonzalez, R. G., Hernandez-Zul, M. I., Quijano Carranza, J. A., & Rico-Garcia, E. (2012). The effect of climate change on plant diseases. *African Journal of Biotechnology*, *11*(10), 2417–2428.

Yang, Y., Tilman, D., Furey, G., & Lehman, C. (2019). Soil carbon sequestration accelerated by restoration of grassland biodiversity. *Nature Communications*, *10*(1), 718. https://doi.org/10.1038/s41467-019-08636-w

Yonow, T., Kriticos, D. J., Kirichenko, N., & Ota, N. (2018). Considering biology when inferring range-limiting stress mechanisms for agricultural pests: A case study of the beet armyworm. *Journal of Pest Science*, *91*(2), 523–538. https://doi.org/10.1007/s10340-017-0938-9

Zhang, H., Xiong, Y., Huang, G., Xu, X., & Huang, Q. (2017). Effects of water stress on processing tomatoes yield, quality and water use efficiency with plastic mulched drip irrigation in sandy soil of the Hetao Irrigation District. *Agricultural Water Management*, *179*, 205–214. https://doi.org/10.1016/j.agwat.2016.07.022

Ziska, L. H., Blumenthal, D. M., Runion, G. B., Hunt, E. R., Jr., & Diaz-Soltero, H. (2011). Invasive species and climate change: An agronomic perspective. *Climatic Change*, *105*(1–2), 13–42. https://doi.org/10.1007/s10584-010-9879-5

# Index

Note: Page numbers in *italics* indicate a figure and page numbers in **bold** indicate a table on the corresponding page.

## W